Bulk-Driven Circuit Techniques for CMOS FDSOI Processes

Friedel Gerfers
Editor

Bulk-Driven Circuit Techniques for CMOS FDSOI Processes

From Circuit Concept to Implementations

Springer

Editor
Friedel Gerfers
Chair Mixed Signal Circuit Design
Technische Universität Berlin
Berlin, Germany

ISBN 978-3-031-85113-1 ISBN 978-3-031-85114-8 (eBook)
https://doi.org/10.1007/978-3-031-85114-8

This Springer imprint is published by the registered company Springer Nature Switzerland AG
The registered company address is: Gewerbestrasse 11, 6330 Cham, Switzerland

I dedicate this book to my three wonderful children, Lea, Jan and Tim, who I will continue to encourage to dream, discover, and enjoy life's journeys.

I would also like to dedicate this book to all my students and project partners, past and present, who have kept my "spirit of inquiry" alive by striving to be the best scientist and researcher to ultimately transform society and improve lives through new and advanced microelectronic technologies.

- Friedel Gerfers -

Preface

In an era where digital systems dominate the technology landscape, the importance of analog signal processing remains paramount. Among the critical components in this domain are Analog-to-Digital Converters (ADCs) and Digital-to-Analog Converters (DACs), which serve as the bridge between the analog and digital worlds. The performance of these converters is crucially dependent on key parameters such as transconductance (g_m), DC gain, device matching, linearity, and timing accuracy to name a few.

Tightening or improving these parameters inherently by design is a significant challenge or even impossible beyond a certain limit using state-of-the-art circuit design techniques. Consequently, as the intrinsic CMOS transistor performance is bounded, for- and/or background calibration schemes are often applied to overcome the MOS device limitations. However, this challenge can be effectively mitigated through the implementation of active or passive bulk-driven circuits, which are feasible within silicon-on-insulator (SOI) CMOS technologies.

Fully Depleted Silicon on Insulator (FD-SOI) CMOS technologies offers superior transistor performance over standard bulk CMOS technology through improved electrical performance, power-efficiency, and scalability.

FD-SOI feature reduced parasitic capacitance, resulting in faster switching speeds and lower power consumption. Its inherent ability to control short-channel effects results in better device performance at smaller transistor nodes. In addition, FD-SOI allows dynamic threshold voltage tuning, enabling ultra-low power operation without sacrificing performance. The isolation provided by the buried oxide layer improves leakage control, making FD-SOI ideal for advanced low-power applications such as mobile devices and IoT compared to traditional bulk CMOS.

Active and passive bulk-driven circuits and systems offer a powerful technology for correcting or fine-tuning the electrical characteristics of FD-SOI CMOS transistors by controlling the local substrate potential. While static bulk-driven circuit techniques typically only take advantage of fixed positive or negative threshold voltage shift, active bulk circuits dynamically adjust the well potential to optimize circuit performance in real time, resulting in improvements in gain, matching, linearity, and therewith overall accuracy.

This is particularly important in precision data converters, where even small enhancements lead to a significant impact on the quality of the converted output signal. On the other hand, passive bulk circuits use fixed settings to improve specific circuit parameters, such as reducing distortion or increasing stability, providing a more straightforward yet effective approach to performance enhancement.

This book delves into the intricate world of analog systems, focusing on the role of active and passive bulk-driven circuits to improve the performance of analog, digital, mixed-signal and mmWave circuits and systems. Readers will be introduced to the fundamental concepts and theory of bulk-driven SOI MOS physics before exploring the specific techniques and benefits of active and passive bulk control. A key emphasis is placed on how these techniques can positively influence essential circuit and system performance metrics.

Through a combination of theoretical explanations and practical implementations, this book aims to provide a comprehensive understanding of how bulk-driven circuits can be leveraged to enhance analog, data converter, or mmWave systems. Real-world examples proven by measurements, coupled with detailed theoretical analyses, demonstrate the practical application of these techniques as well as showing how bulk-driven circuits are utilized to maximize, e.g., energy-efficiency or improve the precision of data converters. Additionally, the design and implementation challenges as well as solutions associated with the realization of bulk-driven circuits are discussed, offering readers valuable insights into the practical implementation aspects of these technologies.

This book serves as a valuable resource for students, engineers, and researchers working in the field of analog, digital, mixed-signal, and mmWave circuit design. It provides a thorough exploration of how active and passive bulk-driven circuits significantly improve the performance of systems and its applications. It is our hope that this work will deepen the readers understanding and spark further innovation in this fascinating area of microelectronics, contributing to the ongoing advancement and benefits of SOI CMOS technology.

Berlin, Germany
December, 2024

Friedel Gerfers

Acknowledgments

This work would not have been possible without the excellent and very supportive cooperation, including technology funding, of *Globalfoundries*, Dresden, Germany. The Chair Mixed Signal Circuit Design and I highly appreciate the great teamwork and dedication of my colleagues at Globalfoundries, especially Claudia Kretzschmar and Rennel Fruto.

In addition, I would like to express my sincere gratitude to Prof. Boris Murrmann (now with the University of Hawai'i at Mānoa, Honolulu, USA) for including me in his outstanding research group at Stanford University and for the scientific insights gained during my stay in California, USA, from 2006 to 2015. I thank Stephen Weinreich, now at Analog Devices, Wilmington, USA, and Prof. Boris Murrmann for their contributions to the second chapter.

I extend my sincere appreciation to Dr. Shai Bonen and Prof Sorin P. Voinigescu, as well as Dr. Oğuzhan Özdamar, Dr. Valentyn Solomko, Dr. Semen Syroiezhin, and Prof. Robert Weigel for their contributions to Chaps. 6 and 10, respectively.

Finally I would like to thank my research staff of the Mixed Signal Circuit Design chair; Dr. Marcel Runge, Julius Edler, Urs Hecht, Helia Ordouei, Enne Wittenhagen, Nima Lotfi, Patrick James Artz, Dr. Philipp Scholz, and Rene Hartmann who believed in their projects even in the face of adversity, and ultimately produced exceptional research results, summarized in the Chaps. 3–5 and 7–9.

Last but not least, I am deeply grateful to the Technical University of Berlin, the Einstein Foundation, the German Research Foundation (*DFG*), and the Federal Ministry of Education and Research (*BMBF*) for providing excellent research infrastructure and funding for projects (BMBF MassiveData6G - 16ES1078K, BMBF 6G-RIC - 16KISK030, DFG DISCO - 509466514, DFG VOLTA - 521607727, BMBF DESign - 16ME0996, PNN - 16ME0980) and laboratory equipment (BMBF FastBIT - 03ZZ05X04 DFG INST 131/874-1, DFG INST 131/849-1, DFG INST 131/848-1 and DFG INST 131/847-1).

Contents

Contributors

U. Alakusu University of Toronto, Toronto, ON, Canada

Patrick James Artz Technische Universität Berlin, Berlin, Germany

A. Bharadwaj University of Toronto, Toronto, ON, Canada

Shai Bonen University of Toronto, Toronto, ON, Canada

G. Cooke University of Toronto, Toronto, ON, Canada

M. S. Dadash University of Toronto, Toronto, ON, Canada

Julius Edler Technische Universität Berlin, Berlin, Germany

Friedel Gerfers Technische Universität Berlin, Berlin, Germany

M. J. Gong University of Toronto, Toronto, ON, Canada

Urs Hecht Technische Universität Berlin, Berlin, Germany

H. Y. Hsu University of Toronto, Toronto, ON, Canada

T. Jaeger University of Toronto, Toronto, ON, Canada

T. Liu University of Toronto, Toronto, ON, Canada

Nima Lotfi Technische Universität Berlin, Berlin, Germany

J. McIntosh University of Toronto, Toronto, ON, Canada

Boris Murmann University of Hawai'i at Mānoa, Honolulu, HI, USA

Helia Ordouei Technische Universität Berlin, Berlin, Germany

Oğuzhan Özdamar Infineon Technologies AG, Neubiberg, Germany

Marcel Runge Technische Universität Berlin, Berlin, Germany

S. Shopov University of Toronto, Toronto, ON, Canada

Valentyn Solomko Infineon Technologies AG, Neubiberg, Germany

Semen Syroiezhin Infineon Technologies AG, Neubiberg, Germany

S. Pati Tripathi University of Toronto, Toronto, ON, Canada

S. P. Voinigescu University of Toronto, Toronto, ON, Canada

Robert Weigel Institute for Electronics Engineering, University of Erlangen-Nuremberg, Erlangen, Germany

Stephen Weinreich Analog Devices, Wilmington, MA, USA

Stanford University, Stanford, CA, USA

Enne Wittenhagen Technische Universität Berlin, Berlin, Germany

L. Wu University of Toronto, Toronto, ON, Canada

A. Zandieh University of Toronto, Toronto, ON, Canada

J. Zhao University of Toronto, Toronto, ON, Canada

Acronyms

A

AC	Alternating Current
ADC	Analog-to-Digital Converter
AMS	Analog Mixed Signal
AUXDAC	Auxiliary Digital-to-Analog Converter
AWG	Arbitrary Waveform Generator
AWGN	Additive White Gaussian Noise

B

BBA	Body Bias Amplifier
BE	Back-End
BER	Bit Error Rate
BOX	Buried Oxide

C

CD	Common-Gate
CDAC	Capacitive Digital-to-Analog Converter
CH	Cherry Hooper
CMFB	Common-Mode Feedback
CMOS	Complementary Metal-Oxide-Semiconductor
OSR	Oversampling Ratio
PRBS	Pseudo Random Binary Sequence
CML	Current-Mode Logic
CT-SDM	Continuous Time Sigma Delta Modulator
CT-SDM	Continuous Time
CS	Common-Source
CV	Coefficient of Variation

D

DAC	Digital-to-Analog Converter

DC	Direct Current
DCA	Digital Communication Analyzers
DDA	Differential-Difference Amplifier
DEMUX	Demultiplexer
DIBL	Drain-Induced Barrier Lowering
DNL	Differential Nonlinearity

E

EML	Electro-Absorption Modulated LASER
ENOB	Effective Number of Bits

F

FBB	Forward Body Bias
FD-SOI	Fully Depleted Silicon On Insulator
FE	Front-End
FEC	Forward Error Correction
FET	Field Effect Transistor
FFE	Feed-Forward Equalization
FIR	Finite Impulse Response
FoM	Figures of Merit

G

GBW	Gain-Bandwidth Product
GF	GlobalFoundries
GF22FDX	GlobalFoundries 22nm FD-SOI
GS/s	Gigasamples per second

H

HBT	Heterostructure Bipolar Transistor
HF	High Frequency

I

IC	Integrated Circuit
IF	Intermediate Frequency
IM3	Third-Order Intermodulation
INL	Integral Nonlinearity
IoT	Internet-of-Things
IP3	Third-Order-Intercept-Point
IQ	In-phase and Quadrature

L

LASER	Light Amplification by Stimulated Emission of Radiation
LSB	Least Significant Bit

LIV	Light-Current-Voltage
LNA	Low Noise Amplifier
LS	Level Shifter
M	
MAG	Maximum Available Gain
MMIC	Monolithic Microwave Integrated Circuit
MMW	Millimeter Wave
MOS	Metal Oxide Semiconductor
MOM	Metal-Oxide-Metal
MSG	Maximum Stable Gain
MZM	Mach-Zehnder-Modulator
N	
NF	Noise Figure
NMOS	N-type Metal Oxide Semiconductor
NRZ	Non-Return-to-Zero
O	
OMA	Optical Modulation Amplitude
OTA	Operational Transconductance Amplifier
P	
P1dB	1 dB Compression Point
PA	Power Amplifier
PAE	Power Added Efficiency
PAM-4	Four-Level Pulse Amplitude Modulation
PDF	Phase-Frequency Detector
PLL	Phase Locked Loop
PMOS	P-type Metal Oxide Semiconductor
PRBS	Pseudo-Random Binary Sequence
PVT	Process, Voltage and Temperature
Q	
QAM	Quadrature Amplitude Modulation
QD	Quantum Dot
R	
RF	Radio Frequency
RL	Return Loss
S	
SAR	Successive Approximation Register

SET	Single Electron Transistor
SFDR	Spurious Free Dynamic Range
SF	Source Follower
SHT	Single Hole Transistor
SIRC	System Impulse Response Correction
SLVT	Super Low-Vth
SNR	Signal-to-Noise Ratio
SPI	Serial Peripheral Interface
SNDR	Signal-to-Noise and Distortion Ratio
SQNR	Signal-to-Quantization-Noise-Ratio
SSB	Single-Sideband
SSF	Super Source Follower
SOTA	State-of-the-Art
SOI	Silicon On Insulator
SWDM	Shortwave Wavelength Division Multiplexing

T

T&H	Track-and-Hold
THD	Total Harmonic Distortion
TI	Time-Interleaved
TIA	Transimpedance Amplifiers
TL	Transmission Line
TX	Transmitter

U

UWB	Ultra-Wideband

V

VCSEL	Vertical-Cavity Surface-Emitting Laser
VGA	Variable Gain Amplifier
VIP3	Third-Order Voltage Intercept Point
VM	Voltage-Mode
VNA	Vector Network Analyzer

Z

ZBB	Zero Body-Biasing

List of Figures

List of Tables

Part I
Analog Circuit Techniques

Chapter 1
Common-Source Amplifier Feedback Using Back-Gate Transconductance

Stephen Weinreich and Boris Murmann

1.1 Introduction

The common-source stage is among the most widely used basic transistor configurations and is a fundamental building block in many voltage amplifiers. It provides voltage-to-current conversion (transconductance) with high input and output impedance resulting in high voltage gain when driving a high-impedance load; however, the gain magnitude is sensitive to process, voltage, and temperature variations, and this poses a problem for precision-gain applications. In order to construct a voltage amplifier with well-defined gain at low supply voltage, a common-source stage may drive a diode-connected load as illustrated in Fig. 1.1a; this simple circuit forms a ratiometric amplifier in which the gain is roughly the ratio of input and load transconductance. However, this amplifier still suffers from gain mismatch and poor linearity. A negative feedback network may instead be wrapped around the common-source stage as in Fig. 1.1b, generating well-defined voltage gain set by the ratio of input and feedback capacitors. This topology also improves linearity but comes at the cost of circuit complexity and area due to the additional feedback elements. Depending on the particular implementation, negative feedback is also likely to reduce the input impedance, which may be unacceptable for certain applications.

S. Weinreich (✉)
Analog Devices, Wilmington, MA, USA

Stanford University, Stanford, CA, USA
e-mail: weinreich@alumni.stanford.edu

B. Murmann
University of Hawai'i at Mānoa, Honolulu, HI, USA
e-mail: bmurmann@hawaii.edu

F. Gerfers (ed.), *Bulk-Driven Circuit Techniques for CMOS FDSOI Processes*,
https://doi.org/10.1007/978-3-031-85114-8_1

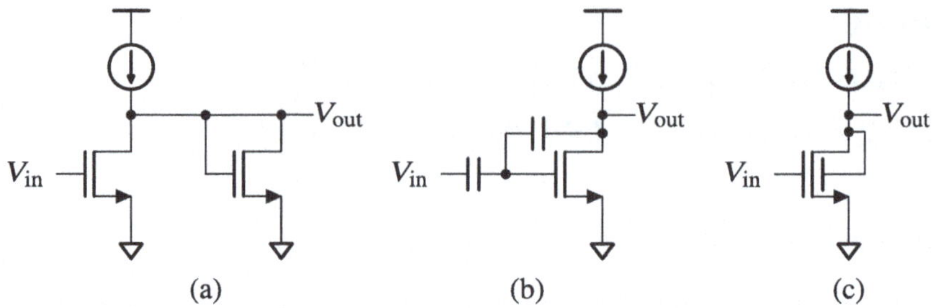

Fig. 1.1 Schematics of (**a**) a conventional common-source amplifier with a diode-connected load, (**b**) a conventional common-source amplifier with negative feedback (gate biasing not shown), and (**c**) a common-source amplifier with back-gate feedback in FD-SOI. Drawn single-ended; similar differential or pseudo-differential topologies also exist

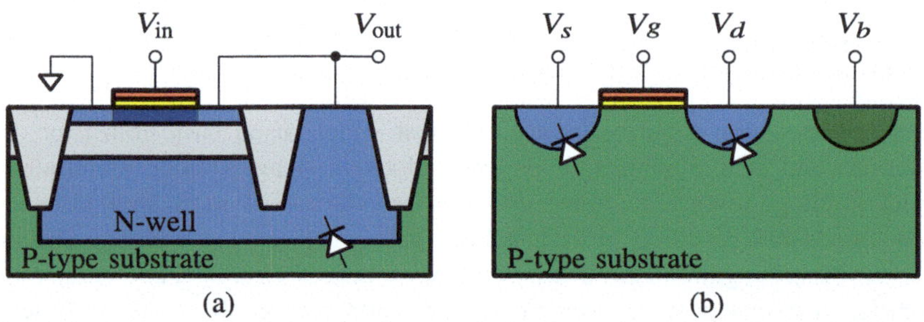

Fig. 1.2 Cross section of (**a**) a flip-well FD-SOI transistor wired as a single-transistor common-source amplifier with back-gate feedback and (**b**) a bulk transistor, with parasitic diodes annotated

This chapter offers an alternative approach, illustrated in Fig. 1.1c: a single-transistor common-source amplifier that exploits the Fully Depleted Silicon-on-Insulator (FD-SOI) back-gate as a feedback injection point without requiring any additional components or reducing input impedance [1]. "Back-gate feedback" offers the simplicity and high input impedance of the ratiometric amplifier but with low mismatch and high linearity provided by the negative feedback, all within a single FD-SOI transistor.

Back-gate feedback is enabled by two unique properties of FD-SOI technology. First, the channel is fully isolated from the substrate by an ultra-thin buried oxide (BOX) layer as illustrated in Fig. 1.2a. As a result, the back-gate voltage can be set independent of the source and drain voltages, up to at least 2–5 V [2, 3]. This is in contrast to bulk technology, illustrated in Fig. 1.2b, where diodes from source and drain to bulk limit forward body voltage tuning to less than a few hundred mV. Additionally, the FD-SOI back-gate exerts significant control over the channel due to the ultra-thin nature of the BOX layer, in contrast to legacy SOI technologies with thick BOX [4].

In the remainder of this chapter, we first describe and characterize the small-signal back-gate transconductance that enables back-gate feedback. Then, we analyze the common-source amplifier with back-gate feedback, followed by prototype amplifier measurements to validate the analysis. Next, we examine the feasibility of this technique in bulk technology before concluding this chapter.

1.2 Back-Gate Transconductance

The back gate is most often modeled as a tuning knob for the device's threshold voltage, offering a threshold adjustment range of up to ±250 mV [5–7]. This suggests a transistor drain current model

$$I_D = f(V_{gs} - (V_{th0} - \gamma V_{bs})) \tag{1.1}$$

where $f(\cdot)$ is some function (exponential, square law, BSIM-IMG model [8], or other), V_{gs} is the front gate[1]–source voltage, V_{th0} is the transistor threshold without any back gate bias, V_{bs} is the back gate–source voltage, and

$$\gamma = -\frac{dV_{th}}{dV_{bs}} \tag{1.2}$$

describes the relative strength of the back gate on the effective threshold voltage

$$V_{th} = V_{th0} - \gamma V_{bs}. \tag{1.3}$$

This large-signal threshold-shifting model implies a static back-gate bias. For example, a static back-gate bias can be used to reduce V_{th} for additional digital processing speed [9–11] or increase V_{th} to enable a low-leakage mode [9, 12, 13]. Static back-gate biasing has also been applied to analog circuits for mismatch compensation [14, 15], filter parameter tuning [16, 17], and enhanced RF performance [18–20], all without disrupting the sensitive signal path.

For back-gate feedback, we instead want to understand the *small-signal* behavior of the back-gate voltage. Just as the front-gate voltage controls charge in the transistor channel with small-signal transconductance g_m, so does the back gate with back-gate transconductance g_{mb}. This is made clear by rearranging (1.1) such that the back-gate voltage appears as a modifier on the front-gate-source voltage rather than a modifier on the threshold voltage. Note that (1.1) and (1.4) are algebraically identical, but (1.4) is more insightful for small-signal modeling.

[1] Usually referred to as just the "gate," we use "front gate" throughout this chapter to explicitly differentiate it from the back gate.

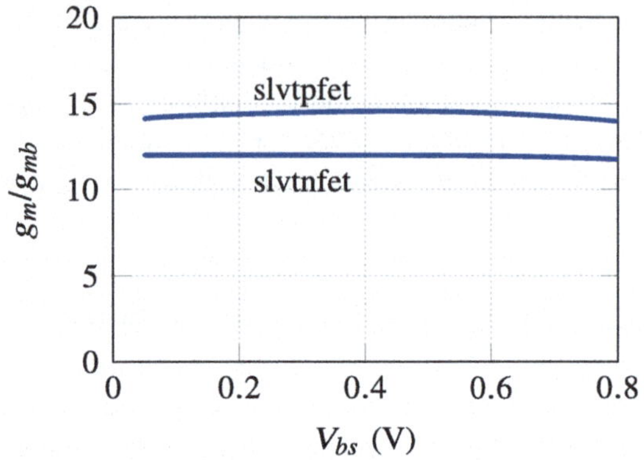

Fig. 1.3 Relative back-gate strength $g_m/g_{mb} = \gamma^{-1}$ with $V_{gs} = V_{bs}$

$$I_D = f((V_{gs} + \gamma V_{bs}) - V_{th0}) \tag{1.4}$$

Given $g_m = dI_D/dV_{gs}$ and (1.4), it is clear that

$$g_{mb} = \frac{dI_d}{dV_{bs}} = \gamma g_m. \tag{1.5}$$

To quantify this effect, γ can be derived as[2]

$$\gamma = \frac{C_{\text{BOX}} \| C_{\text{Si}}}{C_{\text{OX}}} \tag{1.6}$$

where C_{BOX}, C_{Si}, and C_{OX} are the capacitances across the buried (back-gate) oxide, silicon channel, and front-gate oxide, respectively [4]. From (1.6), it can be seen that both an ultra-thin buried oxide layer (large C_{BOX}) and thin silicon channel (large C_{Si}) are necessary for the back gate to exert significant control over the channel. This is satisfied in modern FD-SOI processes where the silicon channel and buried oxide thicknesses are approximately 6–7 nm and 20–25 nm, respectively [6, 21, 22]. Figure 1.3 plots $\gamma^{-1} = g_m/g_{mb}$ across bias point for 22-nm slvt FD-SOI devices.[3] Across all core transistor types, $\gamma^{-1} \approx$ 12–16 in this technology. Thick-oxide IO devices will, by nature of their thicker front-gate oxide, have reduced C_{OX} resulting in weaker front-gate control of the channel and therefore lower γ^{-1}; in this technology, $\gamma^{-1} \approx$ 5–7 for thick-oxide devices.

Given that the back-gate transconductance offers a second, independent small-signal injection point with a well-defined relative strength γ, it enables new design opportunities not practical in bulk technology. In full-duplex wireless systems,

[2] Assuming a long-channel model. For short channels, source and drain coupling reduces γ [8]. In the 22-nm FD-SOI technology used in this work, γ reduces by about 1% at 20-nm gate length.

[3] We plot γ^{-1} rather than γ because, as derived in Sect. 1.3.1, the closed-loop amplifier gain is roughly equal to γ^{-1}.

for example, self-interference cancelation can be realized by subtracting the TX signal directly at the LNA input device without any noise penalty [23]. In op-amps, the back-gate transconductance can inject common-mode feedback without adding any extra device loading [24] and provide feedforward linearization by injecting opposite-phase nonlinear content via the back gate to cancel g_m nonlinearities [25]. The back gate can also be used for linearization by wrapping negative feedback around an amplifier with the back gate as the feedback injection point [1, 26]. As Sect. 1.3 describes, back-gate feedback offers both improved linearity and reduced gain mismatch without any additional circuit components.

1.3 Common-Source Amplifier with Back-Gate Feedback

In order to effectively design the amplifier with back-gate feedback, we must first derive its analog performance. This amplifier, illustrated in Fig. 1.1c, can be more easily understood by comparing it to the similar two-transistor circuit of Fig. 1.1a in which a common-source amplifier drives a diode-connected load. Ignoring output conductance, the voltage gain of this amplifier is set by

$$\frac{v_{\text{out}}}{v_{\text{in}}} \approx \frac{g_m}{g_{ml}} \tag{1.7}$$

where g_m and g_{ml} refer to the input device and diode-connected load, respectively. The amplifier with back-gate feedback has voltage gain

$$\frac{v_{\text{out}}}{v_{\text{in}}} \approx \frac{g_m}{g_{mb}} \tag{1.8}$$

where, again, g_{mb} is the conductance of the back gate.

While (1.7) and (1.8) are similar, in the latter design, the FD-SOI front- and back gates share a single transistor channel, and so the back-gate load connection creates as negative feedback network around the front-gate common-source amplifier. The amplifier with diode-connected load, on the other hand, is open-loop, and this difference has a significant impact on amplifier performance. We will use the small-signal models in Fig. 1.4 to quantitatively derive this performance and benchmark the amplifier with back-gate feedback against the conventional diode-connected load. To avoid confusion, throughout the rest of this chapter we will refer to these two amplifiers as the "back-gate amplifier" and "front-gate amplifier," respectively, noting that this nomenclature describes the load device; both amplifiers use their front gates as inputs.

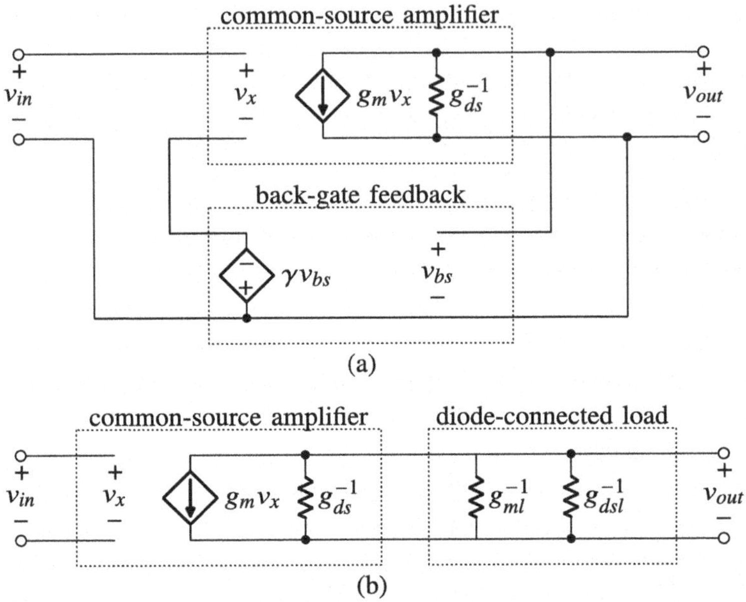

Fig. 1.4 Small-signal models of (**a**) a back-gate amplifier in FD-SOI and (**b**) a conventional front-gate amplifier with diode-connected load. In both cases, $v_{\text{in}} = v_{gs}$ and $v_{\text{out}} = v_{ds}$

1.3.1 Gain

The input common-source amplifiers are identical in both cases with open-loop gain

$$a_0 = \frac{g_m}{g_{ds}}. \tag{1.9}$$

For the back-gate amplifier in Fig. 1.4a, the feedback factor is

$$\beta = \gamma = \frac{g_{mb}}{g_m} \tag{1.10}$$

leading to loop gain

$$T = a_0\gamma = \frac{g_{mb}}{g_{ds}} \tag{1.11}$$

and closed-loop gain

$$A_{cl} = \frac{a_0}{1 + \gamma a_0}. \tag{1.12}$$

Assuming high loop gain, the closed-loop gain simplifies to

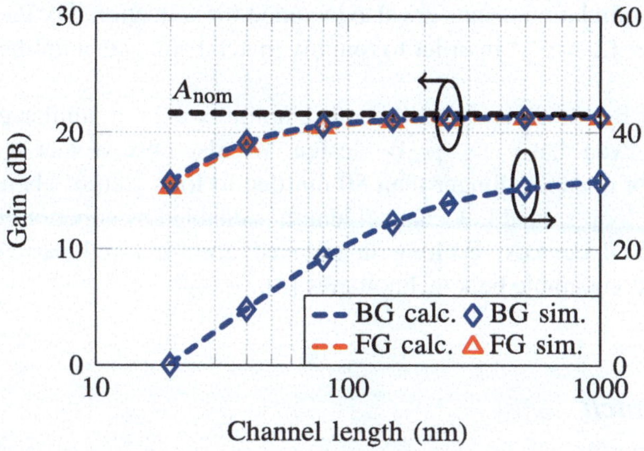

Fig. 1.5 Calculated and simulated gain and loop gain with g_m/I_D = 15 S/A

$$A_{\mathrm{nom}} = \lim_{T\to\infty} A_{cl} = \gamma^{-1} \tag{1.13}$$

which is equivalent to (1.8).

For comparison, we also consider the front-gate amplifier of Fig. 1.4b. This amplifier has gain[4]

$$A_{\mathrm{FG}} = \frac{a_0}{1 + \gamma_{\mathrm{FG}}(a_0 + 1)} \tag{1.14}$$

where we have defined the transconductance ratio between the load and input devices as

$$\gamma_{\mathrm{FG}} = \frac{g_{ml}}{g_m}. \tag{1.15}$$

Provided that $\gamma_{\mathrm{FG}} = \gamma$, (1.12) and (1.14) are nearly equivalent; the extra $+\gamma_{\mathrm{FG}}$ in the denominator of (1.14) accounts for finite output conductance of the load device. In the case where $\gamma_{\mathrm{FG}} a_0$ is large, (1.14) simplifies to

$$A_{\mathrm{nom,FG}} = \lim_{\gamma_{\mathrm{FG}} a_0\to\infty} A_{\mathrm{FG}} = \gamma_{\mathrm{FG}}^{-1} \tag{1.16}$$

which is similar to (1.7).

Figure 1.5 plots gain and loop gain against channel length for amplifiers built of slvtnfet transistors in 22-nm FD-SOI; channel length is varied as a proxy for loop gain. Transistor parameters a_0 and γ are obtained from pre-computed lookup

[4] Subscript FG denotes variables that refer to the front-gate amplifier of Figs. 1.1a and 1.4b.

tables [27], and simulation results are also included for validation. For the front-gate amplifier, $\gamma_{\mathrm{FG}}^{-1} = 12 \approx \gamma^{-1}$ in order to roughly match back-gate amplifier γ^{-1} (see Fig. 1.3).

As predicted by (1.12) and (1.14), both amplifiers exhibit similar gain across channel length given that $\gamma \approx \gamma_{\mathrm{FG}}$ by design. We also observe that $A_{cl} \approx A_{\mathrm{nom}}$ (within 1 dB) for channels longer than 80 nm due to loop gain of about 20 dB or more. This high loop gain at low channel length is enabled by superior electrostatic control in FD-SOI devices resulting in reduced drain-induced barrier lowering (DIBL) than in comparable bulk technologies [3].

1.3.2 Mismatch

While the voltage gain of the two amplifiers is similar, back-gate amplifier feedback attenuates most major random manufacturing variations. To analyze this, consider random variations in open-loop gain Δa_0 and back-gate strength $\Delta\gamma$, resulting in an error in the closed-loop gain ΔA_{cl}. With these variations, closed-loop gain for the back-gate amplifier of Fig. 1.4a becomes

$$A_{cl} + \Delta A_{cl} = \frac{a_0 + \Delta a_0}{1 + (a_0 + \Delta a_0)(\gamma + \Delta\gamma)}. \tag{1.17}$$

Assuming that Δa_0 and $\Delta\gamma$ are small compared to their nominal values, we use superposition and consider each source of mismatch independently.

First, consider the case with ideal feedback such that $\Delta\gamma = 0$. Simplifying (1.17),

$$\left.\frac{\Delta A_{cl}}{A_{cl}}\right|_{\Delta\gamma=0} \approx \frac{\Delta a_0}{a_0} \cdot \frac{1}{1+T} \tag{1.18}$$

as expected for a first-order feedback system with ideal feedback. Even modest loop gain will significantly attenuate mismatch in a_0.

Next, consider the case with an ideal common-source amplifier such that $\Delta a_0 = 0$. Simplifying (1.17),

$$\left.\frac{\Delta A_{cl}}{A_{cl}}\right|_{\Delta a_0=0} \approx -\frac{\Delta\gamma}{\gamma}. \tag{1.19}$$

That is, back-gate feedback does not attenuate feedback factor mismatch, and so to analyze amplifier performance, we must first understand the sources of γ mismatch.

From (1.6), γ is a function of the thickness and dielectric constants of the silicon and front- and back-gate oxides, all parameters with tight processing control. For example, variations in silicon and buried oxide thicknesses are just 0.7%rms and <0.5%, respectively [22]. Since γ mismatch directly impacts A_{cl}, this sets a lower limit on gain matching in the back-gate feedback amplifier. This dependence on feedback factor mismatch is intuitive for any negative feedback system: for example, capacitor mismatch sets a lower bound for the amplifier in Fig. 1.1b.

For the front-gate amplifier of Fig. 1.4b, it can be shown that

$$\frac{\Delta A_{\mathrm{FG}}}{A_{\mathrm{FG}}} \approx \frac{\Delta g_m}{g_m} - \frac{\Delta g_{ml}}{g_{ml}} \tag{1.20}$$

provided that $a_0\gamma_{\mathrm{FG}}$ is large. In this case, the two error terms Δg_m and Δg_{gm} are partially correlated random variables. The correlated portion, whereby g_m and g_{ml} move in the same direction with the same scaling factor, is due to global mismatch and is rejected to first order. Local mismatch, on the other hand, is uncorrelated between the two transistors and so must be considered. Expressing this mismatch in terms of its standard deviation σ, and using Pelgrom's Law to relate the relative mismatch of g_m and g_{ml},

$$\frac{\sigma_{A_{\mathrm{FG}}}}{A_{\mathrm{FG}}} \approx \frac{\sigma_{g_m}}{g_m} \cdot \sqrt{1 + \gamma_{\mathrm{FG}}^{-1}}. \tag{1.21}$$

Whereas loop gain attenuates back gate amplifier device mismatch following (1.18), the front gate amplifier sees no such benefit. In fact, matching degrades relative the standalone common-source amplifier due to the ratiometric design requiring a smaller diode-connected load. As a result, the back gate amplifier has superior gain mismatch.

1.3.3 Linearity

In a typical negative feedback configuration, high loop gain attenuates voltage swing at the open-loop amplifier's input node; this is the mechanism by which negative feedback improves closed-loop linearity. In Fig. 1.1b, for example, the reduction in voltage swing at the transistor's front gate reduces excitement of g_m nonlinearity. The common-source amplifier with back-gate feedback of Fig. 1.1c is unique in that there is no single physical node with reduced swing because feedback summation occurs *within the transistor channel itself* as a result of the shared control between the front and back gates. Instead, the "voltage swing" that is attenuated is $V_{gs} + \gamma V_{bs}$, or the *effective* (but not physical) front-gate–source voltage. This can be seen in the small-signal model in Fig. 1.4a where the transconductor is controlled not by the gate–source voltage v_{in} but rather by v_x, a virtual voltage that combines the effects of both gates; the transistor front-gate swing is unattenuated.

This linearization benefit is predicated on the assumption of a constant feedback factor. Figure 1.3 presents simulated γ^{-1} across inversion level for slvt devices with $V_{gs} = V_{bs}$ and illustrates that γ is nearly constant across the full range. As a result, we assume constant γ for the remainder of this analysis and will validate this assumption with simulations with a BSIM-IMG model [8] and silicon measurements.

Considering up to 3rd order nonlinearities in the open-loop amplifier, the output voltage can be modeled as

$$v_{\text{out},ol} = a_1 v_{\text{in}} + a_2 v_{\text{in}}^2 + a_3 v_{\text{in}}^3 \tag{1.22}$$

with coefficients defined by g_m and g_{ds} nonlinearity; the reader is referred to [1] for details of this derivation. With feedback, the closed-loop gain of the back-gate amplifier is

$$v_{\text{out},cl} = b_1 v_{\text{in}} + b_2 v_{\text{in}}^2 + b_3 v_{\text{in}}^3 \tag{1.23}$$

with coefficients

$$b_1 = A_{cl} \tag{1.24}$$

$$b_2 = \frac{a_2}{(1+T)^3} \tag{1.25}$$

$$b_3 = \frac{a_3(1+T) - 2a_2^2\gamma}{(1+T)^5}, \tag{1.26}$$

where T and A_{cl} are defined in (1.11) and (1.12), respectively. As expected for a negative feedback system, these equations demonstrate that loop gain works to linearize the voltage transfer function. Given that amplifiers with a longer channel length have higher loop gain as seen in Fig. 1.5, they will also exhibit a more pronounced linearization benefit from back-gate feedback. Linearity of the front-gate amplifier is similar to the open-loop amplifier provided that g_{ds} nonlinearity is modified to account for the diode-connected load as detailed in [1].

To quantify linearity, we use the third-order voltage intercept point $\text{VIP}_3 = \sqrt{4a_1/a_3}$, which is the extrapolated input voltage at which the output contains equal parts fundamental and third harmonic. Figure 1.6 plots calculated linearity for the open-loop amplifier, front-gate amplifier, and back-gate amplifier, with g_m and g_{ds} nonlinearities calculated using pre-computed lookup tables across inversion level and channel length [27]. Simulation results are also included for comparison.

Across all inversion levels, the back-gate amplifier provides the highest VIP_3 by 13–27 dB compared to the front-gate amplifier. The largest benefit is at high g_m/I_D when large open-loop amplifier gain enables the highest loop gain. Whereas the short-channel open-loop and front-gate amplifiers offer higher VIP_3 than the long-channel designs due to short-channel effects, this behavior is flipped for the back-gate amplifier: the long-channel back-gate amplifier has the highest linearity because its higher loop gain more than compensates for its lower open-loop linearity. Additionally, the strong correlation between calculated and simulated linearity supports the initial assumption that γ has a negligible effect on nonlinearity.

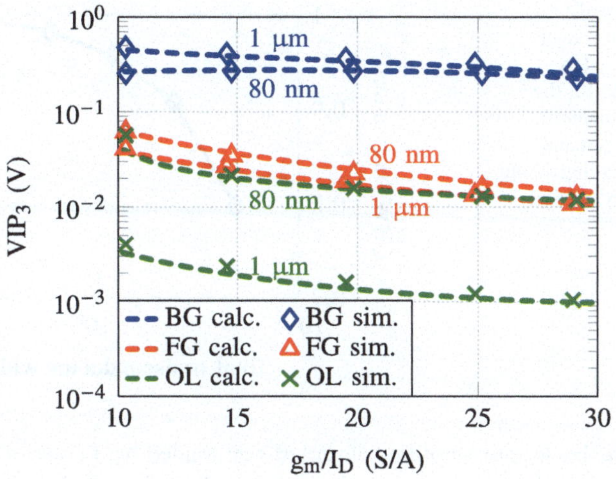

Fig. 1.6 Calculated and simulated VIP_3 for the open-loop (OL), back-gate (BG), and front-gate (FG) amplifiers at 80 nm and 1 µm channel lengths

1.3.4 Power, Noise, Area, and Bandwidth

Back-gate feedback does not explicitly consume additional current or add additional noise compared to the underlying open-loop common-source amplifier because the back-gate transconductance exists whether or not it is utilized. While the large-signal behavior is forward body biasing, thereby reducing the threshold voltage following (1.3), this can be compensated for by slightly reducing the overall size of the transistor to maintain largely back-gate-bias-independent current, noise, and transconductance.

The primary challenges associated with the back-gate feedback technique are area and bandwidth, both a result of the biased well required to construct the circuit as illustrated in Fig. 1.2a. Each amplifier with back-gate feedback requires its own well tap, in contrast to a typical common-source amplifier in which many amplifiers could share a single grounded well tap. On the other hand, the back-gate feedback amplifier can be constructed with just a single transistor finger while the front-gate amplifier relies on the ratio of amplifier-to-load fingers to define gain; a minimum of $A_{\text{nom,FG}} + 1$ fingers would be required for a given gain $A_{\text{nom,FG}}$.

The well capacitance, formed by the reverse-biased well–substrate diode illustrated in Fig. 1.2a, adds an additional capacitive load C_{well} to the amplifier; its impact on gain-bandwidth product is limited provided that $C_L \gg C_{\text{well}}$ given $\text{GBW} \approx g_m/(C_L + C_{\text{well}})$. For large devices, $C_{\text{well}} \approx 0.15 \cdot C_{gg}$ and so the impact is relatively minor; however, the well perimeter capacitance does not scale linearly with device area, and as a result, C_{well}/C_{gg} increases for very small amplifiers. This can be seen in Fig. 1.7, which presents normalized simulated gain-bandwidth

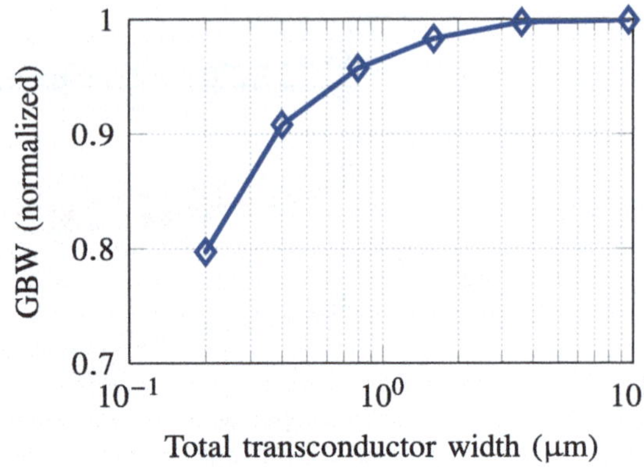

Fig. 1.7 Simulated gain-bandwidth product (GBW; normalized to maximum GBW) for the back-gate amplifier with 1–48 fingers, $L = 80$ nm and $W = 200$ nm per finger, and loaded by a copy of itself

product for the back-gate amplifier of varied size loaded by a copy of itself. Gain-bandwidth product decreases as a result of C_{well} below about 1–2 μm width.

1.4 Prototype Back-Gate Feedback Amplifier Design

A prototype test chip, illustrated in Fig. 1.8, is implemented in 22-nm FD-SOI and uses inverter-based amplifiers to demonstrate back-gate feedback. Comparing Fig. 1.8a with Fig. 1.1c, the inverter-based design offers twice the transconductance efficiency due to the complementary structure. The test chip comprises a total of 128 amplifiers including back-gate amplifiers with varied channel length and size and front-gate amplifiers for benchmarking. All are constructed using slvtnfet/slvtpfet flip-well devices (NFET sits in N-well and PFET sits in P-well).

As illustrated in Fig. 1.8c, the back-gate amplifier PFETs sit in a triple well with the P-well nested in the NFET's N-well; by shorting the wells, the well-to-well parasitic diode can be ignored. Provided that the substrate is grounded and back-gate voltage is positive, as is the case in this structure by design, the N-well–substrate diode is reverse-biased, and so its only impact is capacitive loading on the output as discussed in Sect. 1.3.4. A similar structure is possible with conventional well devices such that the NFET sits in a triple well inside the PFET's N-well.

While the primary purpose of back-gate feedback is to exploit the small-signal back-gate transconductance, this topology inherently applies a static body bias to the transistors, which affects the threshold voltage following (1.3). However, in the inverter-based structure used in this prototype, only the NFET device sees forward body biasing; the PFET device experiences reverse body biasing. In practice, these two effects largely cancel each other out, resulting in a total simulated amplifier transconductance that is within 3% of its pre-body-biased value.

For all measurements, the amplifiers are self-biased off-chip such that the DC voltages at input and output nodes are the same. While post-layout simulations predict that GBW of at least tens of GHz is achievable, capacitive loading from

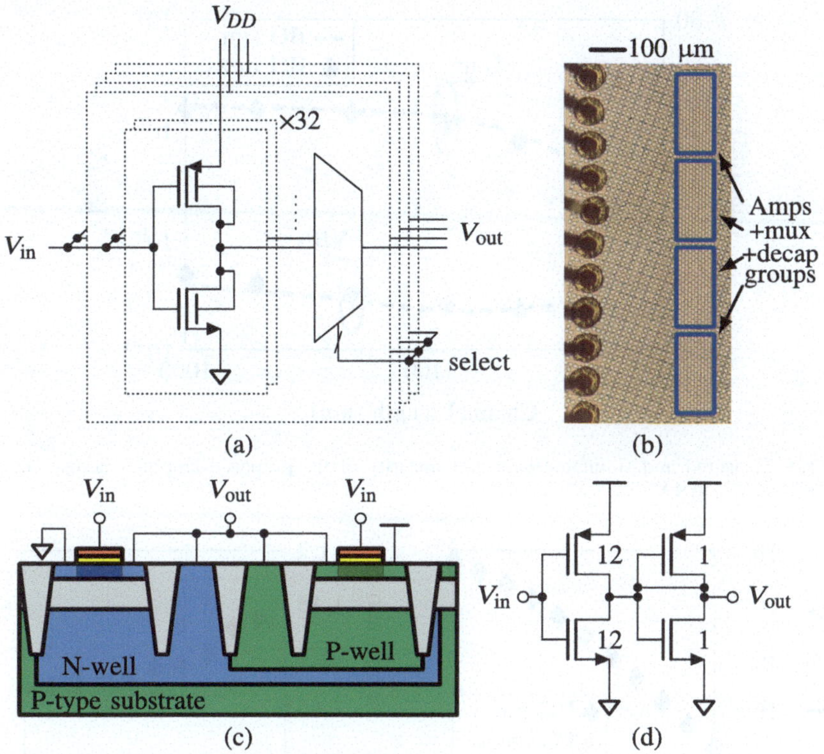

Fig. 1.8 Experimental chip (**a**) schematic with four groups of 32 amplifiers each, with three groups of back-gate amplifiers as shown and one group of front-gate amplifiers; (**b**) annotated die photo; (**c**) cross section showing nested wells (not to scale); and (**d**) schematic of the front-gate amplifiers

the measurement circuitry limits bandwidth and prohibits characterization of back-gate feedback for GBW > 100 MHz. The reader is referred to [28] for additional details of the measurement setup.

1.4.1 Measurement Results

Figure 1.9 presents measured gain and linearity of the back-gate amplifier across channel length at $V_{DD} = 0.8$ V and shows close agreement with simulation results. Across 40–1000 µm channel length, voltage gain varies from 19.0 to 21.9 dB while VIP_3 varies from 0.23 to 0.48 V. As was observed in Figs. 1.5 and 1.6, both gain and linearity improve with increasing channel length as a result of higher intrinsic gain and therefore loop gain.

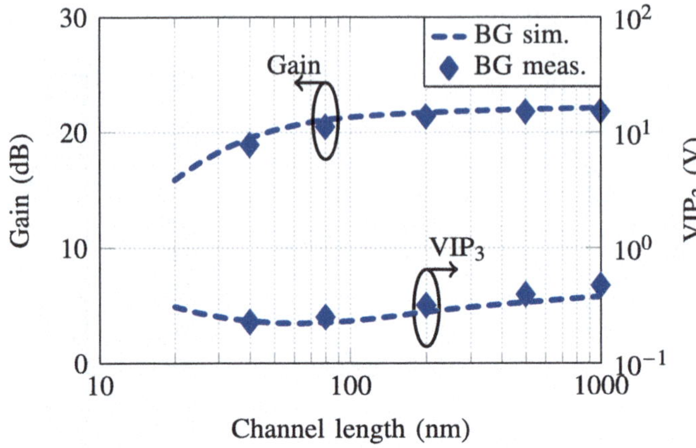

Fig. 1.9 Measured and simulated gain and linearity of the proposed amplifier across channel length at $V_{DD} = 0.8$ V

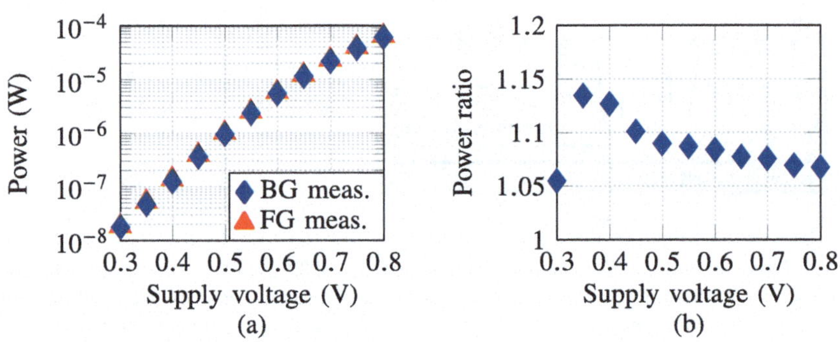

Fig. 1.10 (**a**) Measured power of the self-biased inverter-based amplifiers across supply voltage, per 12-finger amplifier (total 13 fingers for the front-gate amplifier, including its load). (**b**) Ratio of front-gate amplifier to back-gate amplifier power draw. Measurement accuracy at $V_{DD} = 0.3$ V is limited by lab equipment

Unless otherwise noted, the remaining measurements are of back-gate and front-gate amplifiers with 80-nm channel length and 0.2/0.3-µm NFET/PFET widths, with a total of 12 fingers each (plus one finger for the front-gate load). Supply voltage is varied as a proxy for inversion level from about 15 S/A to over 30 S/A.

Power consumption is presented in Fig. 1.10, and front-gate amplifier power is seen to be approximately $13/12 \approx 1.08\times$ that of the back-gate amplifier. This matches expectation given the front-gate amplifier's extra finger for the load device assuming that the biasing conditions are equivalent; as noted previously, body biasing on the back-gate amplifier has only minimal effect.

Measured voltage gain, shown in Fig. 1.11, is 20.5–21.2 dB for the back-gate amplifier and follows the simulated result. For this complementary structure, the nominal gain γ^{-1} of the back-gate amplifier is 22.5 dB while nominal gain for the

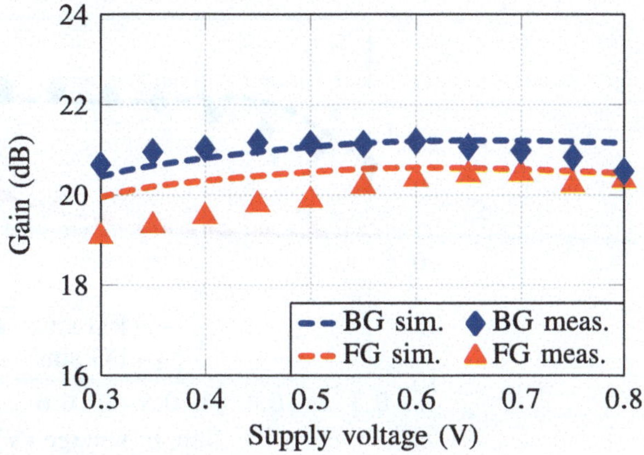

Fig. 1.11 Measured and simulated gain of the self-biased inverter-based amplifiers across supply voltage

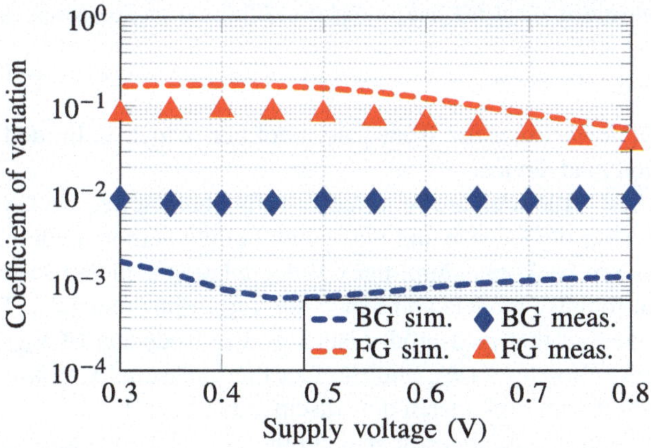

Fig. 1.12 Measured and simulated mismatch of the self-biased inverter-based amplifiers across supply voltage

front-gate amplifier is 21.6 dB. Measured gain is about 1–2 dB lower than these values as a result of finite loop gain, similar to the calculated gain of an NFET-only back-gate amplifier in Sect. 1.3.1.

To quantify gain mismatch, the mean μ and standard deviation σ are derived from measurements of 32 nominally identical amplifiers co-located on the same die. Figure 1.12 presents the coefficient of variation (CV) where $\text{CV} = \sigma/\mu$ is the normalized mismatch. CV for the back-gate amplifier is a nearly constant 0.8–0.9% over V_{DD} but is observed to exceed simulated mismatch by almost 10×. The single-finger back-gate amplifiers, not plotted, have measured CV ≈ 2%, about 5× higher than simulation. Mismatch for the front-gate amplifiers tracks simulation within about a factor of 2, a reasonable discrepancy given the small sample size. While back-gate amplifier gain mismatch is higher than expected from simulation, it is

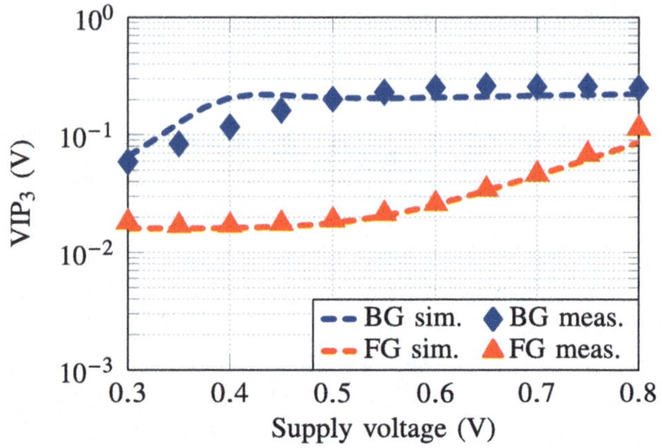

Fig. 1.13 Measured and simulated linearity of the self-biased inverter-based amplifiers across supply voltage

still up to 10× lower than the front-gate amplifier, which is limited by its small diode-connected load device.

Additional test structures are required to fully characterize back-gate amplifier gain mismatch over device size and bias point, but the authors hypothesize that the dominant source of back-gate amplifier CV degradation is most likely γ mismatch, not included in foundry models and hence missing from simulation. This is further supported by the fact that measured mismatch is independent of V_{DD} as expected for γ mismatch following (1.6), and the fact that measured CV is comparable to published FD-SOI wafer mismatch data discussed in Sect. 1.3.2.

Linearity measurements are presented in Fig. 1.13. The back-gate amplifier offers relatively constant high linearity until the supply voltage is low enough to bias the devices in deep subthreshold at a supply of around 0.4–0.5 V. The front-gate amplifier, on the other hand, is more sensitive to supply voltage through moderate and strong inversion. Consistent with the analysis of Sect. 1.3.3, the back-gate amplifier offers 6–20 dB higher linearity than the front-gate amplifier as a result of its negative feedback network. Additionally, strong correlation between measurement and simulation, especially in moderate inversion, validates the assumption of negligible γ nonlinearity from Sect. 1.3.3.

1.5 Feasibility of Body Feedback in Bulk Technology

In bulk technology, the body voltage exerts control over the transistor channel much like the back gate in FD-SOI. In source follower stages, for example, the body effect can be strong enough to demand explicit body biasing in order to restore V_{th}. As a result, one may be motivated to attempt "body feedback" in bulk technology analogous to back-gate feedback in FD-SOI; however, this would not provide the same benefits as back-gate feedback in FD-SOI technology.

First and foremost, the parasitic source- and drain-to-substrate diodes in bulk must remain reverse-biased, limiting the voltage range of this technique. In principle, this could be overcome by either operating in deep subthreshold with restricted voltage swing or by using capacitive coupling to maintain acceptable body biasing, though the latter requires additional circuit complexity to generate and couple the body bias voltage.

Even in that scenario, body feedback would not provide the same linearity and mismatch enhancement as back-gate feedback. This is understood by examining the body voltage sensitivity in bulk

$$\gamma_{\mathrm{bulk}} = -\frac{dV_{th}}{dV_{bs}} = \frac{C_{js}}{C_{\mathrm{OX}}} \tag{1.27}$$

where C_{js} is the depletion layer capacitance and C_{OX} is the front-gate oxide capacitance [29]. Unsurprisingly, (1.29) follows a similar form to (1.6) with the back-gate-to-channel capacitance $C_{\mathrm{BOX}} \| C_{\mathrm{Si}}$ replaced by the body-to-channel capacitance C_{js}. Intuitively, the depletion layer depth and therefore capacitance vary with body–source voltage, indicating some nonlinearity; it is also sensitive to random dopant fluctuations, indicating mismatch. This voltage dependence can be observed by considering the threshold voltage in saturation:

$$V_{th} = V_{th0} + k(\sqrt{2\phi_f - V_{bs}} - \sqrt{2\phi_f}) \tag{1.28}$$

where k is a technology-dependent parameter, V_{th0} is the threshold voltage without body biasing, and ϕ_f is the Fermi level [29].

Combining (1.27) and (1.28), and noting that the body is connected to the output in the body-feedback configuration such that $V_{bs} = V_{\mathrm{out}}$,

$$\gamma_{\mathrm{bulk}} = \frac{k/2}{\sqrt{2\phi_f - V_{\mathrm{out}}}}. \tag{1.29}$$

Unlike FD-SOI γ from (1.6), which is only a function of physical device parameters, γ_{bulk} is a direct function of V_{out}. It is clear from Sect. 1.3 that the closed-loop amplifier is sensitive to feedback factor variation (be it voltage dependence/nonlinearity or mismatch), and therefore, the full benefit of the back-gate feedback technique is limited to FD-SOI technology.

1.6 Conclusion

This chapter described a common-source amplifier configuration in which the FD-SOI back-gate is used a feedback injection point. This achieves negative feedback, with all its benefits, without requiring any extra circuit components or tuning while maintaining a high input impedance. It is enabled by the ultra-thin buried oxide, unique to FD-SOI technology, which allows wide back-gate swing and provides a largely voltage-independent back-gate strength γ. Measurements of a prototype back-gate feedback amplifier in 22-nm FD-SOI were presented, demonstrating that back-gate feedback achieves 5–10$\times$ lower mismatch and 6–20 dB higher linearity compared to a conventional common-source amplifier with a diode-connected load. This amplifier has wide-ranging applications where precise gain and moderate speed and linearity are required.

Acknowledgments The authors of this chapter thank the T.J. Rodgers Stanford Graduate Fellowship for financial support, GlobalFoundries for financial support via the Stanford SystemX FMA program, and GlobalFoundries for providing silicon fabrication through the 22FDX university program.

References

1. Weinreich, S., & Murmann, B. (2022). A single-transistor amplifier with back-gate feedback in 22-nm FD-SOI. *IEEE Solid-State Circuits Letters, 5*, 210–213. https://doi.org/10.1109/LSSC.2022.3198547.
2. Liu, Q., Vinet, M., Gimbert, J., Loubet, N., Wacquez, R., Grenouillet, L., Le Tiec, Y., Khakifirooz, A., Nagumo, T., Cheng, K., Kothari, H., Chanemougame, D., Chafik, F., Guillaumet, S., Kuss, J., Allibert, F., Tsutsui, G., Li, J., Morin, P., …, Sampson, R. (2013). High performance UTBB FDSOI devices featuring 20nm gate length for 14nm node and beyond. In *Technical Digest - International Electron Devices Meeting, IEDM* May 2015. https://doi.org/10.1109/IEDM.2013.6724592
3. Harame, D., McKay, T., & Watts, J. (2017). WSS-1 CMOS FD-SOI technology and benefits for RF. In *2017 IEEE Radio Frequency Integrated Circuits Symposium (RFIC)*. Honolulu, HI.
4. Colinge, J.-P. (2004). *Silicon-on-insulator technology: Materials to VLSI* (3rd ed.) Kluwer Academic Publishers. ISBN: 978-14-61347-95-8.
5. Ong, S. N., Chan, L. H. K., Chew, K. W. J., Lim, C. K., Oo, W. L., Bellaouar, A., Zhang, C., Chow, W. H., Chen, T., Rassel, R., Wong, J. S., Wan, C. W. F., Kim, J., Seet, W. H., & Harame, D. (2019). 22nm FD-SOI technology with back-biasing capability offers excellent performance for enabling efficient, ultra-low power analog and RF/millimeter-wave designs. In *Proceedings of the IEEE Radio Frequency Integrated Circuits Symposium* (pp. 323–326). https://doi.org/10.1109/RFIC.2019.8701768
6. Cathelin, A. (2017). Fully depleted silicon on insulator devices CMOS: The 28-nm node is the perfect technology for analog, RF, mmW, and mixed-signal system-on-chip integration. *IEEE Solid-State Circuits Magazine, 9*(4), 18–26 . https://doi.org/10.1109/MSSC.2017.2745738
7. Raskin, J. P. (2022). Fully depleted SOI technology for millimeter-wave integrated circuits. *IEEE Journal of the Electron Devices Society, 10*, 424–434. https://doi.org/10.1109/JEDS.2022.3165877
8. Kushwaha, P., Pahwa, G., Agarwal, H., Dasgupta, A., Dabhi, C. K., Salahuddin, S., & Hu, C. (2020). *BSIM-IMG103.0.0 Independent Multi-Gate MOSFET Compact Model: Technical Manual*. Technical Report. The University of California, Berkeley. http://bsim.berkeley.edu/models/bsimimg/

9. Asthana, V., Kar, M., Jimenez, J., Noel, J. P., Haendler, S., & Galy, P. (2013). Circuit optimization of 4T, 6T, 8T, 10T SRAM bitcells in 28nm UTBB FD-SOI technology using back-gate bias control. In *European Solid-State Circuits Conference*. https://doi.org/10.1109/ESSCIRC.2013.6649161
10. Zaruba, F., & Benini, L. (2019). The cost of application-class processing: Energy and performance analysis of a linux-ready 1.7-GHz 64-Bit RISC-V Core in 22-nm FDSOI technology. *IEEE Transactions on Very Large Scale Integration (VLSI) Systems, 27*(11), 2629–2640. https://doi.org/10.1109/TVLSI.2019.2926114
11. Bol, D., Schramme, M., Moreau, L., Xu, P., Dekimpe, R., Saeidi, R., Haine, T., Frenkel, C., & Flandre, D. (2021). SleepRunner: A 28-nm FDSOI ULP Cortex-M0 MCU with ULL SRAM and UFBR PVT compensation for 2.6-3.6-uW/DMIPS 40-80-MHz active mode and 131-nW/kB fully retentive deep-sleep mode. *IEEE Journal of Solid-State Circuits, 56*(7), 2256–2269. https://doi.org/10.1109/JSSC.2021.3056219
12. Lallement, G., Abouzeid, F., Daveau, J. M., Roche, P., & Autran, J. L. (2018). A 1.1-pJ/cycle, 20-MHz, 0.42-VTemperature compensated ARM Cortex-M0+ SoC with adaptive self body-biasing in FD-SOI. *IEEE Solid-State Circuits Letters, 1*(7), 174–177. https://doi.org/10.1109/LSSC.2019.2897016
13. Walter, D., Scharfe, A., Oefelein, A., Schraut, F., Bauer, H., Csaszar, F., Niebsch, R., Schreiter, J., Eisenreich, H., & Hoppner, S. (2020). A 0.55V 6.3uW/MHz Arm Cortex-M4 MCU with adaptive reverse body bias and single rail SRAM. In *IEEE Symposium on Low-Power and High-Speed Chips and Systems, COOL CHIPS 2020 - Proceedings*. https://doi.org/10.1109/COOLCHIPS49199.2020.9097639
14. Danilovic, D., Milovanovic, V., Cathelin, A., Vladimirescu, A., & Nikolic, B. (2016). Low-power inductorless RF receiver front-end with IIP2 calibration through body bias control in 28nm UTBB FDSOI. In *Proceedings of the IEEE Radio Frequency Integrated Circuits Symposium (RFIC)*. IEEE (pp. 87–90). https://doi.org/10.1109/RFIC.2016.7508257
15. Mattia, O. E., Sawaby, M., Zheng, K., Arbabian, A., & Murmann, B. (2020). A 10-Gbps continuous-time linear equalizer for mm-wave dielectric waveguide communication. *IEEE Solid-State Circuits Letters, 3*, 266–269. https://doi.org/10.1109/LSSC.2020.3014859
16. Lechevallier, J., Struiksma, R., Sherry, H., Cathelin, A., Klumperink, E., & Nauta, B. (2015). A forward-body-bias tuned 450MHz Gm-C 3rd-order low-pass filter in 28nm UTBB FD-SOI with 1dBVp IIP3 over a 0.7-to-1V supply. In *Digest of Technical Papers - IEEE International Solid-State Circuits Conference* (Vol. 58). https://doi.org/10.1109/ISSCC.2015.7062943
17. Sourikopoulos, I., Frappé, A., Cathelin, A., Clavier, L., & Kaiser, A. (2016). A digital delay line with coarse/fine tuning through gate/body biasing in 28nm FDSOI. In *European Solid-State Circuits Conference* (2016). https://doi.org/10.1109/ESSCIRC.2016.7598263
18. Kumar, A., Debnath, C., Singh, P. N., Bhatia, V., Chaudhary, S., Jain, V., Tual, S. L., & Malik, R. (2016). A 0.065mm2 19.8mW single channel calibration-free 12b 600MS/s ADC in 28nm UTBB FDSOI using FBB. In *European Solid-State Circuits Conference*. https://doi.org/10.1109/ESSCIRC.2016.7598268
19. Rack, M., Nyssens, L., Wane, S., Bajon, D., & Raskin, J. P. (2020). DC-40 GHz SPDTs in 22 nm FD-SOI and back-gate impact study. In *Digest of Papers - IEEE Radio Frequency Integrated Circuits Symposium*. https://doi.org/10.1109/RFIC49505.2020.9218317
20. El-Aassar, O., & Rebeiz, G. M. (2020). Design of low-power sub-2.4 dB mean NF5GLNAs using forward body bias in 22nmFDSOI. *IEEE Transactions on Microwave Theory and Techniques, 68*(10), 4445–4454. https://doi.org/10.1109/TMTT.2020.3012538
21. Carter, R., Mazurier, J., Pirro, L., Sachse, J.-U., Baars, P., Faul, J., Grass, C., Grasshoff, G., Javorka, P., Kammler, T., Preusse, A., Nielsen, S., Heller, T., Schmidt, J., Niebojewski, H., Chou, P.-Y., Smith, E., Erben, E., Metze, C., . . . , Rice, B. (2016). 22nm FDSOI technology for emerging mobile, internet-of-things, and RF applications. In *Proceedings of the IEEE International Electron Devices Meeting (IEDM)*. https://doi.org/10.1109/IEDM.2016.7838029
22. Schwarzenbach, W., Daval, N., Maleville, C., Nguyen, B.-Y., Ecarnot, L., Loubriat, S., Detard, M., Cela, E., Bertrand-Giuliani, C., Chabanne, G., & Maddalon, C. (2019). Advanced FD-SOI and beyond low temperature SmartCut™enables high density 3-D SoC applications.

IEEE Journal of the Electron Devices Society, 7, 863–868. https://doi.org/10.1109/JEDS.2019.2916460
23. Haq, F. U., Englund, M., Antonov, Y., Stadius, K., Kosunen, M., Ryynanen, J., Ostman, K. B., & Koli, K. (2018). Full-duplex wireless transceiver self-interference cancellation through FD-SOI buried-gate signaling. In *Proceedings of the IEEE International Symposium on Circuits and Systems (ISCAS)* (pp. 1–5). https://doi.org/10.1109/ISCAS.2018.8351823
24. Saif, M. A., Dessouky, M., & Aboushady, H. (2021). A 6.4 GHz continuous-time ΣΔ ADC using body-biased feedforward op-amps in 28nm-FDSOI. In *Proceedings of the IEEE International Midwest Symposium on Circuits and Systems (MWSCAS)* (pp. 301–304). https://doi.org/10.1109/MWSCAS47672.2021.9531739
25. Edler, J., Runge, M., & Gerfers, F. (2021). A dynamic body-bias linearization technique enabling wide-band GmC based continous-time sigma-delta converters in 22 nm FD-SOI CMOS. In *Proceedings of the IEEE International Midwest Symposium on Circuits and Systems (MWSCAS)* (pp. 1090–1093). https://doi.org/10.1109/MWSCAS47672.2021.9531807
26. Kuzmicz, W. (2019). A simple ultra-lowpower Opamp in 22 nm FDSOI. In *Proceedings of the 26th International Conference Mixed Design of Integrated Circuits and Systems* (pp. 167–170). https://doi.org/10.23919/MIXDES.2019.8787017
27. Jespers, P. G. A., & Murmann, B. (2017). *Systematic design of analog CMOS circuits: Using pre-computed lookup tables*. Cambridge University Press. https://doi.org/10.1017/9781108125840
28. Weinreich, S. (2022). *Linear periodically time-varying circuits for the Internet of Everything*. PhD Thesis. Stanford University. https://purl.stanford.edu/fv238xg7998
29. Gray, P. R., Hurst, P. J., Lewis, S. H., & Meyer, R. G. (2009). *Analysis and design of analog integrated circuits*. ISBN: 978-04-70245-99-6.

Chapter 2
Transconductance Amplifier Linearization Using Active Back-Gate Input Signal Injection

Julius Edler and Friedel Gerfers

2.1 Introduction

The operational transconductance amplifier (OTA) is a key building block present in a multitude of analog circuit designs. Its core task is to perform a voltage-to-current conversion. The current gain coefficient, representing the ratio between input voltage and output current, is called transconductance, denoted as g_m.

$$g_m = \frac{\Delta I_{\text{out}}}{\Delta V_{\text{in}}} \tag{2.1}$$

Where ΔI_{out} is the change in output current, and ΔV_{in} is the change in input voltage.

2.2 The MOS Transconductor

A Metal-Oxide-Semiconductor (MOS) transconductor is typically realized with a differential pair arrangement, which is a crucial building block in analog circuits. This configuration is widely used to amplify the difference between two voltages while suppressing any components that are common to both inputs. The basic structure of an MOS differential pair consists of two MOS transistors with their sources connected together and typically to a current source. The gates of these transistors receive the differential input signals in the voltage domain, while the drains provide the differential output current. This setup is depicted in Fig. 2.1a.

J. Edler (✉) · F. Gerfers
Technische Universität Berlin, Berlin, Germany
e-mail: edler@tu-berlin.de; friedel.gerfers@tu-berlin.de

F. Gerfers (ed.), *Bulk-Driven Circuit Techniques for CMOS FDSOI Processes*,
https://doi.org/10.1007/978-3-031-85114-8_2

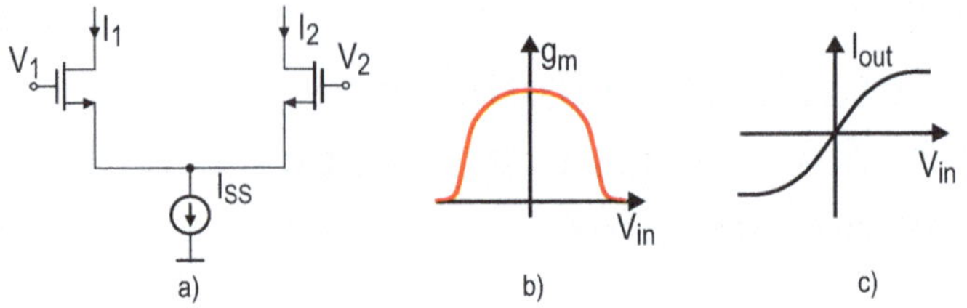

Fig. 2.1 (**a**) Differential pair circuit, (**b**) qualitative transconductance g_m depending on $V_{in} = V_1 - V_2$ and (**c**) the resulting nonlinear output current curve

2.2.1 Transconductance of MOS Differential Pair

The transconductance g_m of a MOS differential pair is the conversion coefficient that translates the differential input voltage V_{in} into a differential output current I_{out}. For a single MOS transistor operating in the saturation region, the transconductance in equillibrium ($V_{in} = 0\,V$) is given by:

$$g_m = \frac{2I_D}{V^*} \tag{2.2}$$

Here, I_D is each transistor's DC drain current and V^* the overdrive voltage ($V_{GS} - V_{TH}$), with V_{GS} being the gate–source voltage and V_{TH} the threshold voltage. In a differential pair, the transconductance determines the gain and noise characteristics [1]. The small-signal differential voltage gain A_v of the pair can be approximated as:

$$A_v = g_m \cdot R_L \tag{2.3}$$

where R_L is the load resistance observed at the output drain node of the differential pair.

2.2.2 Linearity Problem

The circuit is in equilibrium when $V_{in} = 0\,V$, here, both transistors carry the same portion of the tail current I_{SS}, which is provided by the current source. Hence, both transistors provide their maximum amount of small signal output current for a small signal differential input voltage change.

However, once the differential V_{in} is non-zero, the circuit becomes imbalanced, as now one transistor carries more of I_{SS} than the other. This causes g_m to drop until eventually V_{in} exceeds to the range of the overdrive voltage V^*, where one transistor carries most of the current, and the other is essentially switched off. If this is the case, no additional I_{out} is generated for a further increase in V_{in}; hence, the transconductance exceeding this voltage is zero. This suggests that g_m is a function of V_{in} as illustrated by Fig. 2.1b. The resulting $G_m(V_{in})$ function is nonlinear due to the square-law behavior of the MOSFETs and therefore precipitates distortion of the output current as depicted in Fig. 2.1c) and described in detail in [1]. The induced nonlinearity limits the circuit's overall performance in terms of accuracy, and therefore, techniques to mitigate this effect have been introduced.

2.2.3 Linearization Approaches

A multitude of techniques have been developed to deal with the aforementioned linearity problem of MOS transconductors (Fig. 2.2). Firstly, one can simply increase the overdrive V^* (reduce width-to-length ratio W/L) and thereby attain a larger operating range. This though comes at the cost of a reduced transconductance with $g_m = 2I_D/V^*$ and is only practical in a small voltage range, as V^* cannot be increased infinitely. Larger operating ranges are also possible by the addition of a degeneration resistor at the source nodes of the differential pair. This again comes at the cost of transconductance efficiency (lower overall g_m for the same I_D); additionally, the resistors require additional voltage headroom. By using transistors in triode region as degeneration resistors, it is possible to modulate the degeneration factors to achieve a linear transconductor effectively [2]. Other approaches make use of the idea to cancel the caused intermodulation products by the employment of scaled replica transconductors [3, 4]. This improvement in linearity though is limited in full-scale range, since for the techniques to work, the differential pair needs to remain in saturation. Similarly, the tail current source can be replaced by an ohmic resistor, here, by carefully scaling the resistance, a compensation of the nonlinear transconductance can be achieved with an inversly nonlinear current source impedance. Here of course, the process dependency of the resistor needs to be taken into account; therefore, authors of [5] and [6] suggest to use a constant g_m biasing currents. Alternatively, one can try to reduce the amount of gate–source voltage V_{GS} variation seen by the differential pair. In [7], this is achieved by the adoption of a merged input pair of a closed control loop present in a $\Sigma\Delta$-modulator.

An increased overall operating region is possible by adding a second differential pair, which is operated at a DC offset [8, 9]. Thereby both pairs share the same nonlinear transfer curve, and hence, their addition results in a g_m curve, which is linearized in a specific region. Here a drawback is the increased power dissipation since both pairs need to be supplied by the full DC current and contribute noise.

Recently, techniques involving the use of the bulk node have been experimented with. In [10], instead of the gate nodes, the input signal is fed to the transistor bulks, thereby allowing low-voltage operation. Linearization concepts have also been shown, as in [11], where triode region MOS resistor is linearized to increase the operation range of a differential pair. This has been achieved with a combination of source degeneration as in [12, 13] and also [14] where an auxiliary path is employed to generate the bulk voltages needed for a linear transconductor. In the past, these

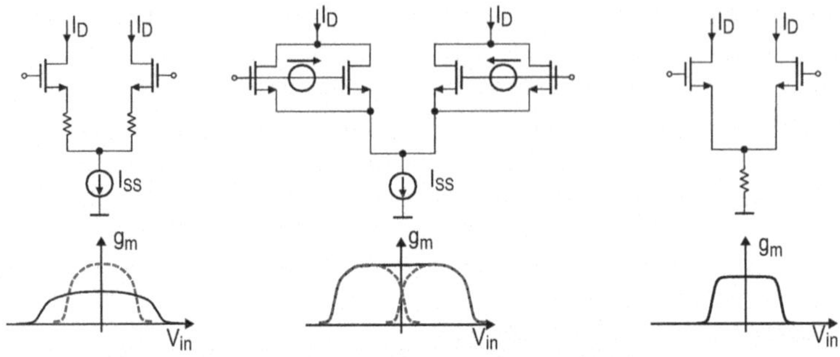

Fig. 2.2 Different lineariazation approaches such as source degeneration, replica offset pairs [8], and tail resistance matching [6]

techniques were limited to weekly nonlinear transconductors, hence relied heavily on source degeneration and thereby came with a significant power penalty. This has been the case of bulk-CMOS technologies, since here, due to the bulk-source diode eventually becoming conductive, the feasible voltage level on this node is limited. In fully depleted silicon-on-insulator (FD-SOI) CMOS technology, this is no longer the case. Here, the source and drain nodes are completely isolated from the bulk node with the depletion layer. Therefore, the term back gate has evolved, since similar as the conventional front gate, larger-scale voltages can be applied to it. This has been shown by Edler et al. [15] and [16], where with a combination of a voltage tracking technique and active back-gate injection of the input signal, linearity can be improved significantly without the employment of any source degeneration. This approach is presented in the following sections.

2.3 Active Back-Gate Input Signal Injection

The bulk is isolated from source and drain in FD-SOI CMOS, thereby the parasitic diodes formed between them are eliminated. This provides an additional degree of freedom in the design process, as now this node can be utilized to act on the MOSFET channel such as a second gate contact, hence the term back gate. Recently, this has been leveraged in manifold ways, namely to provide an second DC bias voltage to compensate for mismatch or to tune the transistors' threshold voltages [17].

In this chapter, a linearization technique for MOS transconductors is proposed. Here, we actively inject the input signal into the back gate by the use of an auxialliary amplifier. By carefully matching the front- and back-gate transfer curves, a linear overall transfer curve is achieved. In the following, this working principle is explained using a block-diagram approach with generic polynomials as in [15]; this is followed by an analysis using large signal equations to determine the effect the technique has on linearity in quantitative terms. Finally, the effects of a gain mismatch and the implications on noise performance are evaluated.

2.3.1 *Working Principle*

The basic circuit structure of this back-gate linearized transconductance amplifier is depicted in Fig. 2.3. As conventional, a differential pair is fed with the differential input voltage $V_{in} = V_1 - V_2$ using the transistors (front-) gate nodes. Additionally, in this implementation, the back gates are injected with a dynamic signal. This voltage is scaled using the auxialliary amplifier with gain A_B. By applying negative feedback, the resulting differential output current ΔI_D depends on the overall G_m curve, which is attained by the subtraction of the front- and back-gate transconductances (g_{m_F} and g_{m_B} respectively).

A linear G_m can be formulated to exhibit no dependence on V_{in}; hence, the derivative $\partial G_m / \partial V_{in}$ equals zero. This can be attained here if both contributions (g_{m_F} and g_{m_B}) derivatives with respect to V_{in} are equal. Therefore, both transfer curves need to be designed to match in that respect. Key parameters here being the DC gain of the auxiliary amplifier and the strength of the nonlinearity in both the front- and back-gate paths.

The block diagram in Fig. 2.3 reveals the two contributors to G_m, which can be seperated into a combination of a front- and a back-gate transfer function H_{FG} and H_{BG} containing the respective transconductance curves g_{m_F} and g_{m_B}. Here, to elaborate on the working principle, we model them using a quadratic polynomial.

$$g_{m_F} = g_{m0}(1 - \alpha V_{in}^2) \tag{2.4}$$

$$g_{m_B} = \gamma \; g_{m0}(1 - \alpha V_{in}^2) \tag{2.5}$$

$$A_B = A_{B_0}(1 - \beta V_{in}^2) \tag{2.6}$$

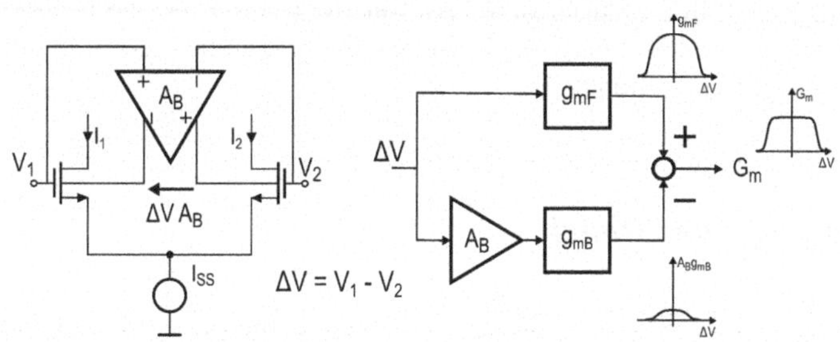

Fig. 2.3 Schematic prototype of the active back-gate linearization scheme and the resulting block diagram

g_{m_0} is the transconductance at zero volt input voltage V_{in} and α models the g_{m} nonlinearity, β the A_{B} amplifiers (nonlinear) transfer curve. The coefficient γ models the threshold voltage shift depending on V_{BB} and hence determines the ratio of the effective transconductance of front- and back gate with

$$\gamma = \frac{g_{\text{mB}}}{g_{\text{mF}}} \tag{2.7}$$

The total G_{m} is the subtraction of both contributors:

$$G_{\text{m}} = g_{\text{mF}} - A_{\text{B}} g_{\text{mB}} = g_{\text{m}_0}(1 - A_{\text{B}}\gamma)(1 - \alpha V_{\text{in}}^2) \tag{2.8}$$

The strength of G_{m}s nonlinearity is defined by the gradient of Eq. (2.8): $\frac{\partial G_{\text{m}}}{\partial V_{\text{in}}}$. First, we analyze the case of an linear amplifier with a constant gain function $A_{\text{B}} = A_{\text{B}_0}$ regardless of input voltage by setting $\beta = 0$. This yields:

$$\left.\frac{\partial G_{\text{m}}}{\partial V_{\text{in}}}\right|_{\beta=0} = 2\alpha g_{\text{m0}} V_{\text{in}}(A_{\text{B}_0}\gamma - 1) \tag{2.9}$$

To achieve linearity, this gradient needs to drop to zero. This condition is met if $A_{\text{B}_0} = 1/\gamma$. Unfortunately, this also eliminates any transconductance since $G_{\text{m}}|_{V_{\text{in}}=0} = g_{\text{m0}}(1-A_{\text{B}_0}\gamma)$ will be zero for this case. Hence, employing an idealized linear A_{B} function cannot yield a practical solution as a certain amount of non-linearity is needed to scale the back gates g_{m} curve gradient. Now, exemplary, we consider both coefficients to be equal $\beta = \alpha$, thereby we get:

$$\frac{\partial G_{\text{m}}}{\partial V_{\text{in}}}|_{\beta=\alpha} = 2\alpha g_{\text{m0}} V_{\text{in}}(2A_{\text{B}_0}\gamma - 1) - g_{\text{m0}}4\gamma A_0\alpha^2 V_{\text{in}}^3 \tag{2.10}$$

We can assume overall a weak nonlinearity, $\alpha \ll 1$; hence, we neglect the cubic term. We note that the gradient drops to zero for $A_{\text{B}_0} = 1/\gamma/2$, which yields the overall $G_{\text{m}}|_{V_{\text{in}}=0}$ to be reduced only by half its original value. Hence, the transconductor is still operating and the nonlinearity is canceled. We can conclude that some amount of nonlinearity in the A_{B} transfer curve is necessary for the circuit to operate. In fact, the less linear A_{B} becomes, the less gain is required, and less G_{m} is lost due to negative feedback.

2.3.2 *Square Law Model*

The preceding section described the basic working principle—the matching of front and back-gate gradients using quadratic polynomials. Now we analyze the circuit topology using square law transistor equations. Although they are valid only in the saturation region and their accuracy is limited in deep submicron technologies, they provide helpful insights by providing quantitative results for the efficacy of the technique in real circuit implementations.

Following the square law equation, the body bias voltages $V_{\mathrm{TH_B}}$ impact on the threshold voltage V_{TH} is modeled in Eq. (2.11) as an additive term $V_{\mathrm{TH_B}}$ to the threshold voltage $V_{\mathrm{TH_0}}$.

$$(V_{\mathrm{GS}} - V_{\mathrm{TH_0}} - V_{\mathrm{TH_B}})^2 = \frac{2I_{\mathrm{D}}}{\mu_{\mathrm{n}} C_{\mathrm{OX}} \frac{W}{L}} \tag{2.11}$$

Now, we replace $K = \mu_{\mathrm{n}} C_{\mathrm{OX}} \frac{W}{L}$ and regroup for the gate source voltage V_{GS}.

$$V_{\mathrm{GS}} = \sqrt{\frac{2I_{\mathrm{D}}}{K}} + V_{\mathrm{TH_0}} + V_{\mathrm{TH_B}} \tag{2.12}$$

With the differential pair depicted in Fig. 2.3, we formulate an expression for the differential input voltage V_{in}.

$$V_{\mathrm{in}} = V_1 - V_2 = (V_{\mathrm{GS1}} - V_{\mathrm{TH_B1}}) - (V_{\mathrm{GS2}} - V_{\mathrm{TH_B2}}) \tag{2.13}$$

Equation (2.12) suggests that a change in the $V_{\mathrm{TH_B}}$ has the same effect as a variation of V_{GS}. The threshold voltage change modeled by $V_{\mathrm{TH_{B1,2}}}$ is linked to the applied back-gate voltage V_{BB} with the back-gate sensitivity coefficient $\gamma \approx 1/12\mathrm{V/V} \approx 83\ \mathrm{mV/V}$ as represented by Eq. (2.14).

$$V_{\mathrm{TH_B1}} - V_{\mathrm{TH_B2}} = \gamma \cdot V_{\mathrm{BB}} \tag{2.14}$$

The differential back-gate voltage V_{BB} is generated by the auxiliary amplifier nonlinear gain curve with $A_{\mathrm{B}} = f(V_{\mathrm{in}})$.

$$V_{\mathrm{BB}} = V_{\mathrm{in}} \cdot A_{\mathrm{B}}(V_{\mathrm{in}}) \tag{2.15}$$

Inserting the equation for the single transistor's V_{GS} Eq. (2.12) into the respective terms for the differential pair Eq. (2.13) yields Eq. (2.16), where the static portion of the threshold voltage $V_{\mathrm{TH_0}}$ is canceled. Hence, only a differential $V_{\mathrm{BB}} = \gamma^{-1}\ (V_{\mathrm{TH_B1}} - V_{\mathrm{TH_B2}})$ needs consideration here.

$$V_{\mathrm{in}} - \gamma\ V_{\mathrm{BB}} = V_{\mathrm{in}}(1 - \gamma\ A_{\mathrm{B}}(V_{\mathrm{in}})) = \sqrt{\frac{2I_{\mathrm{D1}}}{K}} - \sqrt{\frac{2I_{\mathrm{D2}}}{K}} \tag{2.16}$$

Equation (2.16) shows that the addition back-gate voltage-dependent offset yields a voltage degeneration or boosting in accordance to the sign of the feedback. Here, negative feedback is applied, this corresponds to the block diagram in Fig. 2.4. The back-gate amplifier is realized as a $g_{\mathrm{m}}R$ topology by the use of a differential

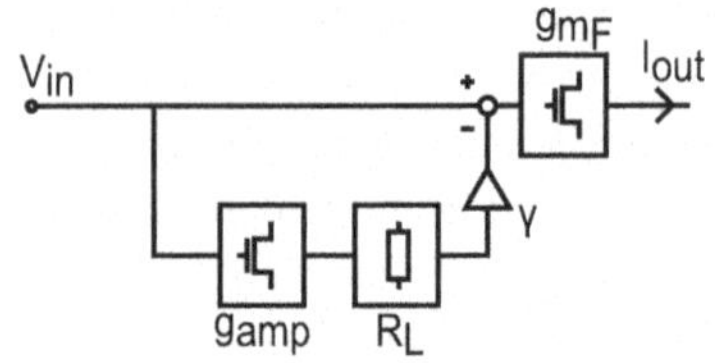

Fig. 2.4 Large signal block diagram of the linearization concept

pair transconductor with small signal transconductance g_{amp} with a load resistance R_L to set the gain.

In order to attain the large signal output current $\Delta I_D = I_{D1} - I_{D1}$, we rearrange Eq. (2.16). This mathematical derivation is discussed in numerous textbooks, as e.g., in [1]. Therefore, here, we compare the coefficients and replace V_{in} with the back-gate-dependent expression $V_{in}(1 - \gamma A_B)$ in the equation for ΔI_D.

$$\Delta I_D = g_{m_0} V_{in}(1 - \gamma\ A_B(V_{in}))\sqrt{1 - \frac{V_{in}^2(1 - \gamma\ A_B(V_{in}))^2}{(2V_{in}^*)^2}} \tag{2.17}$$

In Eq. (2.17), for intuitive analysis, we also substituted the total tail current $I_{SS} = I_{D1} + I_{D2}$ and subsequently the small signal transconductance $g_{m_0} = \sqrt{K I_{SS}}$. The overdrive voltage of the transistors is represented with $V_{in}^* = 2I_D/K$. The linearity of this output current can be analyzed using a Taylor expansion, which yields a higher-order polynomial, which represent the distortion they precipitate. The expansion is performed at $V_{in} = 0$, we thereby attain an approximated function ΔI_{D_t}. We truncate the infinite series to the third order.

$$\begin{aligned}\Delta I_{D_t} =& V_{in} g_{m0}(1 - \gamma A_B(0)) - V_{in}^2 g_{m0} \gamma A_B'(0)\\ &- \frac{V_{in}^3 g_{m0}}{2}\left(\gamma A_B''(0)) - \frac{(\gamma A_B(0) - 1)^3}{4V_{in}^{*2}}\right)\end{aligned} \tag{2.18}$$

In Eq. (2.18), where A_B' and A_B'' are the first and second derivative of A_B respectively, we see, the resulting characteristic is dependent on the main input pair characteristics with g_{m0} and V_{in}^*, but also the back-gate amplifiers gain $A_B(0)$ and its higher-order derivatives. Therefore, we need to formulate the $A_B(V_{in})$ function to attain the required terms. We assume as aforementioned a differential pair with an overdrive voltage V_{amp}^* and a R_L load resistor. Here, we ignore other effects such as channel length modulation:

$$A_B(V_{in}) = g_{amp} \cdot R_L \sqrt{1 - \frac{V_{in}^2}{4V_{amp}^{*2}}} \rightarrow A_B(0) = g_{amp} \cdot R_L \tag{2.19}$$

$A_B(V_{in})$s first derivative needed for Eq. (2.18) is:

$$A'_B(\Delta V_{in}) = -\Delta V_{in} \cdot g_{amp} \cdot R_L \frac{1}{4V_{amp}^{*2}\sqrt{1-\frac{V_{in}^2}{4V_{amp}^{*2}}}} \rightarrow A'_B(0) = 0 \tag{2.20}$$

The second derivative is:

$$A''_B(\Delta V_{in}) = -\frac{g_{amp} \cdot R_L}{4V_{amp}^{*2}\sqrt{1-\frac{\Delta V_{in}^2}{4V_{amp}^{*2}}}} - \frac{\Delta V_{in}^2 g_{amp} \cdot R_L}{4V_{amp}^{*2}\left(1-\frac{\Delta V_{in}^2}{16V_{amp}^{*4}}\right)^{3/2}} \tag{2.21}$$
$$\rightarrow A''_B(0) = -\frac{g_{amp} \cdot R_L}{4V_{amp}^{*2}}$$

Now, inserting the values for $A_B(0)$ and their derivatives from Eqs. (2.19), (2.20) and (2.21) into Eq. (2.18), we find that

$$\begin{aligned} \Delta I_{D_t} &= V_{in} g_{m0}(1-\gamma g_{amp} \cdot R_L) \\ &\quad - \frac{V_{in}^3 g_{m0}}{2}\left(\gamma \frac{g_{amp} \cdot R_L}{4V_{amp}^{*2}} - \frac{(\gamma g_{amp} \cdot R_L - 1)^3}{4V_{in}^{*2}}\right) \end{aligned} \tag{2.22}$$

By substituting the gain for zero input with $A_{B_0} = g_{amp} R_L$ and forming ΔI_{D_t}s derivative, we attain the approximated transconductance $G_{m_t} = \partial \Delta I_{D_t} / \partial V_{in}$.

$$\begin{aligned} G_{m_t} &= g_{m0}(1-\gamma A_{B_0}) \\ &\quad - 3\frac{V_{in}^2 g_{m0}}{2}\left(\gamma \frac{A_{B_0}}{4V_{amp}^{*2}} - \frac{(\gamma A_{B_0} - 1)^3}{4V_{in}^{*2}}\right) \end{aligned} \tag{2.23}$$

By expanding the brackets in Eq. (2.23), we can separate the expression into terms depended on either the front- or back-gate path.

$$\begin{aligned} G_{m_t} = g_{m0}&\left(1 - \frac{3V_{in}^2}{8V_{in}^{*2}}\right) && \Big\} H_{FG} \\ &-g_{m0}\gamma A_{B_0}\left(1 - \frac{3V_{in}^2}{8V_{in}^{*2}}\left(1 + \frac{V_{V_{in}}^{*2}}{V_{amp}^{*2}}\right)\right) && \\ &-g_{m0}\gamma^2 A_{B_0}^2 \frac{3V_{in}^2}{8V_{in}^{*2}} + g_{m0}\gamma^3 A_{B_0}^3 \frac{9V_{in}^2}{8V_{in}^{*2}} && \Big\} H_{BG} \end{aligned} \tag{2.24}$$

As initially formulated in Sect. 2.3.1 , G_{m_t} thereby can be modeled by a separate front-gate function H_{FG} subtracted by the back-gate curve H_{BG} with its dependency of the back-gate sensitivity γ.

$$G_{m_t} = H_{FG} - H_{BG} \tag{2.25}$$

The goal of the circuit topology is to attain a linear transconductance G_m. This is achieved if a change of V_{in} does not yield a deviation of G_m from its small signal value $G_m|_{V_{in}=0V}$. Hence, if the quadratic coefficients of V_{in}^2 in Eq. (2.23) fall to zero, G_{m_t} is linear. This is achieved if the gradients of the front- and back-gate functions match. We therefore form another derivative. This yields:

$$\begin{aligned}\frac{\partial G_{m_t}}{\partial V_{in}} = &-\frac{6V_{in}}{8V_{in}^{*2}}\\ &-g_{m0}\gamma A_{B_0}\left(\frac{6V_{in}}{8V_{in}^{*2}}\left(1+\frac{V_{in}^{*2}}{V_{amp}^{*2}}\right)\right)\\ &-g_{m0}\gamma^2 A_{B_0}^2\left(\frac{6V_{in}}{8V_{in}^{*2}}\right)+g_{m0}\gamma^3 A_{B_0}^3\frac{9V_{in}}{4V_{in}^{*2}}\end{aligned} \tag{2.26}$$

Hence, we set $\frac{\partial G_{m_t}}{\partial \Delta V_{in}} \stackrel{!}{=} 0$ as the linearity condition, this allows to remove common coefficients and yields:

$$0 = 1 - \gamma^3 A_{B_0}^3 + 3\gamma^2 A_{B_0}^2 - 3\gamma A_{B_0} - \gamma A_{B_0}\frac{V_{in}^{*2}}{V_{amp}^{*2}} \tag{2.27}$$

As laid out previously, Eq. (2.27) implies that the fulfillment of the linearity condition relies on the gain A_{B_0} scaled with γ. The different strength of nonlinearity in front- and back-gate path is represented by the ratio of overdrive voltages with $\frac{V_{in}^*}{V_{amp}^*}$. We can now solve the equation to attain solutions for the required gain for certain conditions or the ratio of overdrive voltages for a desired gain. As an example, for $\frac{V_{amp}^*}{V_{in}^*} = 1$, hence, the back-gate amp being a replica design of the main transconductor and setting $\gamma = 1/12 \approx 83\text{mV/V}$, the equation yields the real solution

$$A_{B_{lin}}\big|_{V_{in}^*=V_{amp}^*} = 3.81 \tag{2.28}$$

In Fig. 2.5, the solutions for the optimal gain values depend on the overdrive voltage ratio. As is intuitive, a larger gain is required to compensate for a smaller V_{in}^* since its nonlinear gradient will be stronger than the back-gate paths.

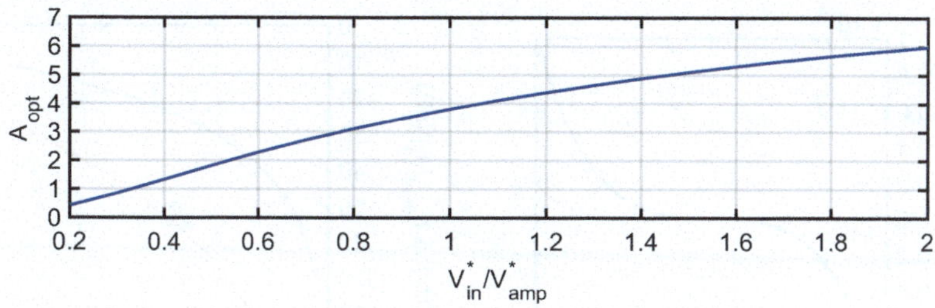

Fig. 2.5 Solution for the gain A_{B_0} to satisfy the linearity condition $\frac{\partial G_{m_t}}{\partial V_{in}} \stackrel{!}{=} 0$

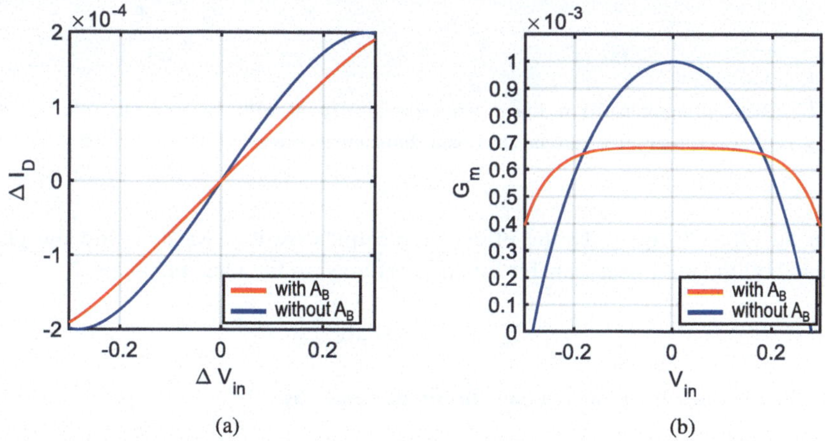

Fig. 2.6 Output current and transconductance with $A_B = 3.81$ and $\frac{V^*_{amp}}{V^*_{in}} = 1$, g_{m0}=1 mS. (**a**) Output current $\Delta I_D(V_{in})$. (**b**) Transconductance $G_m(V_{in})$

2.3.3 *Linearization Efficacy*

In order to asses the attainable linearity with the back-gate input signal injection concept, we analyze the deviation of the resulting $\Delta I_D(V_{in})$ curve from the ideal case. Here, we assume $\frac{V^*_{amp}}{V^*_{in}} = 1$ and $\gamma = 1/12$. On Fig. 2.6a we see the two resulting I/V conversion function for the case with ($A_{B_0} = 3.81$) and without ($A_{B_0} = 0$) back-gate compensation using Eq. (2.17). Its derivative yields the large signal $G_m(V_{in})$ curve on Fig. 2.6b. The successful linearization is apparent. Additionally the overall transconductance is reduced due to the negative feedback via the back-gate path. This reduction can also be derived from Eq. (2.23), here G_m is scaled with the coefficient factor $(1 - \gamma A_{B_0})$. Hence, a higher gain factor in the back-gate path degrades the overall transconductance efficiency while providing the benefit of increased linear range.

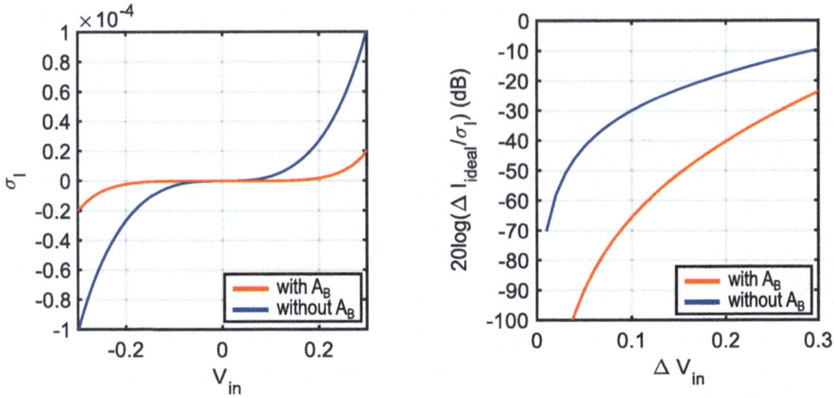

(a) (b)

Fig. 2.7 Residual nonlinearity in linear and logarithmic units with $A_{\mathrm{B}} = 3.81$, $\frac{V^*_{\mathrm{amp}}}{V^*_{\mathrm{in}}} = 1$ and $g_{\mathrm{m0}} = 1$ mS. (**a**) Linear error current. (**b**) Logarithmic error current

In an effort to quantify the effect of the linearization, we compute the error induced by the nonlinear transfer function refereed to the ideal behavior.

$$\Delta I_{\mathrm{ideal}} = V_{\mathrm{in}}\, G_{\mathrm{m}}|_{\mathrm{V_{in}}=0} \tag{2.29}$$

Now the residual error current can simply be found by

$$\sigma_{\mathrm{I}} = \Delta I_{\mathrm{ideal}} - \Delta I_{\mathrm{D}} \tag{2.30}$$

In Fig. 2.7a, the error current is plotted, as can be seen, for a $V_{\mathrm{in}} = V^*_{\mathrm{in}}/2 = 0.1$ V, we attain a reduction of the relative error from -30 dB down to -65 dB. The simplifications performed to yield this result should be noted here, the used large-signal model ignores modeling of channel-length modulation, and the square-law model itself is limited to the saturation region and hence becomes inaccurate when approaching $V_{\mathrm{in}} = V^*_{\mathrm{in}}$. However, in [16], an IM3 reduction of >27 dB has been presented using this technique. Here, total harmonic distortion (THD) can be examined on using the power spectral density (PSD) in Fig. 2.8a. The resulting THD depending on the input amplitude can be seen in Fig. 2.8b.

2.3.4 Gain Variation

The aforementioned analysis concluded that the back-gate amplifiers gain $A_{\mathrm{B_0}}$ needs to be set to a specific value A_{Lin}. This has been derived from the fundamental large signal equations and finally using Taylor expansion to arrive at the optimum

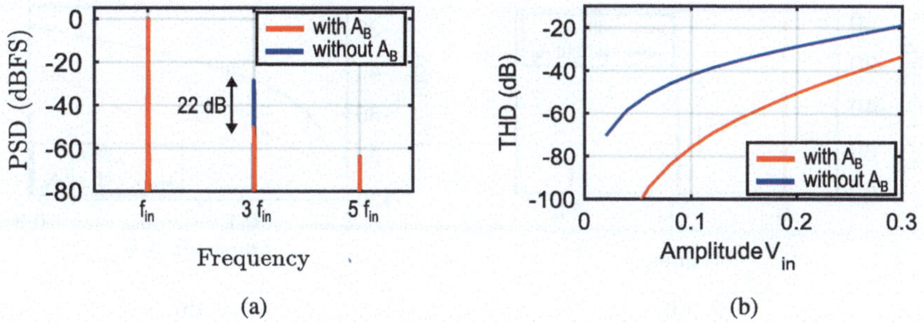

Fig. 2.8 Linearity improvement in the spectral domain, $A_{B_0} = 3.812$, $V_{in}^*/V_{amp}^* = 1$ and $g_{m0} = 1$ mS). (**a**) PSD for input sine wave amplitude 0.2 V. (**b**) THD depending on amplitude

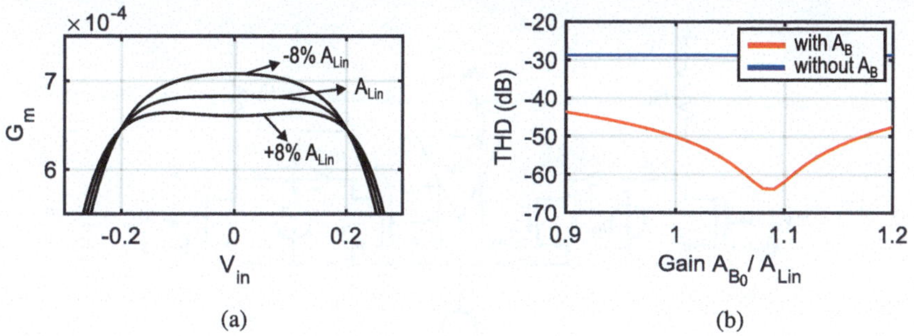

Fig. 2.9 Effect of over- and under compensation of nonlinear distortion with different A_{B_0}. (**a**) $G_m(V_{in})$ for different A_{B_0}. (**b**) THD depending on gain with $V_{in} = 0.2$ V

gain at $V_{in} = 0$. For very low input signals, the linearization is virtually perfect as can be seen on Fig. 2.9b. For larger signals, the linearized function still performs an order of magnitude better; nevertheless, linearity depreciates. This is due to the fact that at some point, the transistors come close to leaving the saturation region. Additionally, the performed Taylor expansion used to arrive at A_{Lin} itself is only accurate at the zero volt input point. Hence, for different input voltage ranges, a different A_{B_0} will be optimal. Figure 2.9a shows the G_m curves with different values for A_B and on Fig. 2.9b the resulting THD for a amplitude equalling $V_{in}^* = 0.2$ V is shown. As can be seen, an increase of the gain from the analytically derived A_{Lin} of 8% yields the best linearity improvement. For this value, the spectral result is shown in Fig. 2.10a, where HD3 now is largely suppressed, so that now HD5 is dominating the THD, which is shown in Fig. 2.10a. Here, THD is slightly worse in the low-voltage regime; however, for larger input amplitudes, the 8% increased gain keeps THD below −60 dB until 0.2 V compared to 0.15 V volt for A_{Lin} in Fig. 2.8b.

Nevertheless, depending on the linearity requirement of the targeted application, an absolute accuracy of the gain A_{B_0} needs to be attained. Variations due to process, voltage, and temperature (PVT) change can significantly affect the gain

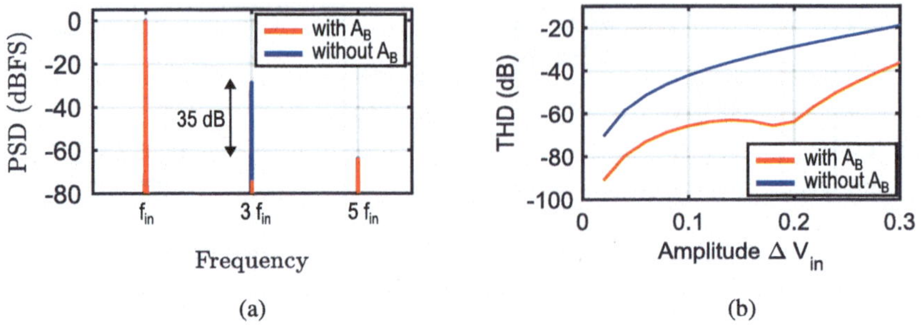

Fig. 2.10 Linearity improvement in the spectral domain with 8% increased $A_{B_0} = 4.12$, $V_{in}^*/V_{amp}^* = 1$ and $g_{m0} = 1$ mS). (**a**) PSD for input sine wave amplitude 0.2 V. (**b**) THD depending on amplitude

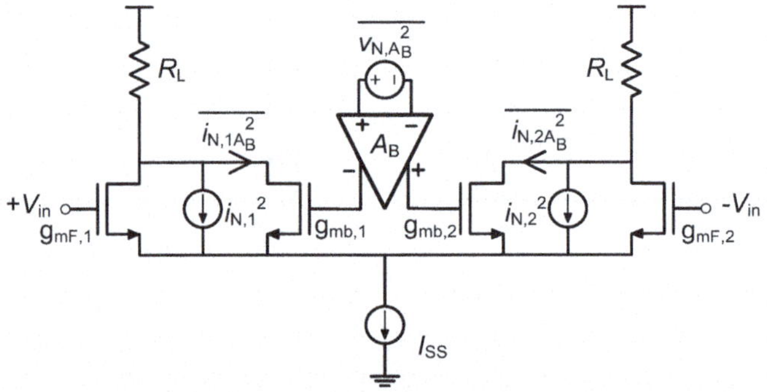

Fig. 2.11 Linearized differential pair circuit including noise sources

curve; therefore, circuit topologies resilient to deviations need to be implemented. Therefore, in Sect. 2.3.6, amplifier architectures mitigating the effect of PVT variations are discussed.

2.3.5 Thermal Noise

Commonly, the transistors' thermal noise contribution is modeled with a current source in parallel to the drain–source nodes [1]. In Fig. 2.11, this is modeled by the currents $\overline{i_{N,1}^2}$ and $\overline{i_{N,1}^2}$ for the respective transistor side. Here, we split up the input pair transistors to model the front- and back-gate transconductors with g_{mF} and g_{m_b} respectively. For analysis, the amplifier's input is disconnected from the main pair, and its noise contribution is modeled by $\overline{v_{N_{A_B}}^2}$. The load resistors R_L are added to close the circuit, and their noise contribution is ignored in the following analysis.

The noise currents are given by

$$\overline{i^2_{\mathrm{N}\,1,2}} = 4kT\gamma_n g_{\mathrm{mF}1,2}\Delta f \tag{2.31}$$

Here, k is the Boltzman constant, T temperature and γ_n the noise factor—here assumed 2/3, Δf is the noise bandwidth. The back-gate amplifier's contribution is modeled with currents $\overline{i^2_{\mathrm{N}_{\mathrm{AB}1,2}}}$ respectively. The noise added due to the actively modulated back-gate is modeled with the noise voltage $\overline{v^2_{\mathrm{N}_{\mathrm{AB}}}}$, which precipitates the noise current $\overline{i^2_{\mathrm{N}_{\mathrm{AB}1,2}}}$ scaled with the gain A_{B} and the back-gate transconductance $g_{\mathrm{mB}1,2}$.

$$\overline{i^2_{\mathrm{N}_{\mathrm{AB}1,2}}} = \overline{v^2_{\mathrm{N}_{\mathrm{AB}}}}\left(A_{\mathrm{B}}\, g_{\mathrm{mB}1,2}\right)^2 \tag{2.32}$$

Since the noise currents are uncorrelated, their power is added to yield a total output noise current $\overline{i^2_{\mathrm{N}_{\mathrm{tot}}}}$. Since $g_{\mathrm{mF}1} = g_{\mathrm{mF}2} = g_{\mathrm{mF}}$ and $g_{\mathrm{mB}1} = g_{\mathrm{mB}2} = g_{\mathrm{mB}}$, and the front- and back-gate transconductances are linked by the sensitivity $\gamma = g_{\mathrm{mB}}/g_{\mathrm{mF}}$, we arrive at

$$\overline{i^2_{\mathrm{N}_{\mathrm{tot}}}} = \overline{i^2_{\mathrm{N}1}} + \overline{i^2_{\mathrm{N}2}} + \overline{i^2_{\mathrm{N}_{\mathrm{AB}1}}} + \overline{i^2_{\mathrm{N}_{\mathrm{AB}2}}} \tag{2.33}$$

$$\overline{i^2_{\mathrm{N}_{\mathrm{tot}}}} = 8kT\gamma_n g_{\mathrm{mF}}\Delta f + \overline{v^2_{\mathrm{N}_{\mathrm{AB}}}}\left(A_{\mathrm{B}}\gamma g_{\mathrm{mF}}\right)^2 \tag{2.34}$$

We here do not consider channel length modulation and the velocity saturation to arrive at this equation, which has g_{mF} as the sole design parameter. A more complete analysis on induced back-gate noise is performed in [18]. Here we assume transistors' noise to be mostly dependent on g_{mF}. To calculate SNR, we compute the overall input referred noise $\overline{v^2_{\mathrm{N_in}}}$. This we attain by dividing both parts of $\overline{i^2_N}$ with their respective transfer function translating $|V_{\mathrm{in}}/V_{\mathrm{out}}|^2$. Thereby we arrive at

$$\overline{v^2_{\mathrm{N_{in}}}} = \frac{4kT\gamma_n\Delta f}{g_{\mathrm{mF}}(1-\gamma\cdot A_{\mathrm{B}})^2} + \overline{v^2_{\mathrm{N}_{\mathrm{AB}}}} \tag{2.35}$$

While the individual transistor's noise currents $\overline{i^2_{N1,2}}$ are not altered by the topology, the effective input referred noise is changed. This is due to the overall reduced effective transconductance. Here, SNR degrades with $(1 - \gamma \cdot A_{\mathrm{B}})$. However, the technique enables the use of larger input signals, which in turn improves SNR. Considering as an example in Fig. 2.10b and a 60 dB HD3 target, V_{in} is limited to 40 mV without, and 160 mV with the linearization, a fourfold increase in linear range. However, since for this case, the gain is set to $A_{\mathrm{B}}(0) \approx 4$, and therefore, the input referred noise only increases by $1/(1 - \gamma \cdot A_{\mathrm{B}}(0)) \approx 1.5$. Hence, in this case, an SNR improvement can be attained by increasing the input signal due to the reduced linearity penalty.

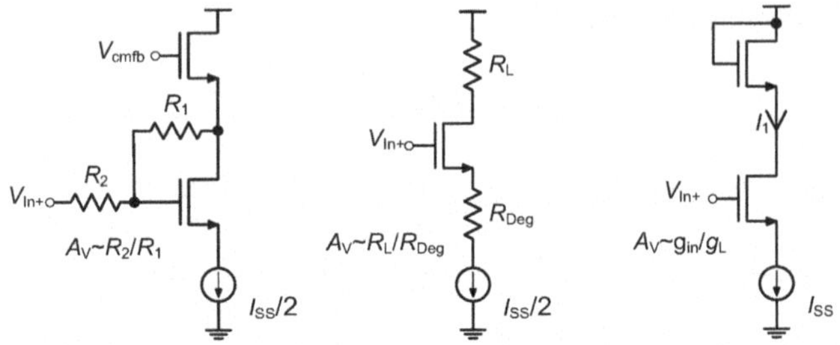

Fig. 2.12 Half circuits of PVT resilient A_{B} amplifier topologies

2.3.6 *Robust Implementations*

For a robust implementation of the concept, the gain A_{B_0} should remain insensitive to PVT variations. The previously described amplifier uses load resistors to set the gain. In integrated CMOS technologies, their resistance can vary up to ±30%. To avoid to let the A_{B_0} deviate by that much, a range of alternative topologies can be implemented. Three different approaches are shown in Fig. 2.12, the first alternative is to set the gain using resistor ratios as in operational amplifiers with a resistive feedback ratio. Here, not their absolute, but their relative value sets the gain of the amplifier, making A_{B_0} much more robust to PVT changes. This though, due to the virtual ground at the input pair, effectively removes the back-gate amplifer's dependence on V_{in} and hence, as discussed in Sect. 2.3.1, would require a very large gain A_{B_0}, thereby reducing the efficiency of the concept. Additionally, the resistive input would remove one of the main benefits of the open-loop OTA configuration, providing a fully capacitive, high-ohmic input load. Alternatively, we can set the gain using a resistive load in combination with resistive source degeneration. Thereby the gain is set by:

$$A_{\mathrm{vb}} = \frac{R_{\mathrm{L}}}{1/g_{\mathrm{m1}} + R_{\mathrm{deg}}} \tag{2.36}$$

If $R_{\mathrm{deg}} \gg 1/g_{\mathrm{m1}}$, the gain is now essentially set as a ratio of resistors; therefore, PVT robustness is improved significantly. This also linearizes the amplifier's transfer curve, and hence, the required gain A_{opt} to achieve compensation will increase. Nevertheless, depending on the required voltage range, linearity, and power requirements, this solution is very viable. Alternatively, we can omit resistors by using a diode-connected load transistor to set the gain. Here, the load is now $1/g_{\mathrm{L}}$; hence, the gain is now $A_{\mathrm{B}} \approx g_{\mathrm{m}}/g_{\mathrm{L}}$. This will set the gain fairly PVT-independent. However, since the diode need to remain in saturation for the circuit to provide gain, voltage swing at the output is limited. This though does not come as disadvantage, as this additional nonlinearity increases the gradient in the back-gate

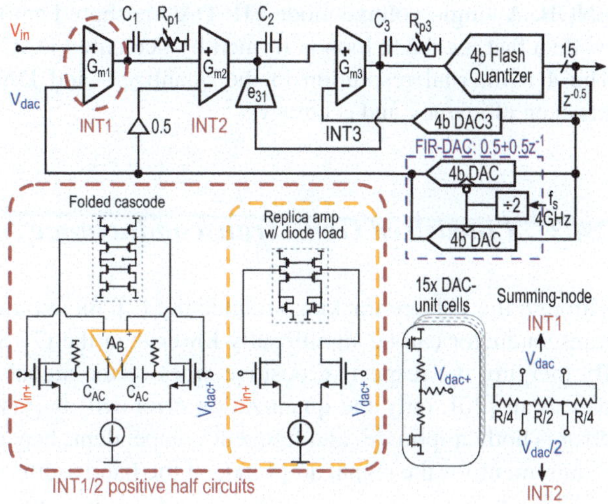

Fig. 2.13 The linearized OTA implemented in a CT-SDM loop filter. (Copyright 2023 The Japan Society of Applied Physics [16])

G_{m} curve and hence enables lower gain values and hence a more efficient use of the main differential pairs G_{m}.

2.4 Design Example

In next-generation mobile receivers, baseband bandwidths exceeding a hundred megahertz are required. Recently, higher-order modulation formats became the norm. Here, with their inherent anti-aliasing and high resolutions >10 b ENOB, continuous-time Sigma Delta ADCs remain a dominant choice to digitize the receive signal. In [19], such an ADC proves to be capable of modulation depths of up to 16384 QAM. Commonly, the CT-SDM loop filter dominates the overall power dissipation of the ADC system. Here, by using a open-loop OTA in GmC configuration, efficiency can be improved by both providing a high-ohmic input load and a favorable noise to power trade off. Here, the active back-gate linearization technique in a differential difference amplifier (DDA) to implement a $G_{\mathrm{m}}C$-based loop filter as published in [16] in GF22FDX technology (Fig. 2.13).

2.4.1 Topology of the Continous Time Sigma Delta Modulator

Here, by employing an open-loop GmC OTA in the CT-SDM loop filter, excellent power efficiency is attained while presenting a purely capacitive input load for preceding receiver stages such as amplifiers and mixers. A third-order loop filter with an optimized zero placement enables an effective resolution of more than 10 b

ENOB. A single voltage mode FIR DAC is shared among the first two integrators, while a fast feedback loop is formed by dedicated DAC around the third integrator. The 4 b internal resolution in the quantizers and DACs offers a good trade-off between efficiency and accuracy.

2.4.2 *Linearized GmC Transconductance Amplifier*

Through the voltage tracking architecture [7], the required signal amplitude for the transconductor can be significantly lower than the CT-SDM's full-scale amplitude. By merging the respective positive sides of the signal together at one differential pair as in [20], only the quantization error will be visible as V_{GS} variation. This error is both amplitude- and frequency-dependent; hence, this V_{GS} variation sets the requirement for the signal amplitude at the bandwidth edge.

The circuit's schematic of the main OTA is depicted on Fig. 2.14a. Since the respective input voltages closely follow each other, their common mode portion is present at the source node of the input pairs. This voltage swing can cause the current sources to become nonlinear due to channel length modulation. This is prevented by the adoption of a regulated cascode topology. The current, finally, is injected in a telescopic cascode output stage. By the employment of auxiliary gain boosters, output impedance is increased, and therefore, a highly linear integration current is ensured.

The transconductor's $G_m(V_{in})$ curve is linearized using the active back-gate input signal injection technique. To ensure good PVT tolerance, the back-gate amplifier's input devices are designed as replicas of the main transconductor. A folded cascode topology is employed to ensure enough voltage headroom for the amplifier's output swing. Here, the gain is set using a diode-connected load transistor to ensure a resilient setting of the amplifier's gain value. The output of the amplifier is AC-coupled. This serves two purposes. Firstly, the DC back-gate bias voltage can be set freely to apply full forward bias and thereby reduce the threshold voltage for the main input pairs. Secondly, low-frequency $1/f$-noise is not injected in the main pair. This does not degrade performance, since as the V_{GS} variation is frequency-dependent, linearization is not needed for DC input voltages where tracking is perfect.

2.4.3 *Characterization Results*

The circuit has been implemented in a 22 nm FD.FSOI CMOS process. The spectral plots in Fig. 2.15 prove the efficacy of the linearization concept. By switching on the amplifier A_B, due to linearization through the back-gate, IM3 linearity improves by >27 dB near the 220 MHz bandwidth edge. The robustness of the concept is shown by Fig. 2.16, here only 3 dB IM3 degradation is observed withing a 10%

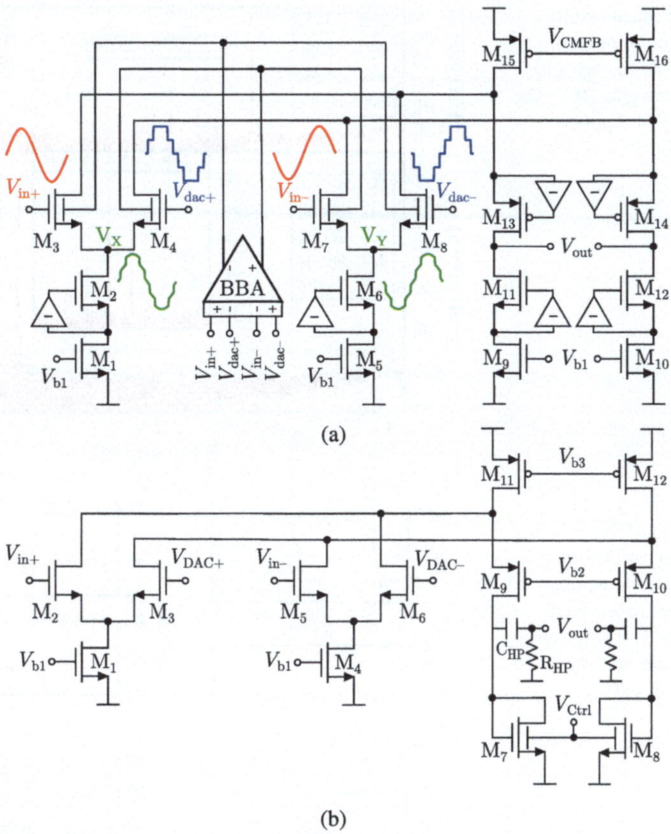

Fig. 2.14 (**a**) Schematic of the GmVC integrator applying active back-gate linearization technique and (**b**) the back-gate amplifier (BBA). (Reprinted from [15], Copyright 2021 IEEE)

gain change. This is accompanied by a chip micrograph, the resilience is proven by IM3 measurements for multiple chip samples.

2.5 Conclusion

The transconductance amplifier linearization using active back-gate input-signal injection can effectively compensate nonlinearity in the differential pairs V/I transfer curve. This is achieved by matching the front- and back-gate g_m curves' gradients to cancel out their input signal dependency. The working principle of the concept is shown with simple polynomials and proven by square-law model analysis where the typology's efficacy can be quantified. Ultimately, a design example in a CT-SDM proves both the robustness and feasibility of the linearization underlined

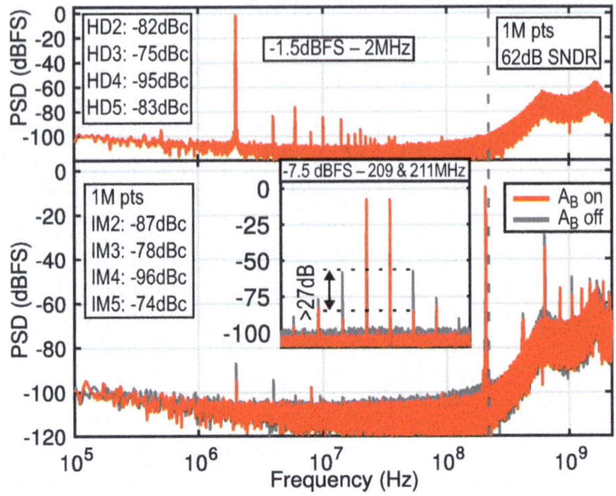

Fig. 2.15 Spectral results underlining the improvement of IM3 with the linearization concept. (Copyright 2023 The Japan Society of Applied Physics [16])

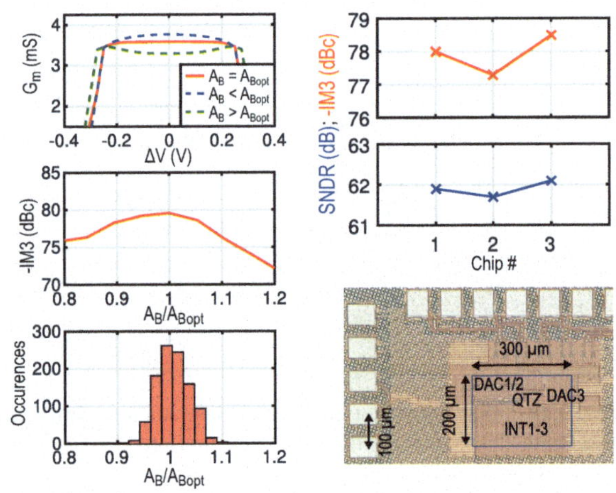

Fig. 2.16 Analysis of the process spread of the gain A_B and its impact on IM3 (simulations) + measured process spread in terms of SNDR and IM3. (Copyright 2023 The Japan Society of Applied Physics [16])

by a measured IM3 improvement of over 27 dB. The concept manages to keep the advantages from simple feed-forward OTA implementations, such as the high-ohmic purely capacitive input load and excellent power efficiency. Alternatively, the concept can also be applied for any other transconductor topologies to mitigate nonlinear distortion precipitated by differential pairs' V/I conversion characteristics.

References

1. Razavi, B. (2001). *Design of analog CMOS integrated circuits second edition*. McGraw-Hill Education. ISBN: 97-80-072524-93-2.
2. Krummenacher, F., & Joehl, N. (1988). A 4-MHz CMOS continuous-time filter with on-chip automatic tuning. *IEEE Journal of Solid-State Circuits, 23*(3), 750–758. https://doi.org/10.1109/4.315

3. Garde, P. (1977). Transconductance cancellation for operational amplifiers. *IEEE Journal of Solid-State Circuits, 12*(3), 310–311. https://doi.org/10.1109/JSSC.1977.1050898
4. Kim, D., Kim, B., & Nam, S. (2015). A transconductor and tunable Gm-C high-pass filter linearization technique using feedforward Gm3 canceling. *IEEE Transactions on Circuits and Systems II: Express Briefs, 62*(11), 1058–1062. https://doi.org/10.1109/TCSII.2015.2456611
5. Akter, M. S., Sehgal, R., & Bult, K. (2020). A resistive degeneration technique for linearizing open-loop amplifiers. *IEEE Transactions on Circuits and Systems II: Express Briefs, 67*(11), 2322–2326. https://doi.org/10.1109/TCSII.2020.2966276
6. Pan, S., & Makinwa, K. A. A. (2021). A 10 fJ·K2 wheatstone bridge temperature sensor with a tail-resistor-linearized OTA. *IEEE Journal of Solid-State Circuits, 56*(2), 501–510. https://doi.org/10.1109/JSSC.2020.3018164
7. Runge, M., Edler, J., Kaiser, T., Misselwitz, K., & Gerfers, F. (2023). An 18-MS/s 76-dB SNDR continuous-time Δ Σ modulator incorporating an input voltage tracking GmC loop filter. *IEEE Journal of Solid-State Circuits, 58*(8), 2288–2299. https://doi.org/10.1109/JSSC.2023.3244718
8. Wang, Z., & Guggenbuhl, W. (1990). A voltage-controllable linear MOS transconductor using bias offset technique. *IEEE Journal of Solid-State Circuits, 25*(1), 315–317. https://doi.org/10.1109/4.50321
9. Lewinski, A., & Silva-Martinez, J. (2006). A high-frequency transconductor using a robust nonlinearity cancellation. *IEEE Transactions on Circuits and Systems II: Express Briefs, 53*(9), 896–900. https://doi.org/10.1109/TCSII.2006.880025
10. Haga, Y., Zare-Hoseini, H., Berkovi, L., & Kale, I. (2005). Design of a 0.8 Volt fully differential CMOS OTA using the bulk-driven technique. In *2005 IEEE International Symposium on Circuits and Systems (ISCAS)* (Vol. 1, pp. 220–223). https://doi.org/10.1109/ISCAS.2005.1464564
11. Arnaud, A., & Miguez, M. (2018). Bulk linearisation of the MOS resistor. *Electronics Letters, 54*(19), 1106–1108. https://doi.org/10.1049/el.2018.5482. Eprint: https://ietresearch.onlinelibrary.wiley.com/doi/pdf/10.1049/el.2018.5482
12. Thomas-Erviti, G., Algueta-Miguel, J. M., De la Cruz Blas, C.A., & Carrillo, J. M. (2017). CMOS transconductor with improved linearity using the bulk of self-cascode transistors. *Electronics Letters, 53*(3), 136–138. https://doi.org/10.1049/el.2016.4107. eprint: https://ietresearch.onlinelibrary.wiley.com/doi/pdf/10.1049/el.2016.4107
13. Gak, J., Miguez, M. R., & Arnaud, A. (2014). Nanopower OTAs with improved linearity and low input offset using bulk degeneration. *IEEE Transactions on Circuits and Systems I: Regular Papers, 61*(3), 689–698. https://doi.org/10.1109/TCSI.2013.2284002
14. Monsurro, P., Pennisi, S., Scotti, G., & Trifiletti, A. (2007). Linearization technique for source-degenerated CMOS differential transconductors. *IEEE Transactions on Circuits and Systems II: Express Briefs, 54*(10), 848–852. https://doi.org/10.1109/TCSII.2007.906203
15. Edler, J., Runge, M., & Gerfers, F. (2021). A dynamic body-bias linearization technique enabling wide-band GmC based continous-time sigma-delta converters in 22 nm FD-SOI CMOS. In *2021 IEEE International Midwest Symposium on Circuits and Systems (MWSCAS)* (pp. 1090–1093). https://doi.org/10.1109/MWSCAS47672.2021.9531807
16. Edler, J., Runge, M., Linnhoff, S., & Gerfers, F. (2023). A 4.4 GS/s 220 MHz ΣΔ ADC with a linearized back-gate controlled GmC filter. In *2023 IEEE Symposium on VLSI Technology and Circuits (VLSI Technology and Circuits)* (pp. 1–2). https://doi.org/10.23919/VLSITechnologyandCir57934.2023.10185281
17. Wittenhagen, E., Runge, M., Lotfi, N., Ghafarian, H., Tian, Y., & Gerfers, F. (2021). Advanced mixed signal concepts exploiting the strong body-bias effect in CMOS 22FDX®. *IEEE Transactions on Circuits and Systems I: Regular Papers, 68*(1), 57–66. https://doi.org/10.1109/TCSI.2020.3023077
18. Radpour, M., & Belostotski, L. (2024). Induced back-gate noise in FDSOI MOSFET. *IEEE Transactions on Circuits and Systems II: Express Briefs, 71*(3), 1022–1026. https://doi.org/10.1109/TCSII.2023.3325809

19. Edler, J., Artz, J. P., Wittenhagen, E., & Gerfers, F. (2024). A wide band high linearity CT-sigma delta converter for receivers reaching 1.75 Gb/s using 16384 QAM. In: *2024 15th German Microwave Conference (GeMiC)*(pp. 233–236). https://doi.org/10.23919/GeMiC59120.2024.10485349
20. Basak, D., Li, D., & Pun, K.-P. (2018). A Gm-C delta-sigma modulator with a merged input-feedback gm circuit for nonlinearity cancellation and power efficiency enhancement. *IEEE Transactions on Circuits and Systems I: Regular Papers, 65*(4), 1196–1209. https://doi.org/10.1109/TCSI.2017.2740501

Chapter 3
A Gain-Enhanced Inverter-Based OTA Employing Active Body-Bias Feedback

Marcel Runge and Friedel Gerfers

3.1 Introduction

The fully depleted silicon-on-insulator (FDSOI) CMOS technology features a buried oxide (BOX) layer that effectively isolates the transistor's source and drain from the substrate or well. This isolation not only eliminates nonlinear junction capacitance [1] but also allows for flexible transistor well architectures, such as flipped-well devices. For instance, when an nMOS device is placed above an nWell, a positive bulk-source voltage (V_{BS}) between 0 and +2 V can be safely applied, resulting in a reduction of the transistor's threshold voltage by up to 140 mV (with a sensitivity of $\approx$70 mV/V). Similarly, for a flipped-well pMOS (pMOS under pWell), the corresponding bulk tuning voltage ranges from 0 to −2 V, with similar V_{BS} sensitivity.

Thanks to the strong influence of the bulk (back gate) on the transistor threshold voltage, process variations in digital circuits can be effectively managed, as demonstrated in [2]. Additionally, [3] explores analog and mixed-signal applications, while [4] achieves >90 dBc linearity by linearizing a current-steering DAC.

Although the aforementioned techniques primarily adjust the transistor threshold voltage (a DC parameter), the back gate also introduces an additional analog input via the bulk-source transconductance. Within this context, [5] employs the back gate in a feedback loop, [6] introduces a feedforward Gm linearization approach, and [7] applies it to transceiver self-interference cancellation. Moreover, [8] implements common-mode feedback through back-gate control, and [9] presents a single-transistor amplifier with back-gate feedback for enhanced linearity.

M. Runge (✉) · F. Gerfers
Technische Universität Berlin, Berlin, Germany
e-mail: marcel.runge@tu-berlin.de; friedel.gerfers@tu-berlin.de

F. Gerfers (ed.), *Bulk-Driven Circuit Techniques for CMOS FDSOI Processes*,
https://doi.org/10.1007/978-3-031-85114-8_3

3.2 Proposed Output Impedance Boosting Technique

This chapter advances the inverter-based amplifier state of the art by exploring a back-gate-enabled gain boosting technique aiming to effectively enhance the transistor's small-signal output resistance r_{OUT} by means of the bulk transconductance g_{mBS}, as depicted in Fig. 3.1.

The circle at the bulk node indicates a flipped-well transistor (here nMOS device above an nWell). The transistor drain is connected to its bulk via the gain A_{b}. This way, a path is formed through the transconductance g_{mBS} compensating the AC current through resistor r_{DS}. The output resistance is derived with

$$r_{\mathrm{OUT}} = \frac{v_{\mathrm{OUT}}}{i_{\mathrm{OUT}}} = \frac{1}{A_{\mathrm{b}} \cdot g_{\mathrm{mBS}} + \frac{1}{r_{\mathrm{DS}}}}. \tag{3.1}$$

If the denominator becomes zero, the output impedance becomes infinite. It follows that:

$$A_{\mathrm{b,OPT}} = -\frac{1}{g_{\mathrm{mBS}} \cdot r_{\mathrm{DS}}}. \tag{3.2}$$

However, tracking the $A_{\mathrm{b,OPT}}$ gain attaining an infinite impedance across process, voltage, and temperature (PVT) effects is not practical. Instead, the goal is to significantly increase r_{OUT} and in turn the inverter small-signal gain as proven by transistor-level simulations for different gains A_{b} and transistor length L. Figure 3.2

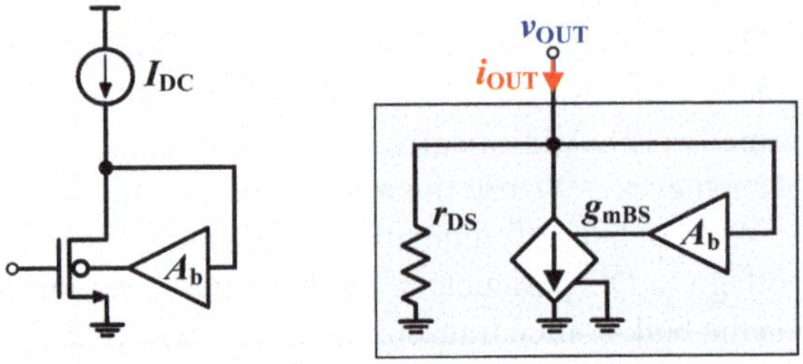

Fig. 3.1 Small-signal model for calculating the output impedance

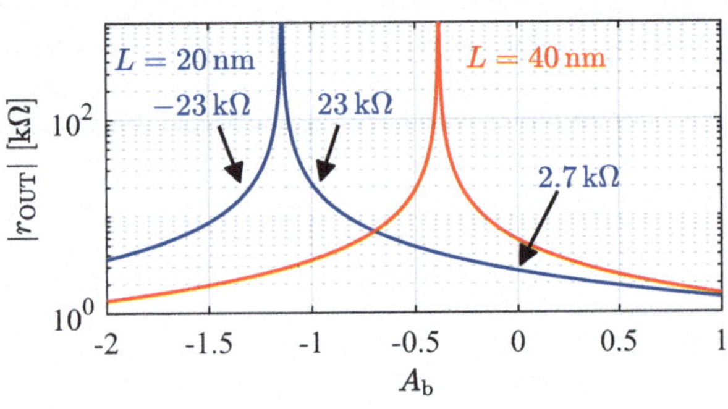

Fig. 3.2 Simulated transistor output impedance $|r_{\mathrm{OUT}}|$ across back-gate gain A_{b} for two channel lengths L © 2024 IEEE. (Reprinted with permission from [10])

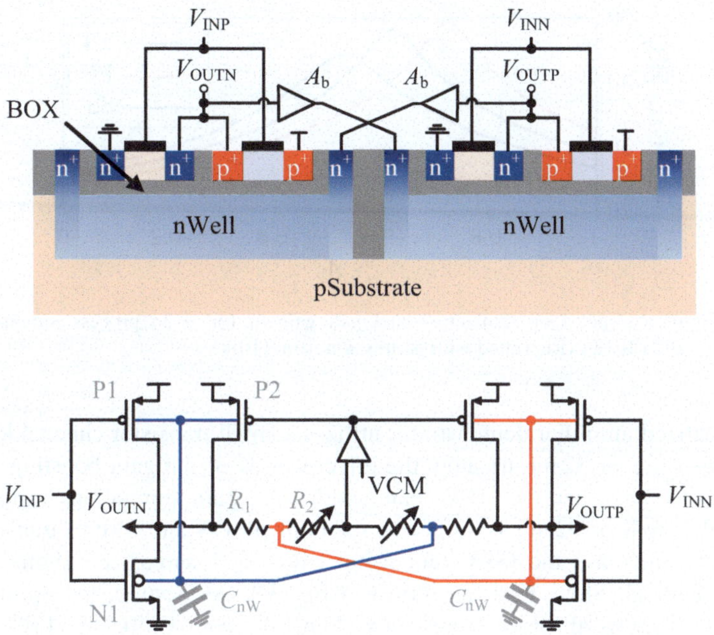

Fig. 3.3 Proposed inverter-based OTA with back-gate gain boosting—modified from © 2024 IEEE. (Reprinted with permission from [10])

depicts r_{OUT} vs the feedback gain A_b for two exemplary channel lengths. Without feedback ($A_b = 0$), the $L = 20$ nm device exhibits an output impedance r_{OUT} of 2.7 kΩ, while for a gain of $A_b = -1$, r_{OUT} increases to 23 kΩ. Around the optimal gain $A_{b,OPT,20nm}$ of -1.15, the output impedance reveals a symmetrical behavior with a $1/x$ dependency. If the feedback gain reduces to $A_b < A_{b,OPT}$, the denominator of r_{OUT} becomes negative, causing sign change of r_{OUT} (phase shift of $-180°$). The r_{OUT} of the $L = 40$ nm device follows the same theory, but due to its larger channel length, it has an inherently higher r_{DS}. Consequently, a smaller negative feedback gain $A_{b,OPT}$ (-0.4) is required as evident from Fig. 3.2. The presented back-gate gain boosting technique works similarly with a pMOS transistor, which further improves the gain enhancement). The introduced concept is applied to an inverter-based OTA, as shown in Fig. 3.3. In order to take advantage of the gain boosting technique for both inverter devices, a regular-well pMOS (pMOS under nWell) and a flipped-well nMOS device (nMOS under nWell) are utilized, sharing the same well. Another constraint to also improve the power efficiency of the proposed gain boosting concept is to limit the gain range of A_b to be smaller than 1 ($|A_b| < 1$), to allow a simple passive gain-control circuitry with

$$A_b = -\frac{R_2}{R_1 + R_2}. \tag{3.3}$$

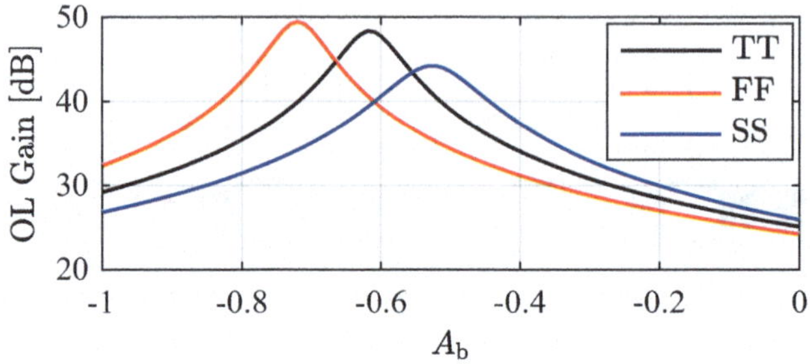

Fig. 3.4 Simulated open-loop gain across back-gate gain A_b for $\pm 3\sigma$ process corners (typical, slow, fast) © 2024 IEEE. (Reprinted with permission from [10])

The realized amplifier demonstrator utilizes a small transistor channel length of $L = 1.6 \cdot L_{min} = 32\,\text{nm}$ to show the effectiveness of the gain boosting concept as well as to maximize the OTA's bandwidth . AC gain simulations validate that a channel length of 32 nm is sufficient for both pMOS and nMOS transistors to significantly increase the OTA gain using $|A_b| < 1$ across ± 3 sigma process corners (**T**ypical, **S**low, **F**ast) as shown in Fig. 3.4. As desired, the optimal gain is $|A_b| < 1$ across all process corners yielding $A_{b,OPT} = -0.7$ for FF, -0.6 for TT, and -0.5 for the SS process corners, respectively. Thanks to the differential configuration, the negative A_b coefficient is achieved by cross-coupling the outputs V_{OUTP} and V_{OUTN} to the opposing bulk node. As a gain of $|A_b| < 1$ is required across PVT, a simple 2-bit tunable resistive voltage divider provides sufficient tuning range. The resistive ratio guarantees independence of A_b from poly sheet resistance variations across PVT. The 2-bit coarse process corner control is realized by an on-chip ring oscillator comprising the same inverter architecture (flipped-well nMOS and regular-well pMOS) to extract the process corner by tracking the oscillation frequency. Process corners TT, SF, FS, 1σ SS, and 1σ FF share the same setting $<01>$ or $A_b = -0.62$, while 2σ SS and beyond demands setting $<10>$ with $A_b = -0.54$, whereas 2σ and 3σ FF utilize $<11>$ with $A_b = -0.7$. The fourth setting $<00>$ shorts resistor R_2, effectively disabling the back-gate gain enhancement ($A_b = 0$). This way, the magnitude of the gain boosting can be effectively shown. The resistance R_1 was chosen to be $3\,\text{k}\Omega$, such that with the 10 fF nWell-substrate diode C_{nW}, a feedback bandwidth of 5.5 GHz is obtained.

3.3 Small-Signal Model

The small-signal model is depicted in Fig. 3.5 taking the nWell capacitance C_{nW} into account. Solving the input to output transfer function yields

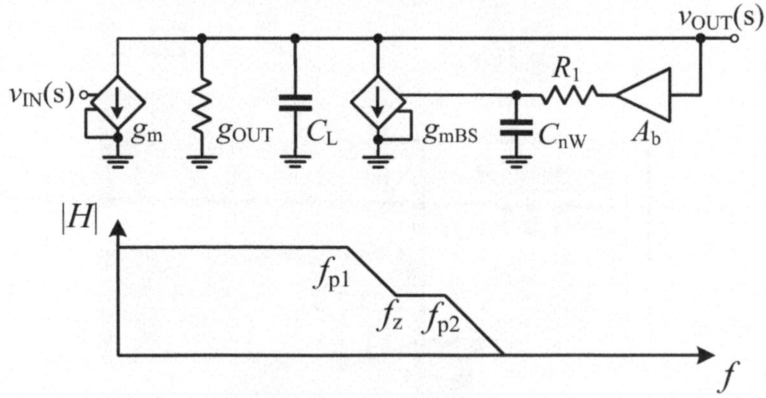

Fig. 3.5 Small-signal model with parasitic nWell capacitance C_{nW}

$$H = -\frac{g_{\mathrm{m}}\left(\omega_0 + s\right)}{C_{\mathrm{L}} s^2 + \left(g_{\mathrm{OUT}} + C_{\mathrm{L}}\omega_0\right) s + \omega_0\left(g_{\mathrm{mBS}} A_{\mathrm{b}} + g_{\mathrm{OUT}}\right)}, \tag{3.4}$$

with

$$\omega_0 = \frac{1}{C_{\mathrm{nW}} \cdot R_1}. \tag{3.5}$$

Hence, the transfer function yields a zero f_{z}, and thanks to negative feedback A_{b} two real poles f_{p1}, f_{p2}.

3.4 Measurement Results

The realized pseudo-differential prototype OTA is fabricated in a 22 nm FDSOI CMOS process, as depicted in Fig. 3.6. The proposed OTA core occupies a core area of only 30 µm × 30 µm. The measurement setup is shown in Fig. 3.7 The demonstrator IC contains both an open- and a closed-loop OTA configuration as shown in Fig. 3.8. The inverter-based OTA operates from a single 1 V power supply, consuming only 3.8 mW. A 10 kΩ input and feedback resistor form the closed-loop configuration. Both designs are connected to a source-follower circuit, which serves as an output isolation buffer to drive the measurement setup. For de-embedding purposes, the source-follower buffer is also placed as a stand-alone component on the same die. Figure 3.9 displays the measured de-embedded OTA open-loop (OL) gain magnitude and phase across the four A_{b} settings. Enabling the proposed back-gate gain boosting technique, the OTA gain is increased from initial 24 dB to a maximum of 42 dB (<01 >). The fabricated OTA features a 28 GHz unity gain frequency for a 120 fF buffer load capacitance. The input and output common-

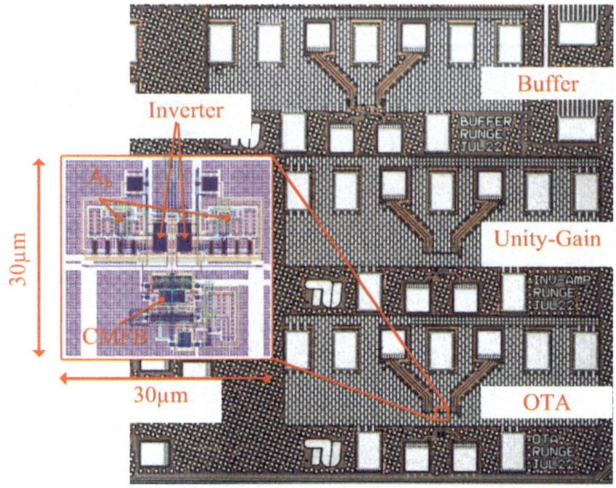

Fig. 3.6 Chip foto © 2024 IEEE. (Reprinted with permission from [10])

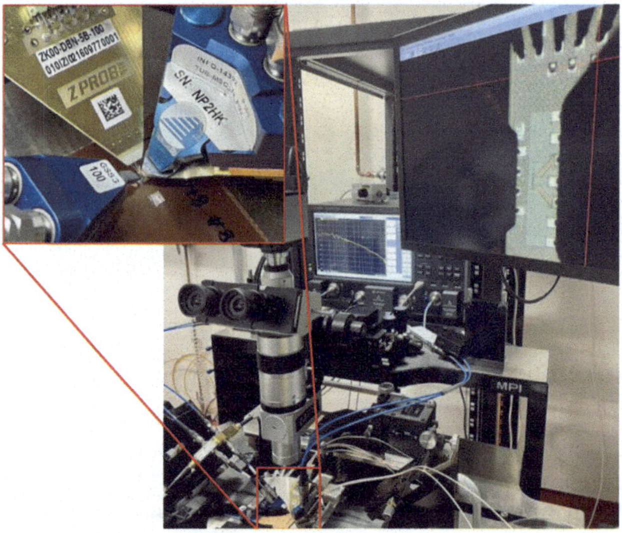

Fig. 3.7 Measurement setup

mode voltage (VCM) sensitivity on the OTA OL gain magnitude is measured in Fig. 3.10. Here, a ±20 mV VCM variation reveals only minor gain variations of 3.5 dB. Figure 3.11 displays OL gain magnitude variations across five samples using the same A_b setting, revealing less than 1 dB gain variations. All samples originate from on the same lot, as there are no cross-wafer samples available. The measured input-referred offset voltage across five samples ranges between ±5 mV. Varying the power supply between 0.9 and 1.1 V, as depicted in Fig. 3.12, maintains an

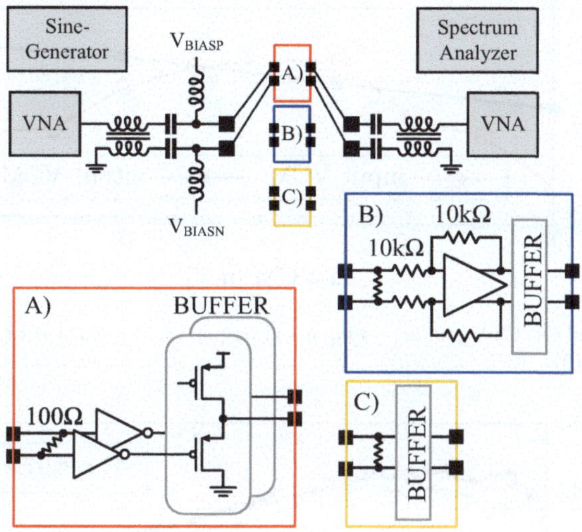

Fig. 3.8 Experimental setup for open-loop , closed-loop , and buffer-only measurements © 2024 IEEE. (Reprinted with permission from [10])

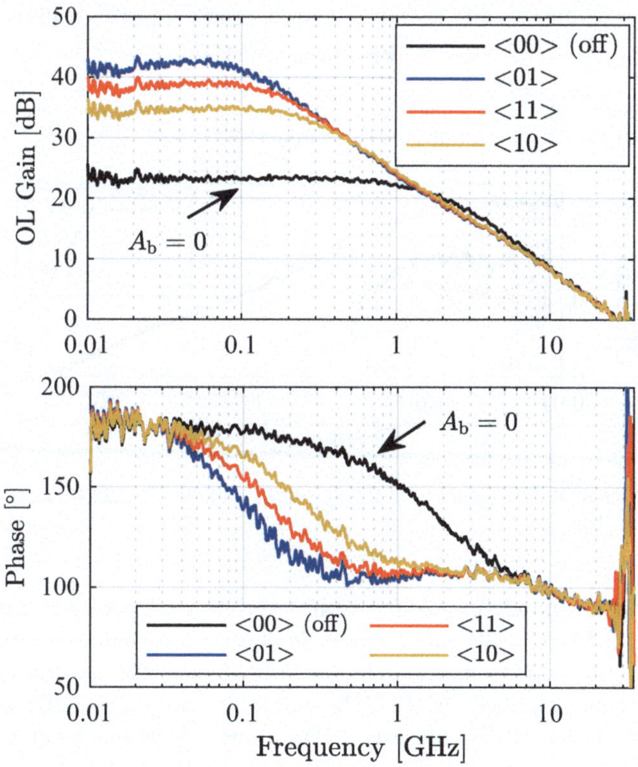

Fig. 3.9 Measured open-loop gain and phase for different back-gate gains A_b © 2024 IEEE. (Reprinted with permission from [10])

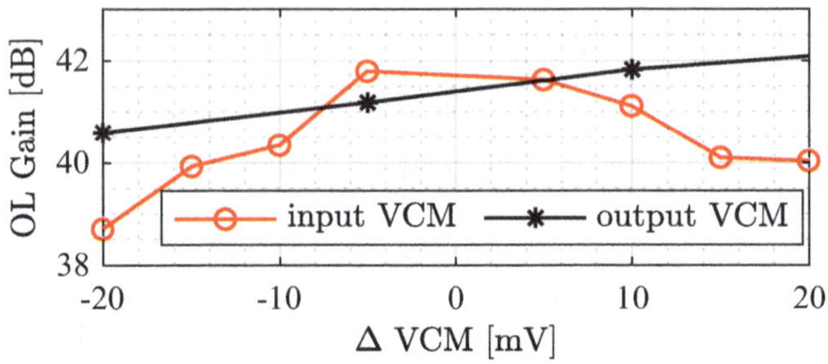

Fig. 3.10 Measured OL OTA gain across input and output VCM © 2024 IEEE. (Reprinted with permission from [10])

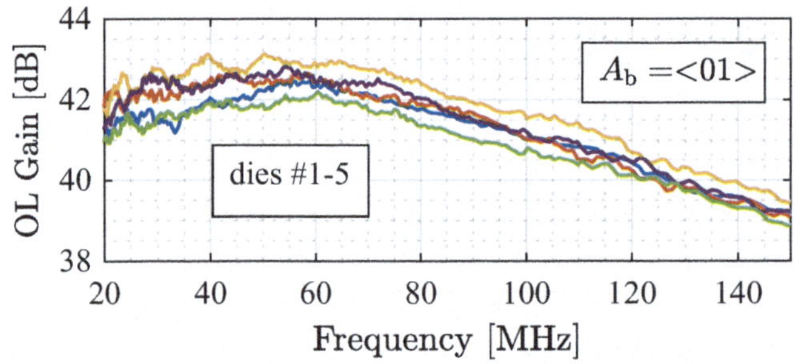

Fig. 3.11 Measured OL OTA gain across five different samples with same back-gate A_b gain setting © 2024 IEEE. (Reprinted with permission from [10])

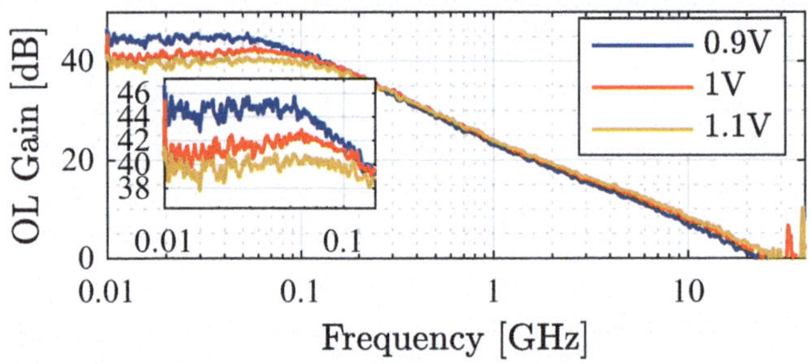

Fig. 3.12 Measured OL OTA gain across supply voltage variation © 2024 IEEE. (Reprinted with permission from [10])

OL OTA gain magnitude exceeding 40 dB. Across a temperature range from 20 to 70 °C (Fig. 3.13) a 5.5 dB gain variation is observed, including a 350 μW (about 10%) power consumption increase. Figure 3.14 depicts the spectral performance of the unity gain amplifier. A 212 MHz sine wave input at −3 dBm (448 mV_{ppd} at OTA input) yields 50 dBc linearity (HD3 limited). The black curve in the plot represents the baseline identifying amplifier independent signals picked up by the measurement setup. The total loss from the sine generator to the spectrum analyzer

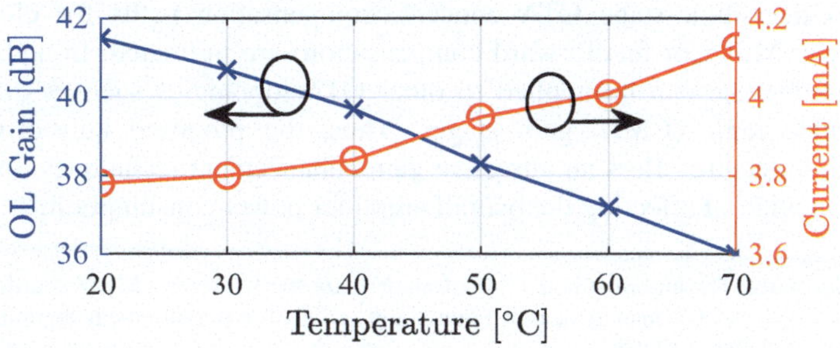

Fig. 3.13 Measured OL OTA gain and power consumption across temperature © 2024 IEEE. (Reprinted with permission from [10])

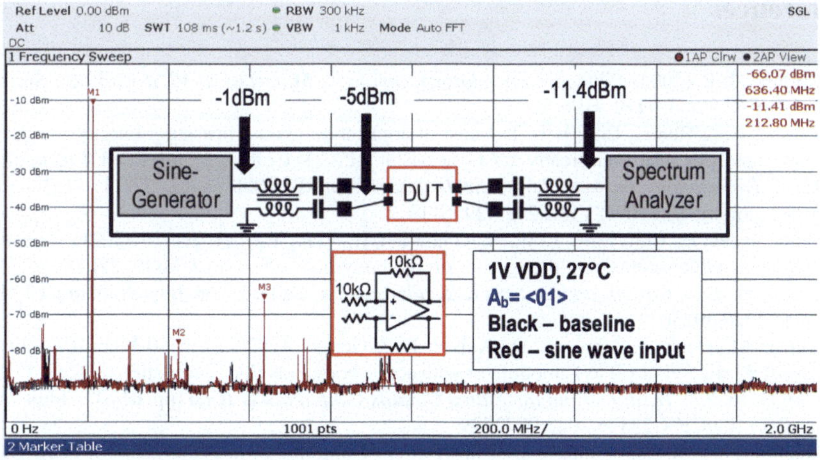

Fig. 3.14 Closed-loop spectral performance measurements with a −3 dBm 212 MHz input signal

Table 3.1 Measured SFDR performance across supply and temperature

SFDR [dBc]								
	12.3 MHz		212 MHz		1.1 GHz		3.2 GHz[a]	
VDD	27 °C	70 °C	27 °C	70 °C	27 °C	70 °C	27 °C	70 °C
0.9 V	49	49	50	49	48	47	62	61
1 V	49	47	50	50	48	47	62	61
1.1 V	49	47	50	50	48	47	61	60

[a] HD3 attenuated by source follower buffer

is 10.5 dB, with the source-follower buffer accounting for approximately 1.5 dB of loss. Table 3.1 summarizes the measured SFDR across temperature and supply voltage variations for four different input frequencies.

In summary, the topology introduced here enables a highly power- and area-efficient realization of OTAs in wireline/wireless filter topologies (e.g. AAF filter).

Thanks to the single-stage OTA concept, compensation paths for closed-loop stability (e.g. Miller or feedforward compensation) are mitigated. In addition, the proposed technique can be applied to other OTA topologies such as differential current mode logic (CML) gain stages. Thus, the proposed bulk-driven gain boosting architecture offers an attractive gain enhancement techniques, improving the OL gain without affecting the overall amplifier power consumption.

Acknowledgments The authors thank Globalfoundries for the University MPW shuttle program. The authors thank the German Research Foundation (DFG) for measurement equipment INST 131/785-1 and 131/786-1 FUGG.

References

1. Colinge, J.-P. (2004). *Silicon-on-insulator technology: Materials to VLSI* (3rd ed.) Springer US. ISBN: 978-1-4419-9106-5.
2. Zaruba, F., & Benini, L. (2019). The cost of application-class processing: Energy and performance analysis of a linux-ready 1.7-GHz 64-bit RISC-V Core in 22-nm FDSOI technology. *IEEE Transactions on Very Large Scale Integration (VLSI) Systems, 27*(11), 2629–2640. https://doi.org/10.1109/TVLSI.2019.2926114
3. Wittenhagen, E., Runge, M., Lotfi, N., Ghafarian, H., Tian, Y., & Gerfers, F. (2021). Advanced mixed signal concepts exploiting the strong body-bias effect in CMOS 22FDX®. *IEEE Transactions on Circuits and Systems I: Regular Papers, 68*(1), 57–66. https://doi.org/10.1109/TCSI.2020.3023077
4. Runge, M., Edler, J., Schmock, D., Kaiser, T., & Gerfers, F. (2022). A 30-MHz BW 74.6-dB SNDR 92-dB SFDR CT ΔΣ modulator with active body-bias DAC calibration in 22nm FDSOI CMOS. In *2022 IEEE Custom Integrated Circuits Conference (CICC)* (pp. 01–02). https://doi.org/10.1109/CICC53496.2022.9772799
5. Kuzmicz, W. (2019). A simple ultra-low power opamp in 22 nm FDSOI. In *2019 MIXDES - 26th International Conference "Mixed Design of Integrated Circuits and Systems"* (pp. 167–170). https://doi.org/10.23919/MIXDES.2019.8787017
6. Edler, J., Runge, M., Linnhoff, S., & Gerfers, F. (2023). A 4.4 GS/s 220 MHz ΣΔ ADC with a linearized back-gate controlled GmC filter. In *2023 IEEE Symposium on VLSI Technology and Circuits (VLSI Technology and Circuits)* (pp. 1–2). https://doi.org/10.23919/VLSITechnologyandCir57934.2023.10185281
7. Haq, F. U., Englund, M., Antonov, Y., Stadius, K., Kosunen, M., Ryynänen, J., Östman, K. B., & Koli, K. (2018). Full-duplex wireless transceiver self-interference cancellation through FD-SOI buried-gate signaling. In *2018 IEEE International Symposium on Circuits and Systems (ISCAS)* (pp. 1–5). https://doi.org/10.1109/ISCAS.2018.8351823
8. Saif, M. A., Dessouky, M., & Aboushady, H. (2021). A 6.4 GHz continuous-time ΣΔ ADC using body-biased feedforward op-amps in 28nm-FDSOI. In *2021 IEEE International Midwest Symposium on Circuits and Systems (MWSCAS)* (pp. 301–304). https://doi.org/10.1109/MWSCAS47672.2021.9531739
9. Weinreich, S., & Murmann, B. (2022). A single-transistor amplifier with back-gate feedback in 22-nm FD-SOI. *IEEE Solid-State Circuits Letters, 5*, 210–213. https://doi.org/10.1109/LSSC.2022.3198547
10. Runge, M., Kellermann, J., Scholz, P., Wittenhagen, E., & Gerfers, F. (2024). A single-stage inverter based OTA using back-gate gain boosting in 22nm FDSOI CMOS. In *2024 IEEE European Solid-State Electronics Research Conference (ESSERC)* (pp. 53–56). https://doi.org/10.1109/ESSERC62670.2024.10719410

Chapter 4
Ultra-Low-Power Voltage-Mode VCSEL-Driver with Back-Gate Bias Tuning

Urs Hecht, Helia Ordouei, and Friedel Gerfers

4.1 Introduction

Since the rise of modern long-range communication systems, innovation and advancements in this field, such as mobile services, radar, Internet-of-Things (IoT), and optical communication systems, rely heavily on the availability of high-speed data transfer. While this development also calls for even faster networking equipment, the energy efficiency of the devices becomes equally important. Low power efficiency leads to several problems on all architectural levels, from thermal management when dealing with high integration to higher operating costs to the toll on the environment. Transmitters (TX) utilizing directly modulated vertical-cavity surface emitting laser (VCSEL) diodes are well-known to be much more energy-efficient at converting electrical signals into optical signals than comparable optical sources or modulators like Mach–Zehnder Modulators (MZMs) or Electro-absorption Modulated LASER (EML) -based systems. However, the gain in power efficiency for VCSEL-based transmitter systems is much smaller than expected, since the driver's power draw surpasses the VCSEL's by several orders of magnitude. State-of-the-art (SOTA) VCSEL drivers achieve data rates up to 71 Gbit/s at an efficiency of 13.4 pJ/bit [1] and 56 Gbit/s at an efficiency of 0.71 pJ/bit [2]. This chapter showcases an optimization of the TX architecture and transistor sizing specifically for the VCSEL driving application, utilizing back-gate trimming and advancing the state-of-the-art energy efficiency of optical transmitter systems. Additionally, the use of Short Wavelength Division Multiplexing (SWDM) is shown, enabling to quadruple the data rate per fiber for up to 224 Gbit/s transmission.

The chapter is organized as followed: in Sect. 4.2, the new voltage-mode architecture is introduced and compared to previous current-mode designs, Sect. 4.3 shows

U. Hecht (✉) · H. Ordouei (✉) · F. Gerfers
Technische Universität Berlin, Berlin, Germany
e-mail: hecht@tu-berlin.de; h.ordouei@tu-berlin.de; friedel.Gerfers@tu-berlin.de

F. Gerfers (ed.), *Bulk-Driven Circuit Techniques for CMOS FDSOI Processes*,
https://doi.org/10.1007/978-3-031-85114-8_4

the implementation of the driver in CMOS, Sect. 4.4 explains the measurement setup and showcases the measurement results, Sect. 4.5 explains the SWDM experiment, and in Sect. 4.6, the results are summarized.

4.2 Voltage- and Current-Mode Architectures

VCSEL drivers are often implemented using a current-mode logic (CML) -based topology, like in [1, 3]. CML topologies enable very high data rates, since they only use fast NMOS/NPN-based current switches. Equalization (utilizing either continuous time linear equalization and/or feed-forward equalization (FFE)) can be easily implemented with CML. Furthermore, CML offers a well-controlled output impedance, which has huge benefits for copper-based wireline applications. The bandwidth of CML circuits can be enhanced by using shunt peaking techniques like inductors or T-coils [4]. Despite all these benefits, the CML topology has two significant flaws: increased power supply voltage ($V_{DD,H}$) to keep all current sources in saturation and high current consumption (current sharing). This can be explained with the schematic in Fig. 4.1a. The entire modulation current I_{Mod} is drawn to $V_{DD,H}$, while the TX output is in "1" state, resulting in low energy efficiency. Adding FFE taps results in further efficiency degradation, since the additional FFE tap currents are also wasted through the R_{Dummy} path. The VCSEL impedance is a strong function of the modulation current and frequency and can vary by up to 280% as measurements indicate in Fig. 4.2a, meaning that the well-controlled output resistance R_{Load} of CML drivers is not required or even of any benefit. Same applies to the differential design of CML, as VCSELs are single-ended devices.

The schematic shown in Fig. 4.1b depicts one implementation of the recently proposed push–pull topologies like [2, 6]. To improve power efficiency (compared to CML), the load resistance was replaced with a switched current source I_{Push} and the dummy path was removed. This enables individual tuning of rise- and fall times, helping with VCSEL nonlinear resonant behavior [7]. Despite the improvements, power efficiency is still not optimal: In the "0" state, a large part of the source I_{Bias} current is drawn to ground by I_{Pull}, wasting current. Contrary to being called "Voltage-Mode" (VM) these push–pull architectures still exhibit a quite large output impedance since the switches act as cascodes. The parallel connection (at the output node) of the bias current source as well as additional FFE taps shown in [2] would not function with purely voltage-mode low impedance taps.

Overcoming energy efficiency bottlenecks, the first true VM VCSEL driver for operation up to 60 Gbit/s error-free bit error rate (BER) ($<10^{-12}$) was published in [5]. A simplified schematic of the concept is shown in Fig. 4.1c. Switches are used to connect the VCSEL either to the power supply V_{DD} or to a controlled voltage V_{Ref}, relying on the current-voltage (IV) behavior of the VCSEL (shown in Fig. 4.2c) to define the "0" and "1" TX signal level. The huge benefit is that the whole TX output stage current flows through the VCSEL without wasting crowbar currents or

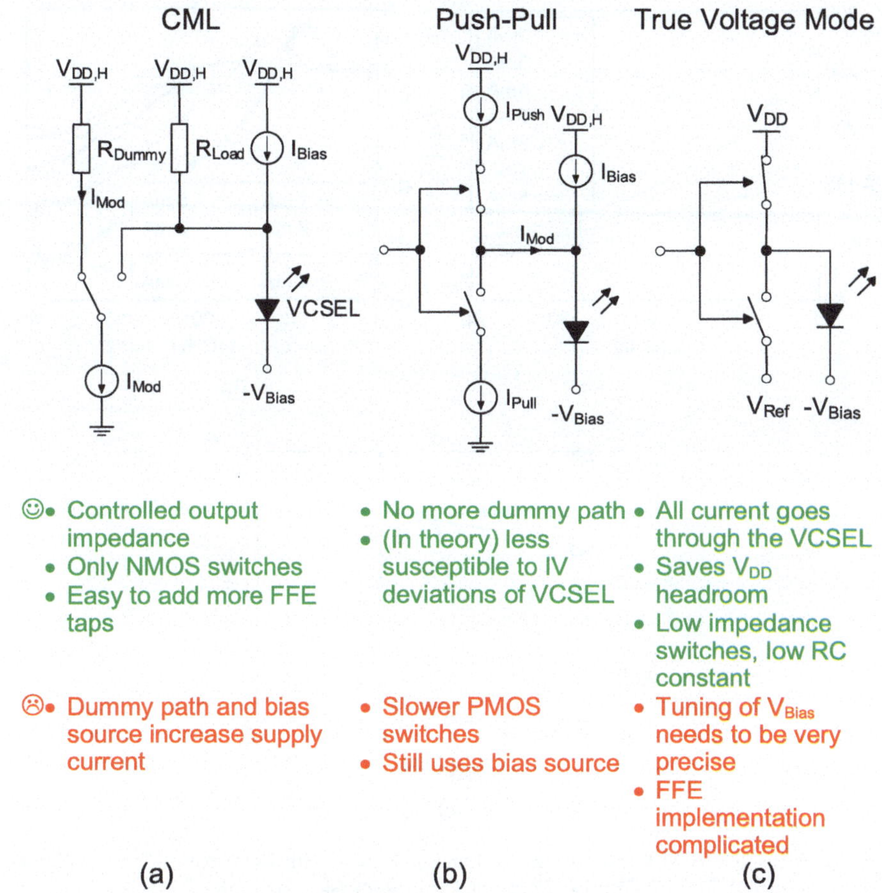

Fig. 4.1 CML vs. Push-Pull vs. True Voltage-Mode (VM) output stages © 2023 IEEE. (Reprinted with permission from [5])

demanding higher supply voltage levels to keep the current sources in saturation region. The voltage-mode architecture comes with two drawbacks: firstly, FFE implementation is complicated since you cannot use simple current summing to add FFE taps. Another way to implement an FFE is by adding two additional reference voltages per tap and digital logic that makes sure no two switches are closed at the same time. However, it has been shown that NRZ VCSEL transmission is feasible up to 50 Gbit/s without any equalization [8]. The nonlinear VCSEL current-dependent peaking/bandwidth variations shown in Fig. 4.2b have a far less-severe impact for NRZ modulation than PAM-4 since the overshoots do not cross through adjacent eyes. As can be seen in Sect. 4.4, the unpleasantly looking deformations of the eye resulting from nonlinear behavior do not necessarily have an impact on bit error rate. Secondly, the voltage-mode approach requires precise VCSEL supply voltage

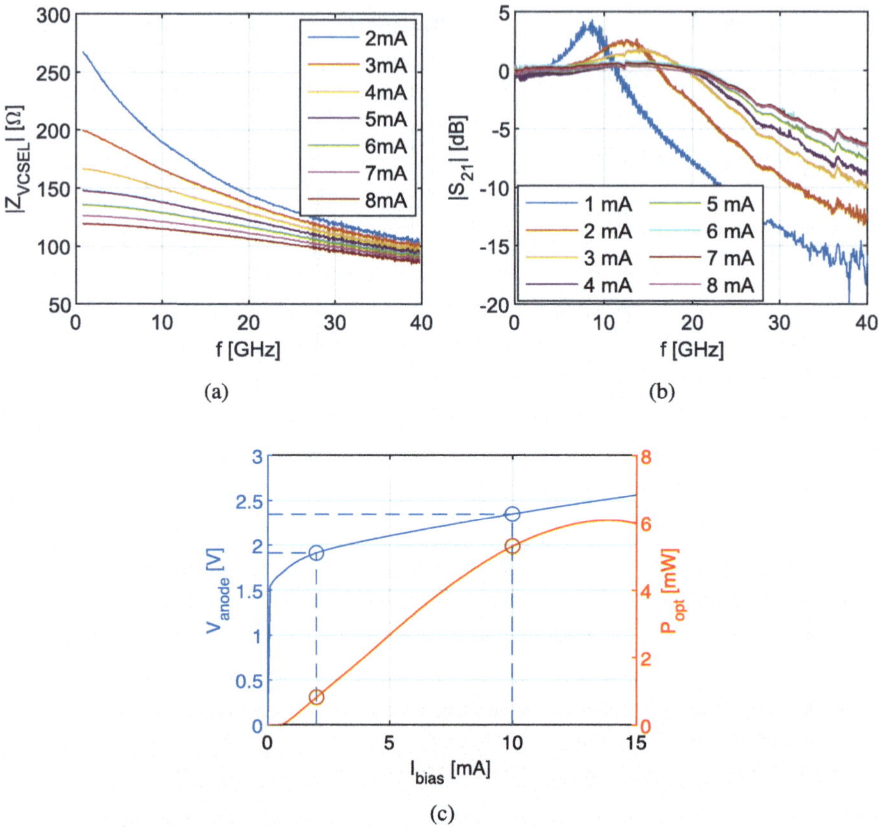

Fig. 4.2 Measured VCSEL behavior (**a**) AC input impedance, (**b**) electro-optical $|S_{21}|$, (**c**) Light-Current-Voltage (LIV) © 2023 IEEE. (Reprinted with permission from [5])

tuning. This is, however, also true for CML and push–pull TX approaches because of their limited swing.

The simplest way to implement the proposed topology is an inverter-based driver, shown in Fig. 4.3a, using ultra-fast nm-CMOS switches connected to V_{DD} and V_{Ref}. CMOS switches offer very low output impedance and, hence, low RC time constants at the TX output node. To generate V_{Ref}, however, a power-hungry low output-impedance reference voltage generator is required. To mitigate that need, the CMOS inverter architecture was refined by replacing the NMOS switch with a tuneable NMOS-based source-follower (SF) structure, shown in Fig. 4.3b. The SF configuration enables tuning of the effective transistor drive strength through parallel switched connection as well as varying the threshold voltage through the back-gate, directly enabling tuning of the "0" level. Simulations of both topologies driving the same electro-optical VCSEL model (based on [9]) are shown in

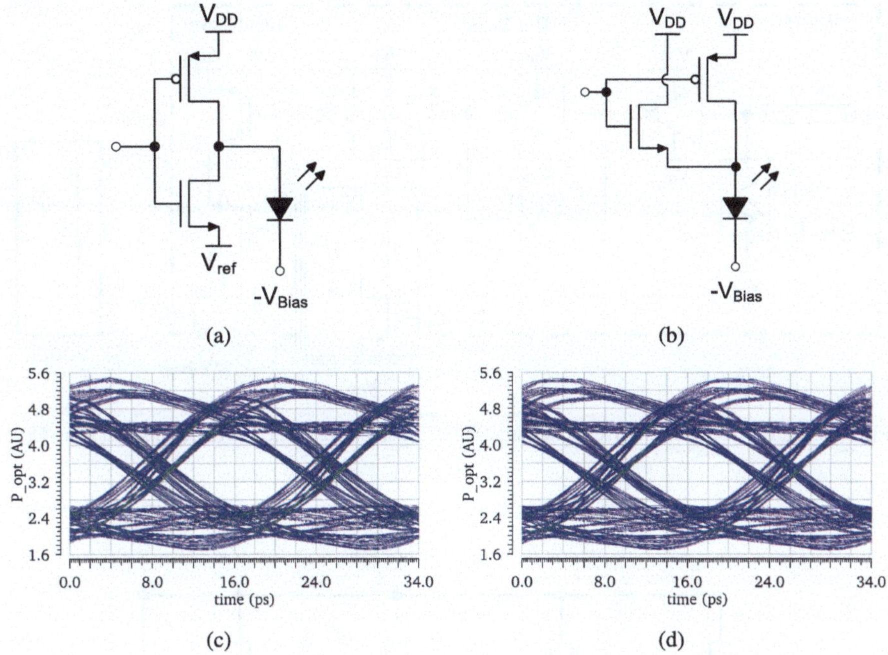

Fig. 4.3 True VM simulated 60 Gbit/s eye diagrams and implementations (**a**, **c**) NMOS switched reference based, (**b**, **d**) NMOS SF based © 2023 IEEE. (Reprinted with permission from [5])

Fig. 4.3c–d and reveal no significant performance difference, but the SF topology enables easier tuning, better power efficiency, and easier implementation.

4.3 Transmitter Design

An overview of the driver schematic is shown in Fig. 4.4. The driver is divided into three parts: the input network, the pre-driver, and the output driver. Special care has to be taken when designing the input network. The required input impedance has to be 50 Ω for the whole bandwidth of the NRZ signal; otherwise, the signal is distorted leading to reduced eye openings. This is achieved through a carefully tuned planar 50 Ω transmission line connecting input pad and a four-bit ($D_{R0} \ldots D_{R3}$) tuneable load resistance. The schematic of the load resistance is given in Fig. 4.5. Forward biasing is used to make the CMOS switches as small as possible, minimizing the parasitic capacitances, which helps with broadband matching. A version with actively driven bulk wells showed to significantly reduce leakage currents, but was omitted due to the additional area required (since every transistor would need its own well).

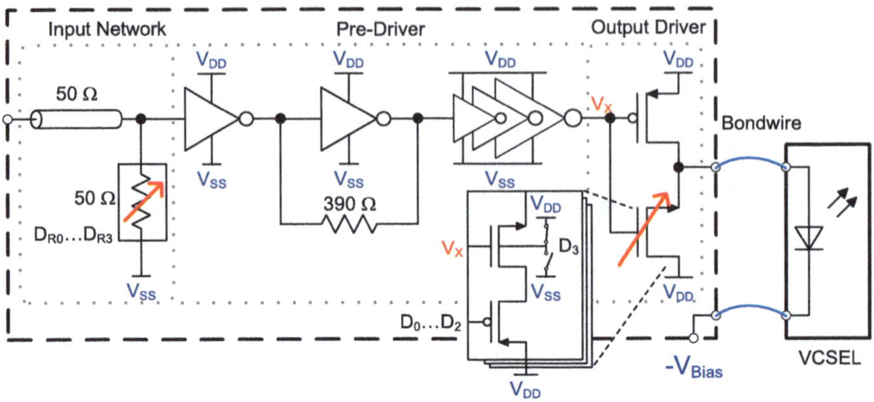

Fig. 4.4 True VM VCSEL driver schematic © 2023 IEEE. (Reprinted with permission from [5])

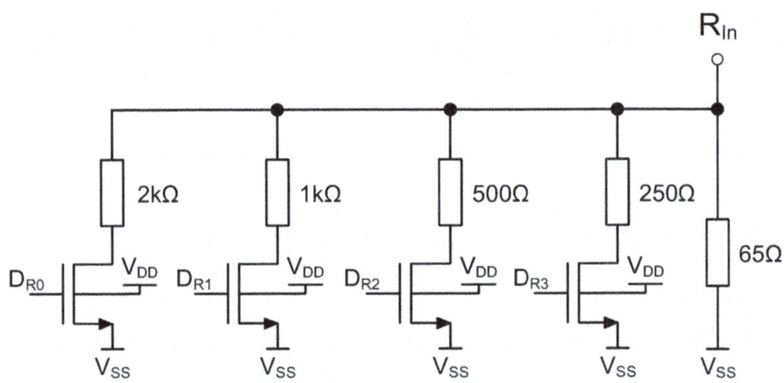

Fig. 4.5 Trimmable input resistance schematic

The pre-driver consists of an inverter chain, amplifying and buffering the input signal. The second inverter is using a self-biased resistive feedback configuration, enhancing the bandwidth and keeping the biasing point at an optimal level. All transistors are forward-biased (NMOS well connected to V_{DD} and PMOS well to V_{SS}), reducing the threshold voltages of the transistors and, hence, increasing the bandwidth of the chain. The main innovation of the driver lies in the voltage-mode output driver. The bias point of the VCSEL can be influenced by changing the external bias voltage V_{Bias}, changing the "1" bias current, as well as tuning the drive strength and threshold voltage of the NMOS source follower.

As can be seen in Fig. 4.6, three-bit course trimming, $D_0 \ldots D_2$ is achieved by switching the drive strength of the NMOS source follower through PMOS switches. One constantly "ON" NMOS, is connected parallel to the switched NMOS devices to prohibit the VCSEL to completely turn off and possibly damaging the output stage transistors. A fourth bit (D_3) is used to control the NMOS SF bulk voltage

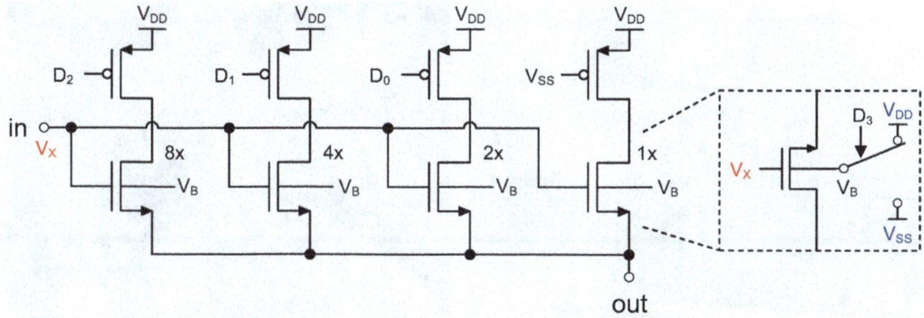

Fig. 4.6 Trimmable source follower schematic

V_B between V_{DD} and V_{SS}. For simplicity, this was done with one bit, but it would be also possible to control the bulk voltage more precisely, enabling a fine-tuned optical output power adjustment. Comparable solutions are used in voltage controlled oscillators, where a discrete amount of load capacitors can be switched in parallel, and fine-tuning is done through the input voltage of a varactor [10]. Schematic-level optimization of the driver was performed using a precise electro-optical VCSEL model (based on [9]). A comparison of the simulated 60 Gbit/s eye diagram (in Fig. 4.8d) with the measurement result shown in Fig. 4.8c shows good correlation in eye opening and shape, bandwidth, as well as overshoots, proving the value of this method.

4.4 Measurement Setup & Results

The 22 nm FD-SOI technology is used to implement and fabricate the driver. The driver is measured with the setup shown in Fig. 4.7. 30–60 Gbit/s PRBS-11 data are generated by a Keysight M8194A arbitrary waveform generator (AWG). Internal pre-emphasis of the AWG was used to compensate for nonlinear losses of the input GSG probe, the 67 GHz cable, as well as the AWG output stage. A separate channel is used to synchronize the sampling scope to the AWG. The trimming bits of the driver are set through a serial interface. VCSEL and TX IC are connected through two bondwires, 200 μm long and 25 μm in diameter. A lensed 1.5 m OM3 multi-mode fiber pigtail is used to couple the optical output signal of the VCSEL and feed it into the Keysight DCA-X sampling oscilloscope. The oscilloscope was set to use the internal compensation of the receiver module (called SIRC); otherwise, no additional equalization was used.

Because of the nonlinear behavior of the VCSEL [7] and the driver purposely lacking equalization, BER cannot be calculated by the common method for AWGN channels of fitting one Gaussian distribution per level. Instead, the total interference histogram from the oscilloscope was exported, and multiple superimposed Gaussian distributions were fitted.

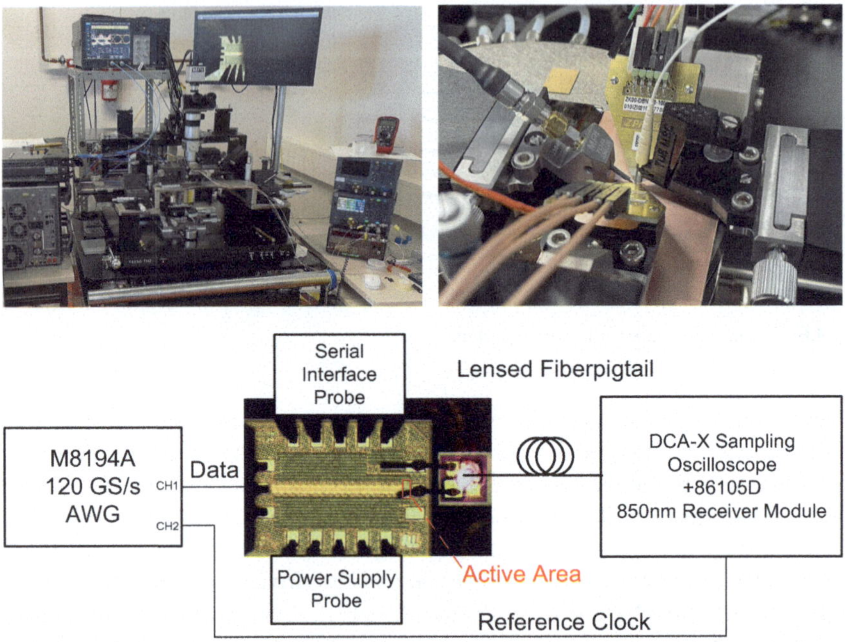

Fig. 4.7 Measurement setup © 2023 IEEE. (Reprinted with permission from [5])

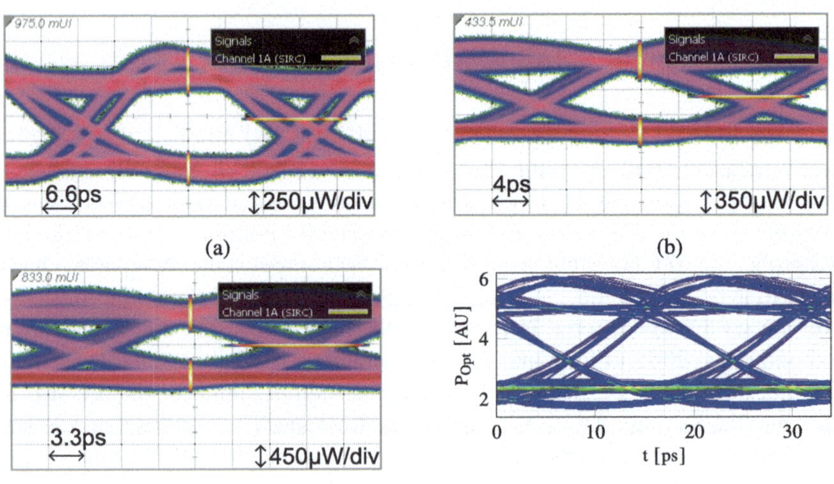

Fig. 4.8 (**a**) 30 Gbit/s, (**b**) 50 Gbit/s, (**c**) 60 Gbit/s measured eye diagram, (**d**) 60 Gbit/s simulated eye diagram © 2023 IEEE. (Reprinted with permission from [5])

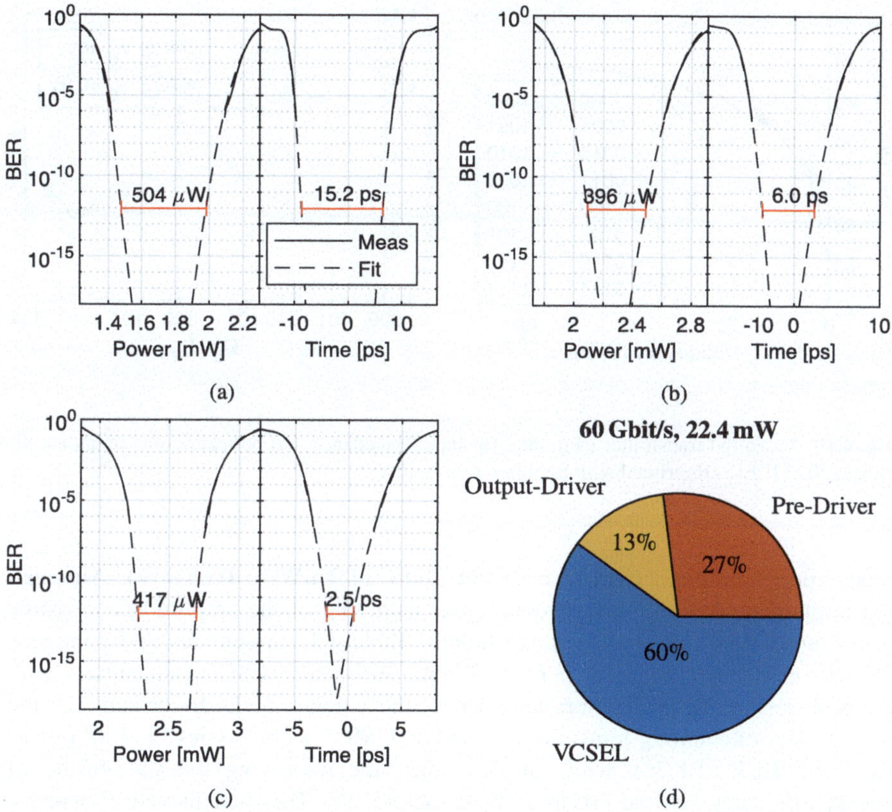

Fig. 4.9 (**a**) 30 Gbit/s, (**b**) 50 Gbit/s, (**c**) 60 Gbit/s bathtub curves, (**d**) 60 Gbit/s power breakdown © 2023 IEEE. (Reprinted with permission from [5])

Figure 4.8a–c show the measured eye diagrams for 30 Gbit/s to 60 Gbit/s with marked slicing points and Fig. 4.9a–c the vertical and horizontal bathtub curves at those slicing points using the discussed method. The vertical BER $< 10^{-12}$ eye openings range from 504 μW to 396 μW, and the horizontal eye openings range from 15.2 ps to 2.5 ps. State-of-the-art values are reached at a data rate of 60 Gbit/s with best-in-class energy efficiency below 0.37 pJ/bit. To increase the efficiency at lower data rates, the supply voltage (as well as the VCSEL bias current) was adjusted down to 0.65 V/4.5 mA when transmitting 30 Gbit/s (Fig. 4.8a). The high efficiency numbers are reached through the VM output stage, drawing only 13% of the total power consumption (as can be seen by Fig. 4.9d). The effective trimming can be seen in Fig. 4.10. The transmitter achieves good input matching with $S_{11} < -17$ dB for up to 30 GHz. Low-level trimming offers a range of the optical

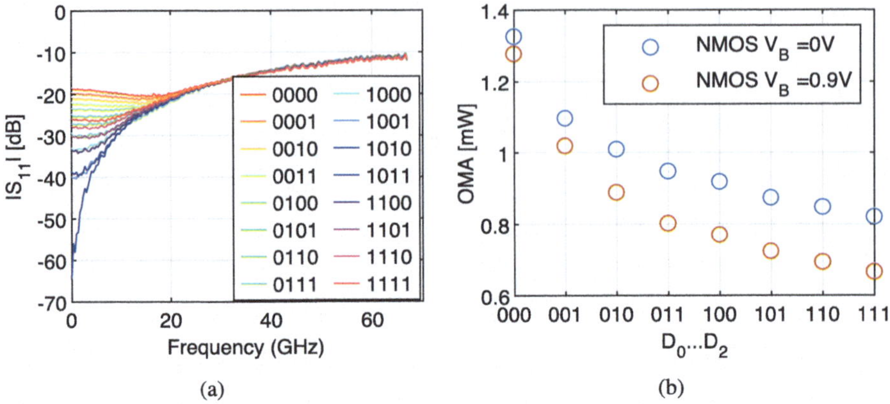

Fig. 4.10 Measured transmitter trimming: (**a**) input impedance, (**b**) optical output signal amplitude © 2023 IEEE. (Reprinted with permission from [5])

modulation amplitude (OMA) of about 200% (1.3 mW to 0.65 mW). Although not implemented in this driver, continuous trimming of the amplitude is possible using more than 1 bit bulk voltage tuning. Table 4.1 summarizes and compares the performance of recently published state-of-the-art VCSEL transmitters with the VM design. To the authors knowledge, the voltage-mode driver achieves the highest data rate among transmitters based on CMOS technologies [2, 3, 6] (error-free with BER$<10^{-12}$), while at the same time improving the state-of-the-art energy efficiency beyond 50 Gbit/s by a factor 1.9x. This is achieved by using a high-bandwidth low-impedance, simple to implement voltage-mode approach and purposefully omitting any equalization in the driver. The novel source-follower approach with back-gate tuning enables easy trimming. Occupying a very small active area (0.001 mm^2), the driver is predestined for SWDM applications. Hence, in the following section, an SWDM setup is measured.

4.5 SWDM Experiment

SWDM enables higher speed by sending multiple signals over one fiber. Contrary to changing the modulation (i.e., going from NRZ to PAM-4), the efficiency is less degraded because data rate and power draw scale by the number of signals over the fiber. The SWDM measurement setup is shown in Fig. 4.11 and is an extension of the previous setup. The 850 nm output of the VCSEL+driver combination was muxed with the 880 nm output of a packaged VCSEL, acting as an aggressor, through a Huber+Suhner Color Cube. The aggressor VCSEL is driven at 56 Gbit/s from the AWG and biased at 4.5 mA through a bias-tee. Connected through 0 m to 100 m

Table 4.1 VCSEL driver state-of-the-art comparison © 2023 IEEE. (Reprinted with permission from [5])

Metric	[1]	[6]	[2]	This Work
Data rate [Gbit/s]	**64–71**	35	32–56	30–60
Process	130 nm (BiCMOS)	64 nm (CMOS)	28 nm (CMOS)	22 nm (FDSOI)
Supply voltages [V]	4	3	0.7–1.15; − 2	0.65–0.9; − 2
VCSEL terminal	Cathode	Anode	Anode	Anode
Output type	CML	Push-pull current	Push-pull current	Voltage-mode
VCSEL biasing	Current source	Current source	Current source	Voltage-based
Inductors	Shunt	Series	Active	**None**
Equalization	2-tap fixed delay FFE	None	Pulse shaping 3-tap UI FFE	None
VCSEL bias current [mA]	8.2	6	5	**4–5.5**
OMA [mW]	**2**	1	1 to 1.3	0.7 to 1.3
ER [dB]	2.1	1.8	2.1	**2.4**
VCSEL BW	26 GHz @ 8.2mA	20 GHz	14 GHz @ 3 mA 25 GHz @ 5 mA	25 GHz @ 4mA **28 GHz @ 6 mA**
Active area [mm^2]	0.06	0.009	0.0024	**0.001**
Efficiency [pJ/bit]	13.4 (71 Gbit/s)	0.65 (35 Gbit/s)	**0.32 (30 Gbit/s)** 0.57 (50 Gbit/s) 0.71 (56 Gbit/s)	0.43 (30 Gbit/s) **0.38 (50 Gbit/s)** **0.37 (60 Gbit/s)**

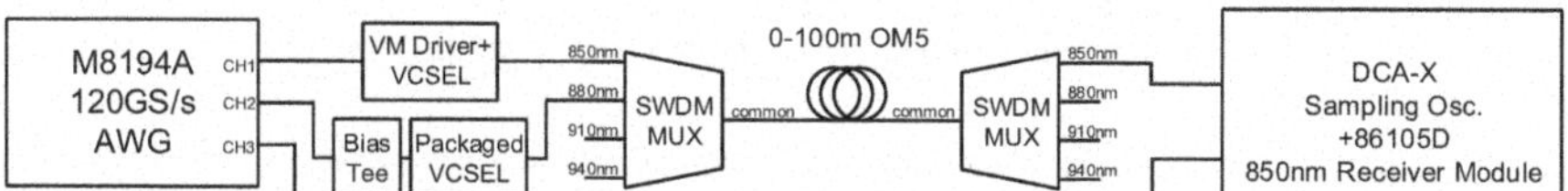

Fig. 4.11 SWDM experiment measurement setup [11]

of OM5 fiber, a second Color Cube acts as an SWDM demux separating the wavelengths. BER of the 850 nm signal is analyzed with the methods described in Sect. 4.4.

Figure 4.12 shows the SWDM measurement results at a bit rate of 56 Gbit/s and for a fiber length of 0 m to 100 m. Similar to before, the driver reaches an efficiency of 0.4 pJ/bit. The system is able to reach error-free transmission for lengths of fiber of up to 50 m, resulting a projected total data rate of 224 Gbit/s per fiber.

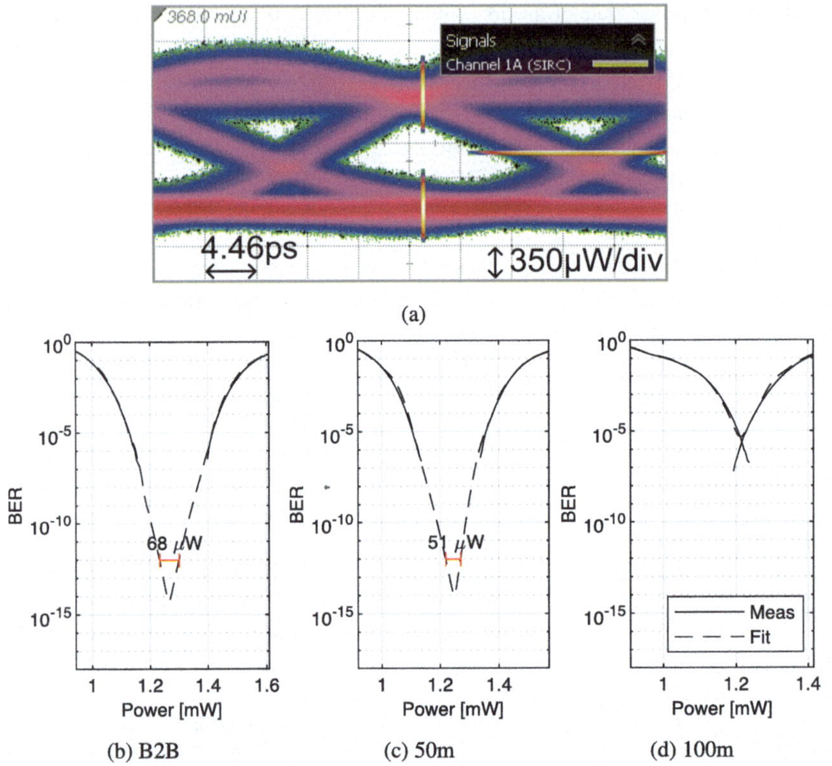

Fig. 4.12 SWDM measurements at 56 Gbit/s (**a**) back-to-back (B2B) eye diagram with marked slicer values, (**b–d**) vertical bathtubs [11]

For lengths of 100 m, forward error correction (FEC) would be needed for error-free transmission.

4.6 Conclusion

This chapter showcases a voltage-mode low-impedance VCSEL driving approach, utilizing the back-gate of the NMOS source follower for fine trimming. In transmission experiments, the driver achieves 30–60 Gbit/s at an efficiency of down to 0.37 pJ/bit, improving the state-of-the-art energy efficiency beyond 50 Gbit/s by a factor of 1.9x.

The use of SWDM is demonstrated to increase the data rate per fiber without significantly effecting the efficiency. Using SWDM, a projected total data rate of 224 Gbit/s per fiber at an efficiency of 0.4 pJ/bit is reached for fiber lengths of up to 50 m.

References

1. Kuchta, D. M., Rylyakov, A. V., Doany, F. E., Schow, C. L., Proesel, J. E., Baks, C. W., Westbergh, P., Gustavsson, J. S., & Larsson, A. (2015). A 71-Gb/s NRZ modulated 850-nm VCSEL-based optical link. *IEEE Photonics Technology Letters, 27*(6), 577–580. ISSN: 1041-1135. https://doi.org/10.1109/LPT.2014.2385671
2. Mansuri, M., Inti, R., Kennedy, J., Qiu, J., Hsu, C. M., Sharma, J., Li, H., Casper, B., & Jaussi, J. (2021). A scalable 32–56 Gb/s 0.56–1.28 pJ/b voltage-mode VCSEL-based optical transmitter in 28-nm CMOS. *IEEE Journal of Solid-State Circuits, 57*(3), 757–766. https://doi.org/10.1109/JSSC.2021.3124289
3. Khafaji, M., Pliva, J., Zoldak, M., Henker, R., & Ellinger, F. (2017). A 42 Gbps VCSEL driver with adjustable 2-tap feed-forward equalizer in 14 nm SOI CMOS. In *2017 12th European Microwave Integrated Circuits Conference (EuMIC)* (pp. 179–182). https://doi.org/10.23919/EuMIC.2017.8230689
4. Razavi, B. (2015). The bridged T-coil [a circuit for all seasons]. *IEEE Solid-State Circuits Magazine, 7*(4), 9–13. https://doi.org/10.1109/MSSC.2015.2474258
5. Hecht, U., Ordouei, H., Ledentsov, N., Scholz, P., Kurth, P., Ledentsov, N., & Gerfers, F. (2023). True voltage-mode NRZ VCSEL transmitter enabling 60 Gbit/s at 0.37 pJ/bit in 22 nm FDSOI. In *ESSCIRC 2023-IEEE 49th European Solid State Circuits Conference (ESSCIRC)* (pp. 365–368). https://doi.org/10.1109/ESSCIRC59616.2023.10268797
6. Choi, H.-S., Hwang, J., Jeong, G.-S., Kim, G., & Jeong, D.-K. (2018). A 35-Gb/s 0.65-pJ/b asymmetric push-pull inverter-based VCSEL driver with series inductive peaking in 65-nm CMOS. *IEEE Transactions on Circuits and Systems II: Express Briefs, 65*(12), 1824–1828. https://doi.org/10.1109/TCSII.2018.2870181
7. Hecht, U., Ledentsov, N., Scholz, P., Agustin, M., Schulz, P., Ledentsov, N. N., & Gerfers, F. (2019). Non-Linear PAM-4 VCSEL equalization and 22 nm SOI CMOS DAC for 112 Gbit/s data transmission. In *2019 12th German Microwave Conference (GeMiC)* (pp. 115–118). https://doi.org/10.23919/GEMIC.2019.8698167
8. Ledentsov, N., Kottke, C., Chorchos, L., Shchukin, V. A., Yu. Makarov, O., Bou Sanayeh, M., Kalosha, V. P., Jungnickel, V., Freund, R., Turkiewicz, J. P., & Ledentsov, N. N. (2020). Up to 600 Gbit/s data transmission over 100 m of single multi-mode fiber using $4 \times \lambda$ 850 – 940 nm VCSELs. In *2022 Optical Fiber Communications Conference and Exhibition (OFC)* (pp. 1–3). https://doi.org/10.1364/OFC.2022.Tu2D.3
9. Belfiore, G., Khafaji, M., Henker, R., & Ellinger, F. (2016). A compact electro-optical VCSEL model for high-speed IC design. In *2016 12th Conference on Ph.D. Research in Microelectronics and Electronics (PRIME)* (pp. 1–4). https://doi.org/10.1109/PRIME.2016.7519508
10. Kurth, P., Misselwitz, K., Scholz, P., Hecht, U., & Gerfers, F. (2022). A 0.007 mm^2 48–53 GHz low-noise LC-oscillator using an ultra-compact high-Q resonator. In *2022 14th German Microwave Conference (GeMiC)* (pp. 104–107). ISBN: 978-3-9820397-2-5
11. Hecht, U., Ordouei, H., Ledentsov, N., Scholz, P., Kurth, P., Titkov, I. E., Ledentsov, N. N., & Gerfers, F. (2023). A 0.4 pJ/bit NRZ voltage mode VCSEL driver for up to 224 Gbit/s SWDM links. In *Optical Fiber Communication Conference (OFC) 2023*. Optica Publishing Group. https://doi.org/10.1364/OFC.2023.M4E.1. https://opg.optica.org/abstract.cfm?URI=OFC-2023-M4E.1

Part II
Cryo/Quantum Computing

Chapter 5
Maximizing the Figures of Merit, Temperature Range, and Optimizing the Algorithmic Design Methodologies Based on Constant Current Density Bias for FDSOI Analog-Mixed-Signal, Digital, mm-Wave, and Fiberoptic Circuits from the Back-Gate Voltage

S. Bonen, S. Pati Tripathi, A. Bharadwaj, J. McIntosh, J. Zhao, A. Zandieh, M. S. Dadash, S. Shopov, M. J. Gong, L. Wu, U. Alakusu, T. Jaeger, G. Cooke, T. Liu, H. Y. Hsu, and S. P. Voinigescu

5.1 Introduction

In this chapter, we will focus on several applications of the back-gate voltage control in: (i) optimally biasing analog mixed signal (AMS) and CMOS logic circuits for maximized performance over a wide range of temperatures from 2 K to 400 K, (ii) broadband linear electro-optical interface circuits; (iii) mm-wave radar sensors, and (iv) cryogenic classical control circuits for quantum processors operating in the 60–200 GHz range.

We will start in Sect. 5.2 by deriving the threshold voltage expression as a function of the backgate voltage to capture its temperature dependence. Next we will describe the equations that link the drain–source current, I_{DS}, and the analog-mixed-signal (AMS) figures of merit (FoMs): transconductance, g_m, g_m/I_{DS}, intrinsic voltage gain, A_v, cutoff frequency, f_T, maximum frequency of oscillation, f_{MAX}, maximum available power gain, MAG, and the minimum noise figure, NF_{MIN} to the back-gate voltage, V_{BG}, in the triode, saturation, and subthreshold regions of operation. With the support of experimental data collected on two different commercial FDSOI CMOS technology nodes [1, 2], whose cross sections are depicted in Fig. 5.1, we will illustrate how, by plotting them versus the drain current

S. Bonen (✉) · S. Pati Tripathi · A. Bharadwaj · J. McIntosh · J. Zhao · A. Zandieh · M. S. Dadash · S. Shopov · M. J. Gong · L. Wu · U. Alakusu · T. Jaeger · G. Cooke · T. Liu · H. Y. Hsu · S. P. Voinigescu
University of Toronto, Toronto, ON, Canada
e-mail: shai.bonen@mail.utoronto.ca; sorinv@ece.utoronto.ca

F. Gerfers (ed.), *Bulk-Driven Circuit Techniques for CMOS FDSOI Processes*,
https://doi.org/10.1007/978-3-031-85114-8_5

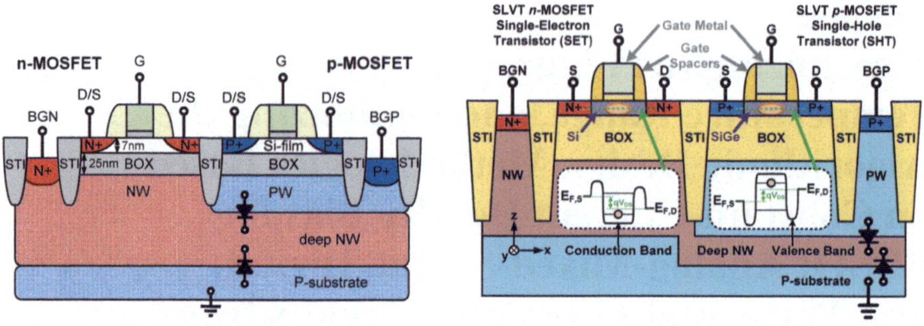

(a) (b)

Fig. 5.1 Cross section through production (**a**) 28-nm [3] and (**b**) 22-nm FDSOI CMOS technologies showing n- and p-MOSFETs in reverse-doped wells. (**b**) also illustrates how quantum dots are formed in the MOSFET channels at cryogenic temperatures below 70 K [4]

density, I_{DS}/W, the dependence on the threshold voltage, V_t, and on V_{BG} of all of these FoMs largely disappears, pointing to a generally applicable design principle for all CMOS AMS, mm-wave, and broadband linear front-end fiberoptic circuits.

In Sect. 5.3, we will develop constant current density biasing techniques taking advantage of the back-gate voltage to maximize the linearity and linear output voltage swing in all circuits. The specific mm-wave building block circuit topologies, design considerations, and performance comparison are discussed in Sect. 5.4. Circuit design examples are provided in Sect. 5.5, followed by a discussion of the significance of these techniques and experimental results in Conclusions.

Throughout this chapter, we will use the following notation N_fx$L W_f$ to describe the size of a particular MOSFET, where N_f is the number of gate fingers, L is the gate length, and W_f is the gate finger width.

5.2 FDSOI MOSFET Fundamentals

5.2.1 Threshold Voltage

Unlike in older bulk CMOS nodes where V_t varies with the square root of the bulk-to-source voltage, V_{BS}, in advanced-node planar bulk MOSFETs, V_t depends linearly on both V_{BS} and drain-to source voltage, V_{DS}, according to [5]

$$V_t = V_{t0} + \alpha V_{SB} - \eta V_{DS} \tag{5.1}$$

where $\eta \approx 0.1$ V/V captures the drain-induced barrier lowering (DIBL) effect and α is the body bias factor. The latter is practically constant in advanced planar bulk CMOS nodes due to the steep retrograde doping profile in the channel [5] and is very small in FinFETs:

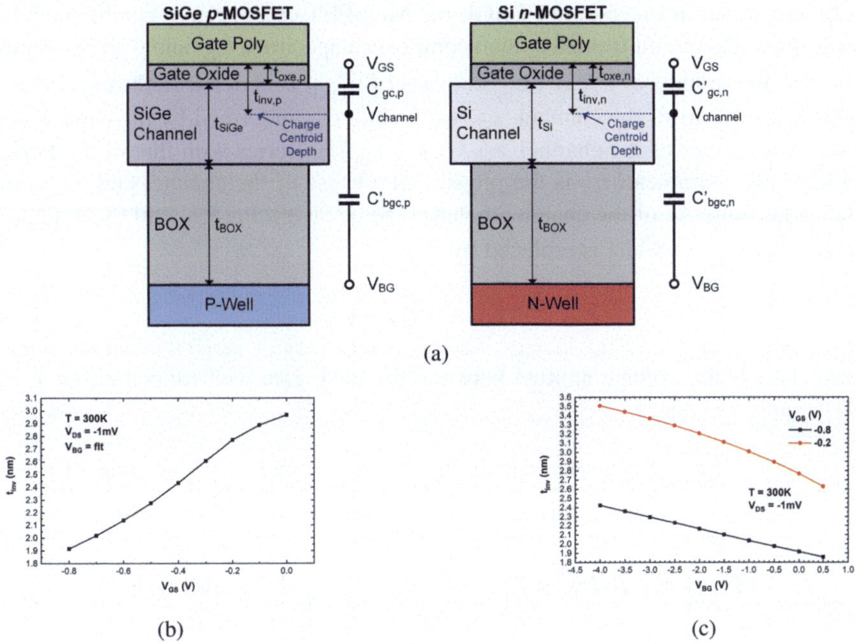

Fig. 5.2 (**a**) Cross section used to define t_{inv} in p-MOSFET and n-MOSFET. Sentaurus-simulated t_{inv} as a function of (**b**) V_{GS} with floating (flt) back gate and (**c**) V_{BG} at two different V_{GS} values in an FDSOI p-MOSFET

$$\alpha = \frac{C_{dep}}{C_{OXE}} = \frac{\epsilon_S}{\epsilon_{OX}} \frac{t_{OXE}}{W_{d\,\max}} \approx \frac{3t_{OXE}}{W_{d\,\max}}. \tag{5.2}$$

In (5.2), C_{dep} and W_{dmax} describe the depletion capacitance and the maximum depletion width in the semiconductor (Si or SiGe) channel, respectively, while t_{OXE} and C_{OXE} (in units of F/cm^2) are the effective top gate oxide thickness and the effective top gate oxide capacitance per area, respectively. They account for the multi-dielectric $HfO_2/SiON/SiO_2$ gate oxide stack, but not for the charge centroid offset, t_{inv}, from the top gate oxide/channel interface due to charge quantization in the channel [5]. The charge centroid position is illustrated in Fig. 5.2. Since, when operated in the deep subthreshold region or close to V_t, the channel is practically depleted of charge, the channel charge can be approximated by a metallic plate placed at t_{inv} below the top gate/channel interface. In this case, (5.2) can be modified to include the impact of t_{inv} :

$$\alpha = \frac{\epsilon_S}{\epsilon_{OX}} \frac{t_{OXE}}{W_{d\,\max}} + \frac{t_{inv}}{W_{d\,\max}} \approx \frac{3t_{OXE} + t_{inv}}{W_{d\,\max}} \tag{5.3}$$

As illustrated in Fig. 5.2, in FDSOI, the MOSFET cross section can be modeled as an equivalent circuit formed by two composite capacitors, C'_{gc} and C'_{cb}, connected in series, and in units of F/cm^2. C'_{gc} consists of C'_{OXE} in series with the capacitance due to the top part of the channel, $\varepsilon_{\mathrm{ch}}/t_{\mathrm{inv}}$, while C'_{cb} is formed by the capacitance of the bottom part of the channel, $\varepsilon_{\mathrm{ch}}/\left(t_{\mathrm{ch}}-t_{\mathrm{inv}}\right)$, in series with that of the buried oxide, $\mathrm{C_{BOX}}$. Parameter t_{ch} is the physical thickness of the channel and $\varepsilon_{\mathrm{ch}}$ is the relative permittivity of the channel material (Si or SiGe). In (5.1) and (5.2), C'_{dep} is replaced by C'_{cb}, and V_{BS} is replaced by

$$V_{\mathrm{BS}}=V_{\mathrm{BG}}-\left(\varphi_{\mathrm{BG}}-\varphi_{\mathrm{channel}}\right) \tag{5.4}$$

where V_{BG} is the voltage applied between the back-gate well (doped either p- or n-type) and the source,

$$\phi_{\mathrm{channel}}=\frac{kT}{q}\ln\left[\frac{N_A(\text{ channel })}{n_i}\right] \text{ or } \phi_{\mathrm{channel}}=\frac{-kT}{q}\ln\left[\frac{N_D(\text{ channel })}{n_i}\right] \tag{5.5}$$

$$\phi_{BG}=\frac{kT}{q}\ln\left[\frac{N_A(BG\text{ well })}{n_i}\right] \text{ or } \phi_{BG}=\frac{-kT}{q}\ln\left[\frac{N_D(BG\text{ Bell })}{n_i}\right] \tag{5.6}$$

The " + " sign applies when the doping in the channel or back-gate well is p-type (N_{A}), while the "−" sign applies when the doping is n-type (N_{D}). By plugging (5.4) into (5.1), the FDSOI threshold voltage dependence on the back gate becomes:

$$V_t=V_{t0}-\frac{C'_{cb}}{C'_{gc}}\left(V_{BG}-\Delta\phi_{BC}\right)-\eta V_{DS} \tag{5.7}$$

and, by expanding the capacitance formulae:

$$V_t=V_{t0}-\frac{t_{OXE}+\frac{\epsilon_{OX}}{\epsilon_{ch}}t_{inv}}{t_{BOX}+\frac{\epsilon_{OX}}{\epsilon_{ch}}\left(t_{ch}-t_{inv}\right)}\left(V_{BG}-\Delta\phi_{BC}\right)-\eta V_{DS} \tag{5.8}$$

where

- $\Delta\phi_{BC}=\phi_{BG}-\phi_{\mathrm{channel}}$ is the potential difference between the back-gate well and the channel,
- V_{to} is determined primarily by the metal–semiconductor work function at the top gate/channel interface through the flat band voltage term, V_{FB}. Since in FDSOI and advanced-node FinFET technologies the channel is undoped with $N_{\mathrm{A}}<10^{16}\ \mathrm{cm}^{-3}$ and $t_{\mathrm{ch}}<10$ nm, the second and third terms in (5.9) are in the mV range. V_{t0} is only weakly dependent on C_{OXE}, channel doping, and channel thickness and does not depend on the doping level in the back-gate well or on the voltage applied to the back gate,

$$V_{t0} = V_{FB} + \frac{qN_A t_{ch}}{C_{OXE}} + \frac{qN_A t_{ch}^2}{2\epsilon_{ch}} \tag{5.9}$$

In (5.9), which is obtained by solving Poisson's equation analytically in the direction perpendicular to the gate, $\phi_S = \frac{qN_A t_{ch}^2}{2\epsilon_{ch}}$ and $\xi_S = \frac{qN_A t_{ch}}{\epsilon_{ch}}$ are the expressions of the top gate surface potential and surface electric field, respectively.

Equation (5.8) can be recast as

$$V_t = V_{t0} + \alpha \Delta\phi_{BC} - \alpha V_{BG} - \eta V_{DS} \tag{5.10}$$

where

$$\alpha = \frac{t_{OXE} + \frac{\epsilon_{OX}}{\epsilon_{ch}} t_{inv}}{t_{BOX} + \frac{\epsilon_{OX}}{\epsilon_{ch}} (t_{ch} - t_{inv})} \tag{5.11}$$

By assuming $t_{\rm inv} = 2.8$ nm (as simulated in Fig. 5.1 at $V_{\rm GS} = -0.2$ V $\approx V_{\rm t}$ with floating back-gate), $\varepsilon_{\rm OX} = 4$, $\varepsilon_{\rm ch} = 12$, $t_{\rm BOX} = 20$ nm, $t_{ch} = 6$ nm, and $t_{\rm OXE} = 1$ nm, we calculate $\alpha = 91.8$ mV/V. As shown in Fig. 5.1, over all bias conditions, $t_{\rm inv}$ varies between 1.8 nm and 3.5 nm, which leads to a maximum variation of α with $V_{\rm BG}$ and $V_{\rm GS}$ from 104 mV/V to 74.7 mV/V.

The parameter $\Delta\varphi_{\rm BC}$ is tuned by highly doping the back-gate well *p*-type or *n*-type to provide MOSFETs with different $V_{\rm t}$ flavors. Since the impact of $\Delta\varphi_{\rm BC}$ is attenuated by α, the $V_{\rm t}$ values achievable in this manner differ by at most 100 mV.

Except for $t_{\rm inv}$, which is bias-dependent as shown in Fig. 5.26b and c, all the parameters appearing in the expression of α are bias- and temperature-independent. Therefore, the body-bias factor is practically independent of temperature from 2 K to 400 K and has only a weak dependence on $V_{\rm BG}$ and $V_{\rm GS}$, as evidenced by measurements from several groups [6–8]. The $V_{\rm t}$ variation with temperature is due to the $V_{\rm to}$ and $\Delta\varphi_{\rm BC}$ terms in (5.10).

5.2.2 *I-V Characteristics*

Even without considering the impact of the back gate, because of the thin FDSOI channel, the top gate has better control of the channel charge and current flow between source and drain when compared to a planar bulk MOSFET. This is illustrated in Fig. 5.3, where the measured output characteristics of planar bulk (dashed lines) and FDSOI *p*- and *n*-MOSFETs with identical layout geometry, manufactured by the same foundry, are compared [3]. In this case, the back gate of the FDSOI transistors was left floating. As can be seen, the FDSOI MOSFETs exhibit lower output conductance, g_0, which manifests itself in flatter output characteristics and results in higher intrinsic gain, $A_{\rm v} = g_{\rm m}/g_0$.

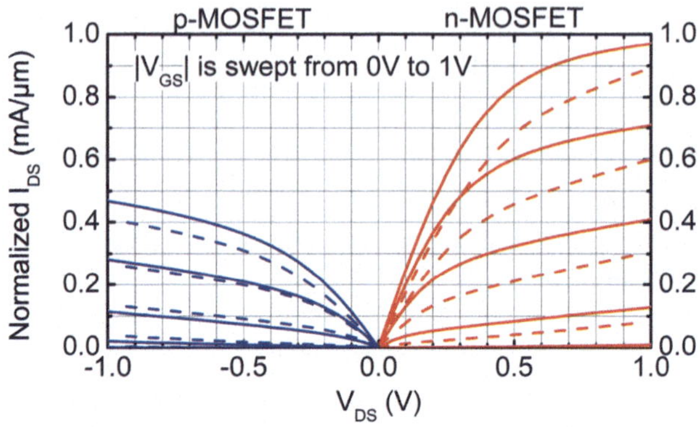

Fig. 5.3 Measured normalized (to total transistor gate width, W) output characteristics of production planar bulk (dashed lines) and FDSOI (solid lines) 28-nm MOSFETs at 300 K [3]

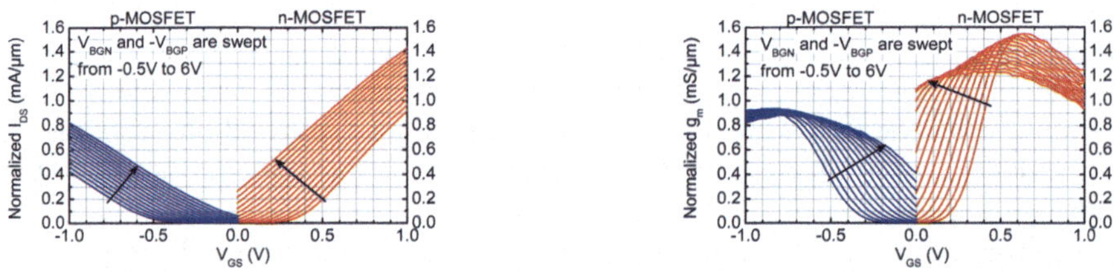

Fig. 5.4 Measured (**a**) normalized to gate width transfer characteristics and (**b**) normalized transconductance as a function of V_{GS} and V_{BG} for 28-nm FDSOI p- and n-MOSFETs at 300 K and $V_{DS} = 1$ V, $L = 24$ nm, $W_f = 500$ nm and $N_f = 40$ [3]

The wide tuneability of all AMS and high-frequency (HF) FoMs of the transistor from V_{BG} was recognized and exploited early on [3], when the first commercial 28-nm FDSOI CMOS technology [1] was introduced. Figures 5.4, 5.5, and 5.6 reproduce the measured normalized transfer characteristics, $I_{DS}(V_{GS})/W$, and the FoMs g_m, f_T, f_{MAX}, and MAG at 60 GHz, as a function of V_{GS} and V_{BG} in saturation at $|V_{DS}| = 1$ V [3] for the 28-nm FDSOI p- and n-MOSFETs in Fig. 5.3. These plots show body factors $\alpha_n \approx 92.8\,\text{mV/V}$ and $\alpha_p \approx 86.2\,\text{mV/V}$ for the n- and p-MOSFETs, respectively, in line with the calculations presented earlier.

The resistive, capacitive, and inductive parasitics associated with the wiring above the transistor up to the top metal were not removed in the f_T, f_{MAX}, and MAG measurements. Thus, Figs. 5.5, 5.6, and 5.7 reflect the real performance of the fully wired transistor in a mm-wave circuit. The expected (from [9]) yet still striking behavior of these HF FoMs when plotted vs. the drain current density, I_{DS}/W, in Figs. 5.6b and 5.7, confirms that, for any given current density, the f_T, f_{MAX}, and MAG values vary by less than 20% as the threshold voltage is changed by hundreds of mV when $|V_{BG}|$ varies from -0.5 V to $+6$ V. The variation is larger for the

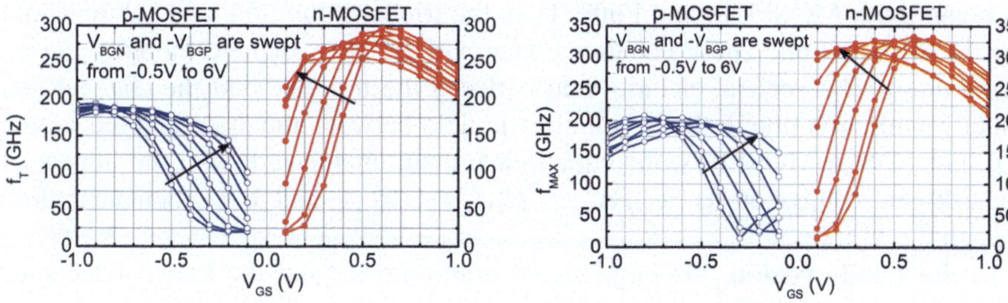

(a) (b)

Fig. 5.5 Measured (**a**) f_T and (**b**) f_{MAX} as a function of V_{GS} and V_{BG} for 28-nm FDSOI p- and n-MOSFETs at 300 K and $V_{DS} = 1$ V, $L = 24$ nm, $W_f = 500$ nm and $N_f = 40$ [3]

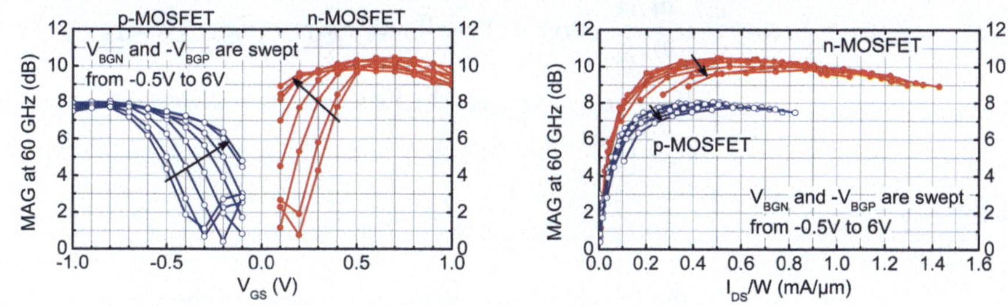

(a) (b)

Fig. 5.6 Measured MAG at 60 GHz as a function of (**a**) V_{GS} and V_{BG}, and (**b**) of I_{DS}/W for 28-nm FDSOI p and n-MOSFETs at 300 K and $V_{DS} = 1$ V, $L = 24$ nm, $W_f = 500$ nm and $N_f = 40$ [3]

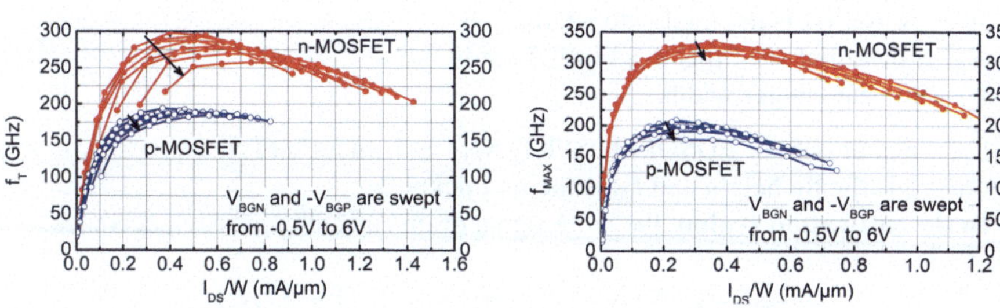

(a) (b)

Fig. 5.7 (**a**) f_T and (**b**) f_{MAX} as a function of I_{DS}/W and V_{BG} from the same data presented in Fig. 5.5 [3]

n-type than for the p-type MOSFET, is highest around the peak values of g_m and f_T, and accurately tracks the dependence of t_{inv} on V_{BG}.

The measured characteristics discussed earlier are well-captured and explained by the short-channel piece-wise linear $I_{DS}\,(V_{GS}, V_{DS})$ and $g_m\,(V_{GS}, V_{DS})$ equations for planar bulk MOSFETs and FinFETs in the triode, saturation, and subthreshold regions of operation. These equations, summarized next, include mobility degradation due to the vertical electric field between the top gate and the channel and velocity saturation due to the lateral electric field between the drain and the source [10]. For FDSOI MOSFETs, the equations remain unchanged, with the impact of the back-gate voltage on I_{DS}, g_{m}, A_{v}, $g_{\mathrm{m}}/I_{\mathrm{DS}}$, f_{T}, f_{MAX}, and MAG being captured through the V_{t} term.

In the triode region, the large signal drain current and the transconductance expressions are given by

$$i_{DS}\left(v_{GS}, v_{DS}\right) = \frac{\mu_{nso} C_{OXE} \frac{W}{L}\left[v_{GS} - V_t - \frac{m}{2} v_{DS}\right] v_{DS}}{1 + \theta\left(v_{GS} - V_t\right) + \frac{\mu_{nso} v_{DS}}{2 L v_{\mathrm{nsat}}}} \tag{5.12}$$

$$\begin{aligned} g_m &\stackrel{\mathrm{def}}{=} \frac{\partial i_{DS}}{\partial v_{GS}} \lfloor v_{DS} = V_{DS} \text{ fixed}, v_{BG} = V_{BG} \text{ fixed} \rfloor \\ &= \frac{\mu_{nso} C_{OXE} N_f N_{fin} W_f V_{DS}}{L\left[1 + \theta\left(V_{GS} - V_t\right)\right]^2} \end{aligned} \tag{5.13}$$

where

- $m = 1 + \alpha \approx 1.1$,
- $\mu_{\mathrm{nso}}/\mu_{\mathrm{pso}}\left[\mathrm{cm}^2/\mathrm{V}^*\mathrm{s}\right]$ is the low-field electron/hole mobility in the channel,
- $v_{\mathrm{nsat}}/v_{\mathrm{psat}}\,[\mathrm{cm/s}]$ is the saturation velocity of electrons/holes in the channel material,
- $C_{\mathrm{OXE}}\left[F/\mathrm{cm}^2\right]$ is the effective top gate oxide capacitance (may include impact of average t_{inv} for higher precision),
- $\theta[1/\mathrm{V}]$ captures the mobility degradation due to vertical electric field,
- $W = N_{\mathrm{f}} N_{\mathrm{fin}} W_{\mathrm{f}}$ is the total gate width,
- N_{fin} is the number of fins in a FinFET, and
- W_{f} is the gate finger width.

For a planar bulk or FDSOI MOSFET, $N_{\mathrm{fin}} = 1$. In a FinFET, $W_{\mathrm{f}} = 2h_{\mathrm{ch}} + t_{\mathrm{ch}}$, where h_{ch} is the fin height and t_{ch} is the fin thickness.

In the saturation region, the large signal drain current expression becomes:

$$i_{DSat} = \frac{\mu_{\mathrm{nso}} C_{OXE} N_f N_{fin} W_f \left(v_{GS} - V_t\right)^2 \left(1 + \lambda v_{DS}\right)}{2mL\left[1 + \theta'\left(v_{GS} - V_t\right)\right]} \tag{5.14}$$

where

$$\theta' = \theta + \frac{\mu_{\mathrm{nso}}}{2mLv_{\mathrm{nsat}}} \tag{5.15}$$

- $\lambda[1/\mathrm{V}]$ is the channel length modulation parameter,

- θ'[1/V] captures both mobility degradation due to vertical field and velocity saturation due to lateral field, which are indistinguishable when the transistor operates in the saturation region.

$$g_{msat} = \frac{\partial i_{DS}}{\partial v_{GS}}|_{v_{DS},v_{BG}=\text{fixed}}$$
$$= \frac{\mu_{nso} C_{OXE} N_f N_{fin} W_f}{mL} \frac{(V_{GS} - V_t)(1 + \lambda V_{DS})}{[1 + \theta'(V_{GS} - V_t)]^2} \left[1 + \frac{\theta'}{2}(V_{GS} - V_t)\right] \quad (5.16)$$

with

$$V_{Dsat} = \frac{V_{GS} - V_t}{m} \frac{[1 + \theta(V_{GS} - V_t)]}{1 + \left[\theta + \frac{\mu_{nso}}{2mLv_{\text{nsat}}}\right](V_{GS} - V_t)} \quad (5.17)$$

Finally,

$$g_{mb} \stackrel{\text{def}}{=} \frac{\partial i_{DS}}{\partial v_{BG}}|_{v_{DS},v_{GS}=\text{ fixed}} = \alpha g_m \approx 0.1 g_m \quad (5.18)$$

The parameter θ can be easily extracted from DC measurements in the triode region by plotting $1/g_{\text{m}}$vs. $(V_{\text{GS}} - V_{\text{t}})^2$ when the transistor is biased at $V_{\text{DS}} = 1$ mV to 5 mV, to ensure that the lateral electric field is $\approx$1 mV/10 nm $\approx 10^3$ V/cm, low enough to be significantly below the onset of velocity saturation at the drain end of the channel. The classical long-channel, "square" law equation

$$I_{DSat} \approx \frac{\mu_{nso} C_{OXE} N_f N_{fin} W_f (V_{GS} - V_t)^2 (1 + \lambda V_{DS})}{2mL} \quad (5.19)$$

that describes the relationship between the drain current and the gate–source voltage is valid only at low $V_{\text{GS}} - V_{\text{t}}$, corresponding to current densities $I_{\text{DS}}/W <$ 0.15 mA/μm[10], as can be observed in Fig. 5.4. This region is easily identified in Fig. 5.4b as the region where g_{m} is a linear function of V_{GS}:

$$g_{msat} \approx \frac{\mu_{nso} C_{OXE} N_f N_{fin} W_f}{mL} (V_{GS} - V_t)(1 + \lambda V_{DS}) \quad (5.20)$$

However, for most of the V_{GS} range, where $I_{\text{DS}}/W \geq 0.3$ mA/μm [10], the dependence of the drain current on V_{GS} simplifies to a linear law, g_{m} reaches its peak value and remains approximately constant as V_{GS} and the drain current density continue to increase. In this part of the saturation region, the MOSFET behaves like a very linear device since its small signal gain does not change with the drain current or the gate–source voltage:

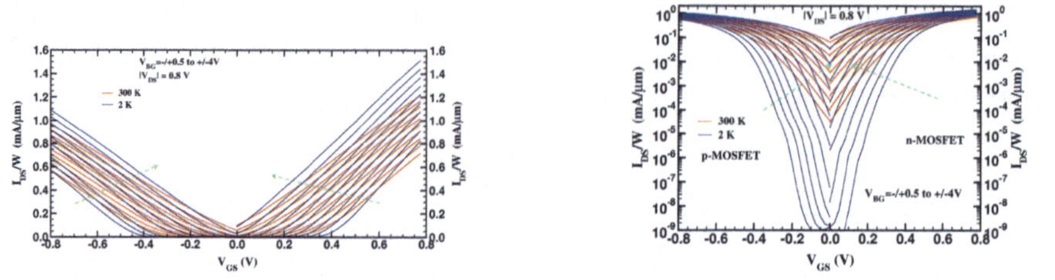

(a) (b)

Fig. 5.8 Measured I_{DS}/W vs. V_{GS} as a function of V_{BG} for 22 nm FDSOI 40 × 20 nm × 590 nm *p*- and 40 × 20 nm × 720 nm *n*-MOSFETs with 2× gate pitch and double gate contact at 2 K and 300 K biased at $|V_{DS}| = 0.8$ V and $V_{BG} = +/-0.5$ V to $-/+4$ V, 0.5 V steps: (**a**) linear *y*-axis showing the saturation region and (**b**) logarithmic *y*-axis highlighting the subthreshold region

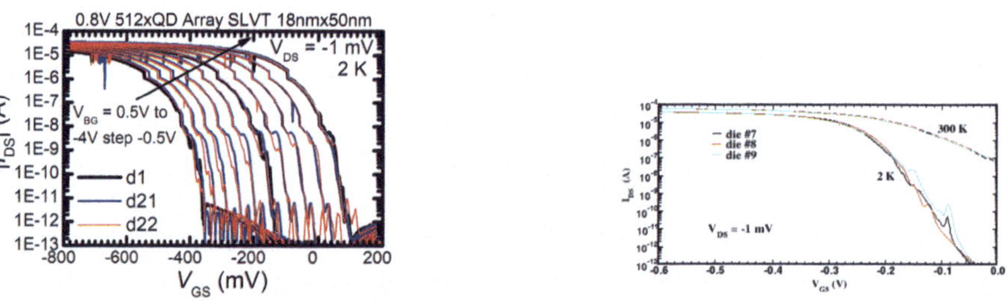

(a) (b)

Fig. 5.9 Transfer characteristics of 512 minimum-size *p*-type MOSFETs connected in parallel, measured in the subthreshold triode region at $V_{SD} = 1$ mV across 3 dies: (**a**) 22-nm FDSOI technology, as a function of V_{BG} at 2 K and (**b**) 3-nm FinFET technology at 2 K and 300 K

$$I_{DSat} \approx \frac{\mu_{nso} C_{OXE} N_f N_{fin} W_f \left(V_{GS} - V_t\right)\left(1 + \lambda V_{DS}\right)}{2mL\theta'} = g_{mpsat}\left(V_{GS} - V_t\right) \tag{5.21}$$

$$g_{mpsat} \approx \frac{\mu_{nso} C_{OXE} N_f N_{fin} W_f \left(1 + \lambda V_{DS}\right)}{2mL\theta'} \tag{5.22}$$

The distinct "square law" (close to V_t) and "linear law" sections of the saturation region can be clearly observed in the measured transfer and $g_m - V_{GS}$ characteristics of both 28-nm FDSOI (in Fig. 5.4) and 22-nm FDSOI (in Figs. 5.8a and 5.9, respectively) MOSFETs.

In the subthreshold region, the drain current expression is given by [5]:

$$i_{DS} = \frac{\mu_{nso} C_{OXE} N_f N_{fin} W_f (m-1)}{L}\left(\frac{k_B T}{q}\right)^2 e^{\frac{q(v_{GS} - V_t)}{mk_B T}}\left(1 - e^{\frac{-qv_{DS}}{k_B T}}\right) \tag{5.23}$$

where k_B is Boltzmann's constant. In (5.23), the back-gate voltage appears in the exponential term through V_t. The expression of the exponential term in $V_{GS} - V_t$ can be obtained using the capacitive voltage divider equivalent circuit shown in Fig. 5.2a. Here, $v_{channel}$ is an approximation of the surface potential, φ_S:

$$v_{channel} = \phi_S = \frac{C'_{gc,p/n}}{C'_{gc,p/n} + C'_{bgc,p/n}} v_{GS} + \frac{C'_{bgc,p/n}}{C'_{gc,p/n} + C'_{bgc,p/n}} V_{BG} \quad (5.24)$$

As observed in the measured data of 22-nm FDSOI MOSFETs for V_{BG} in the $+/-0.5$ V to $-/+4$ V range shown in Fig. 5.8, the entire $I_{DS} - V_{GS}$ characteristics, including the subthreshold region in Fig. 5.8b, can be shifted to lower or higher V_{GS} values by more than 0.4 V over this range of back-gate voltages. As the temperature decreases from 300 K to 2 K, V_t, the maximum drain–source current, $I_{ON} = I_{DS}$ ($V_{GS} = 0.8$ V, $V_{DS} = 0.8$ V), and the subthreshold slope increase while the drain–source leakage current (I_{DS} at $V_{GS} = 0$ V) decreases by several orders of magnitude. For each V_{BG} value, there is a V_{GS} value at which the drain current density remains constant as the temperature changes from 2 K to 300 K. These constant-with-temperature V_{GS} and drain current values have also been observed in FinFETs and planar bulk MOSFETs. What is unique to FDSOI MOSFETs is that they can be adjusted to some extent from the back-gate voltage. This feature can be exploited to optimize the performance of cryogenic control circuits in quantum processors (QPs).

Notes

1. The DC operating point equations are obtained by replacing the large signal current and voltages i_{DS}, v_{GS}, v_{DS} in (5.12), (5.14), and (5.23) with the DC current and voltages: I_{DS}, V_{GS}, V_{DS}.
2. For the p-MOSFET equations, v_{GS} is replaced by v_{SG}, v_{DS} by v_{SD}, i_{DS} by i_{SD}, and V_{tn} by $|V_{tp}|$ in the n MOSFET equations aforementioned.
3. To accurately account for the impact of the bias dependence of the inversion charge offset, t_{inv}, C_{oxe} should be replaced by C_{gc} in all the drain–source current and transconductance equations: (5.12)–(5.14), (5.16), and (5.19). This is important to explain why the peak transconductance value decreases by approximately 15% as the back gate is forward-biased, pushing the charge centroid closer to the channel-to-buried-oxide interface from the top-gate/channel interface, allowing for a more uniform distribution of the charge in the vertical cross section of the channel.
4. By setting $m = 1, \theta = 0$, and $\theta' = 0$ (i.e., $v_{nsat} \to \infty$), (5.12)–(5.14), (5.16) and (5.19) default to their long-channel versions, which ignore the subthreshold current, velocity saturation due to the lateral electric field at the drain end of the channel, and mobility degradation due to the vertical electric field at large V_{GS} in both the triode and the saturation regions.
5. Since channel strain is applied in advanced MOSFETs to enhance mobility, the latter becomes layout-geometry-dependent via W_f, L, and, through N_f, the width of active area occupied by the MOSFET. Therefore, the finger width and gate

length are normally fixed and given to the circuit designer, and a large number of dummy gates are placed on both sides of the active gates of a transistor to minimize mobility variation along the width of the active region due to decreasing strain for gates placed farther from the edges of the active region.

6. At 2 K, humps can be observed in the subthreshold region at low current densities in Fig. 5.8b. At large V_{DS}, these can be explained by a combination of charge confinement and quantization in two or even three dimensions in the thin channel, forming a 2D or 1D (depending on channel width) electron or hole gas [10], and the presence of a temperature-independent tunneling current between source and drain at small gate lengths [11], which dominates over the classical diffusion current at cryogenic temperatures. As shown in Fig. 5.9, for narrow width (50 nm) and minimum gate length (18 nm) FDSOI and 3-nm FinFET MOSFETs operating in the triode region at low V_{DS}, these humps in the subthreshold region evolve into large peaks and deep valleys. They are the result of Coulomb blockade effects and sequential tunneling through the quantized energy levels of the quantum dot (QD) formed in the channel below the gate [4, 12]. At temperatures below 50 K, the presence of gate oxide spacers as narrow as 3 nm is sufficient for potential barriers to be formed between the source and drain contact regions and the channel. These potential barriers confine the electron- or hole charge in the third dimension [12]. Another cause of the strong current "oscillations" (i.e., peaks and valleys) in the subthreshold region can be a single random dopant atom accidentally present in the otherwise undoped channel below the gate. Statistically, at a doping level of 10^{14} cm^{-3} in the 6 nm $\times$ 50 nm $\times$ 18 nm volume of the FDSOI QD, a dopant atom could be found in one of every 1852 QDs. If the channel doping is 10^{16} cm^{-3}, a dopant atom is expected to be present in every second QD. Small channel width, channel length, and gate-oxide spacer width variations, as well as random dopants can cause fluctuations in the V_{GS} coordinate and value of the Coulomb current peaks at cryogenic temperatures of the 512 QDs connected in parallel in Fig.5.9, without affecting room temperature characteristics, as shown in Fig. 5.9b. Finally, it is noted that, even at 2 K, there is excellent match in the exponential regions between Coulomb peaks across the three 22-nm FDSOI and across the three 3-nm FinFET dies.

5.2.3 Low-Frequency AMS FoMs

The measured g_m/W and g_m/I_{DS} in the saturation and subthreshold regions at $V_{DS} = \pm 0.8$ V of production 22-nm FDSOI p- and n-MOSFETs are plotted in Fig 5.10 as a function of V_{GS} for back-gate voltages in the $+/-0.5$ V to $-/+4$ V range at both 2 K and 300 K. As in the case of the transfer characteristics in Fig. 5.8, V_t and the entire $g_m - V_{GS}$ and $g_m/I_{DS} - V_{GS}$ curves can be shifted from the back gate by more than 0.4 V to the desired V_{GS} value that best suits a

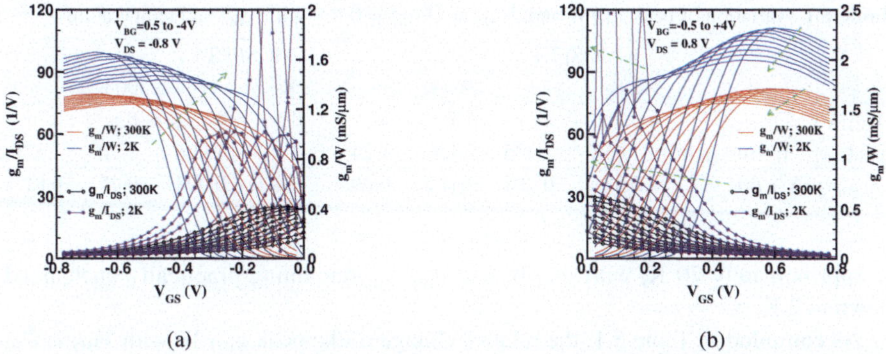

Fig. 5.10 Measured g_m/W (red and blue lines) and g_m/I_{DS} (black and violet lines and symbols) vs. V_{GS} as a function of V_{BG} and temperature for 22-nm FDSOI (**a**) 40 × 20 nm × 590 nm p-and (**b**) 40 × 20 nm × 720 nm n-MOSFETs with 2× gate pitch and double gate contact. $V_{DS} = 0.8$ V and $V_{BG} = 0.5$ V to -4 V, -0.5 V steps

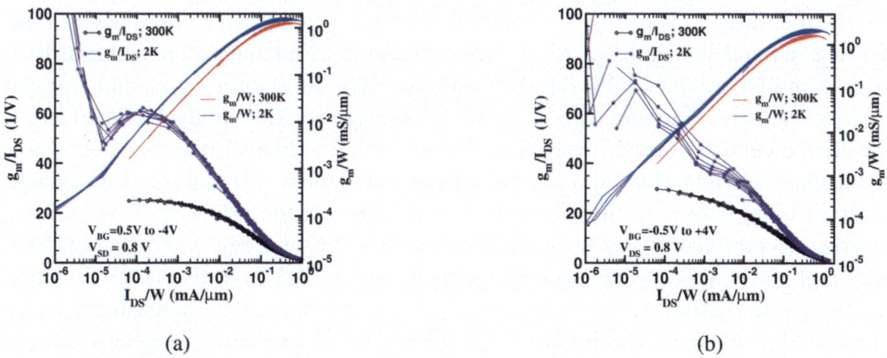

Fig. 5.11 Measured g_m/W (red and blue lines) and g_m/I_{DS} (black and violet lines and symbols) in Fig. 5.10 replotted vs. I_{DS}/W

particular circuit design strategy and circuit performance. However, when g_m/W and g_m/I_{DS} are plotted versus current density in Fig. 5.11, the g_m/W vs. I_{DS}/W and g_m/I_{DS} vs. I_{DS}/W curves corresponding to different V_{BG} values overlap in both the saturation and the subthreshold regions, re-enforcing the idea that CMOS circuit design based on setting the current density rather than V_{GS} or $V_{GS} - V_t$ is largely insensitive to V_t variation [10]. Moreover, Fig. 5.11 shows that the drain–source current density at which the normalized transconductance reaches its peak at each V_{BG} value remains largely unchanged over temperature from 2 K to 300 K. This provides critical insight into how to design AMS circuits that are robust to process variation and have the highest FoMs at any temperature. The peak normalized transconductance and g_m/I_{DS} values in saturation at $|V_{DS}| = 0.8$ V increase at 2 K compared to 300 K. The peak g_m improves by 25% and 35%, respectively, for the p MOSFET and n-MOSFET, respectively. As reported by several groups [6–8], this improvement is the result of higher mobility and saturates

Table 5.1 Variation of peak g_m/W and $J_{p,gm}$ at $|V_{DS}| = 0.8$ V with V_{BG} and temperature

	p-MOSFET				n-MOSFET			
	2 K		300 K		2 K		300 K	
V_{BG}(V)	0.5	−4	0.5	−4	−0.5	4	−0.5	4
peak- g_m/W (mA/μm)	1.658	1.435	1.326	1.24	2.314	1.996	1.709	1.571
$J_{p,gm}$ (mA/μm)	0.371	0.708	0.412	0.696	0.429	0.993	0.471	0.81

at approximately 70 K, with g_m/W and g_m/I_{DS} remaining practically unchanged down to 2 K.

As compiled in Table 5.1, the relative change of the peak g_m/W with V_{BG} is 8% and 15% at 300 K and 2 K, respectively, and is well explained by the change in t_{inv} with V_{BG} simulated in Fig. 5.2c. At the same time, the $J_{p,gm}$ of the p-MOSFET can be increased by 70% and 90% at 300 K and 2 K, respectively, by forward biasing the back gate from 0.5 V to $-$ 4 V. The corresponding increase of the n-MOSFET $J_{p,gm}$ is 70% and 130%, respectively.

The measurements in Fig. 5.11 point to a unique property of FDSOI MOSFETs. While at a fixed V_{DS} and V_{BG} $J_{p,gm}$ does not change with temperature and transistor size, as in planar bulk MOSFETs [10] and FinFETs, in FDSOI, $J_{p,gm}$ and I_{ON} can be increased by changing V_{BG} to provide stronger forward body bias and move the charge centroid farther away from the top-gate-to-channel interface and closer to the channel-BOX (buried oxide) interface. The strongest signature of the charge centroid being closer to the channel-BOX interface is the "shoulder" that can be clearly observed in the $g_m/W - V_{GS}$ blue curve at 2 K and $V_{BG} = 4$ V in Fig. 5.8b. Although the peak-g_m value is degraded by up to 10–15% from its maximum when the back gate is strongly forward-biased, g_m remains flat for an increasingly wider range of V_{GS} values. The higher $J_{p,gm}$ allows for the design of higher-linearity, high-frequency, and high-bandwidth, large-power circuits. Similarly, higher I_{ON} increases the speed and drive of digital circuits [10] without having to increase the power supply voltage, V_{DD}, thus reducing DC and dynamic power consumption. The g_m/I_{DS} ratio is an important figure of merit used in low-frequency analog circuit design. In low-voltage, low-power circuits, the MOSFET is typically biased in the subthreshold region, where, as in a bipolar transistor, the drain current is due to diffusion rather than drift. It is in this region that the MOSFET g_m/I_{DS} ratio approaches that of a bipolar transistor, with the added benefit of a lower turn-on voltage and, in the case of FDSOI, tuneability from V_{BG}. As can be seen in Fig. 5.11 (black and violet curves), when the MOSFET is biased at the lowest current densities, g_m/I_{DS} reaches its maximum value of $\approx$25 V^{-1} and $\approx$30 V^{-1}, respectively, for p- and n-MOSFETs, respectively, at 300 K, and $\approx$70–100 V^{-1} at 2 K. These values are similar to those measured in FinFETs in Fig. 5.12, and approach those of the SiGe HBT for which measurements are collected in Fig. 5.13a. The maximum value of g_m/I_{DS} obtained from (5.23) is q/mk_BT. In FDSOI MOSFETs and FinFETs, $m \approx 1.1$, while $m = 1$ in the case of the SiGe HBTs. Excepting the region of heavy injection at large current densities where it decreases with

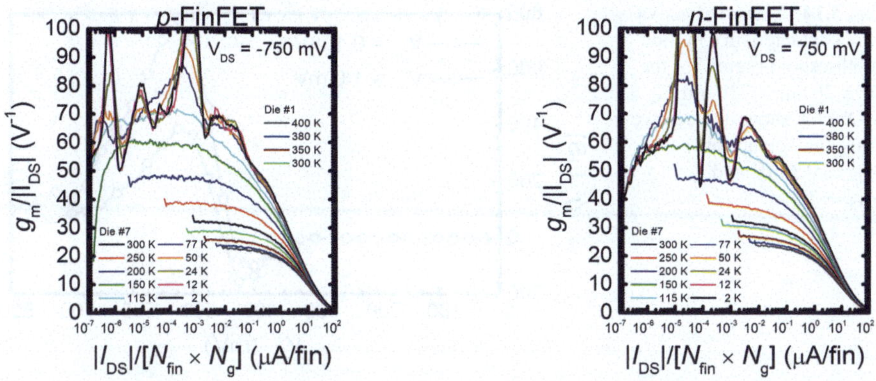

(a) (b)

Fig. 5.12 Measured g_m/I_{DS} vs. $I_{DS}/N_f N_{fin}$ as a function of temperature in 3-nm (**a**) p- and (**b**) n-FinFETs at $|V_{DS}| = 0.75$ V

(a) (b)

Fig. 5.13 (**a**) Measured g_m/I_C vs. I_C and (**b**) comparison of the measured peak value of g_m/I_C with theory as a function of temperature for SiGe HBTs at $V_{BC} = 0$ V

increasing current, for SiGe HBTs $g_m/I_C = q/k_B T$. It remains constant over many decades of current, with a value of 38 V^{-1} at 300 K, which is well matched by the measurements in Fig. 5.13b. However, as reported in [13] and immediately apparent in Fig. 5.13b, the slope of the exponential dependence of I_C on V_{BE} no longer follows the expected theory below 100 K. The same unexpected behavior has been observed for the subthreshold slope in all types of MOSFETs. In the latter case, it has been empirically attributed to the broadening of the conduction and valence bands at cryogenic temperatures [14]. However, most recently, it has been demonstrated that the saturated slope can be explained by the dominance of the tunnelling current and modelled analytically [15].

The plots in Figs. 5.10, 5.11, 5.12, and 5.14 also show that g_m/W and g_m/I_{DS} become negative and experience large fluctuations in the deep subthreshold region at 2 K due to the phenomena already covered when explaining the transfer characteristics in Figs. 5.8 and 5.9. These fluctuations or oscillations in I_{DS}, g_m and g_m/I_{DS} are universal in thin channel devices such as 16 nm [16], 7 nm, 5 nm [17], and

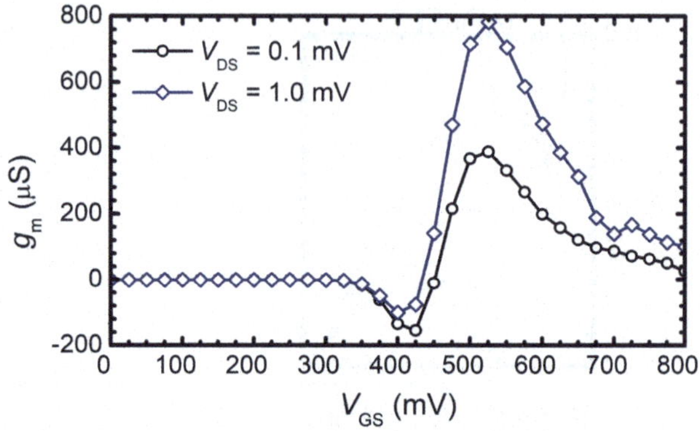

Fig. 5.14 Measured g_m vs. V_{GS} characteristics in the subthreshold region for the 40 × 20 nm × 720 nm MOSFET showing negative transconductance at $V_{DS} = 0.1$ mV and $V_{DS} = 1$ mV

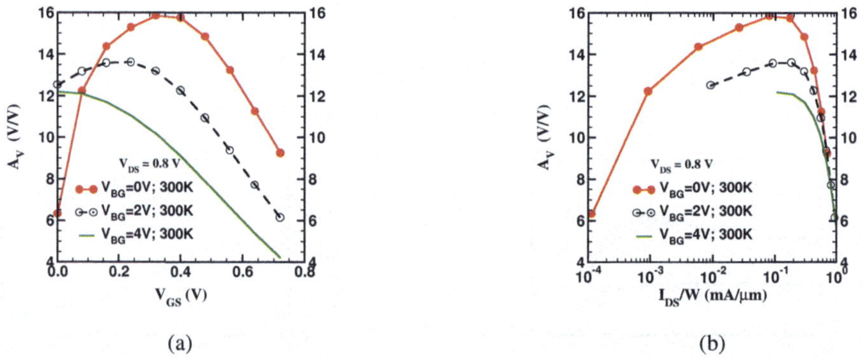

Fig. 5.15 Measured (**a**) A_V vs. V_{GS} and (**b**) A_V vs. I_{DS}/W at 300 K as a function of V_{BG} for 22-nm FDSOI 40 × 20 nm × 720 nm n-MOSFET with 2× gate pitch and double gate contact. $V_{DS} = 0.8$ V

3 nm FinFETs (Fig. 5.12), not restricted to FDSOI. Therefore, biasing MOSFETs in the deep subthreshold region in AMS circuits is not recommended at cryogenic temperatures unless the Coulomb blockade and tunnelling effects are well-captured by the design kit model to avoid operation in the negative transconductance region at low current densities. Unfortunately, unlike in SiGe HBTs where its value is maximized over many decades of current, in MOSFETs, the g_m/I_{DS} ratio is a strong function of current density and reaches its maximum only at the very lowest drain current densities, below 0.1μA/μm, precisely where fluctuations in g_m and g_m/I_{DS} occur.

For opamp design, another important figure of merit is the intrinsic voltage gain, $A_V = g_m/g_o$, plotted in Figs. 5.15 and 5.16 for the same 22-nm FDSOI short channel p- and n-MOSFETs as in Figs. 5.8, 5.10 and 5.11. In these plots, g_m and g_o (MOSFET output conductance), and therefore A_v, are extracted from the measured S-parameters in the 0.5–5 GHz range. Figure 5.15a shows that the $A_V - V_{GS}$ characteristics shift with V_{BG} just like the transconductance characteristics. The peak voltage gain increases from 12 to 16 when V_{BG} changes from 4 V to 0 V for

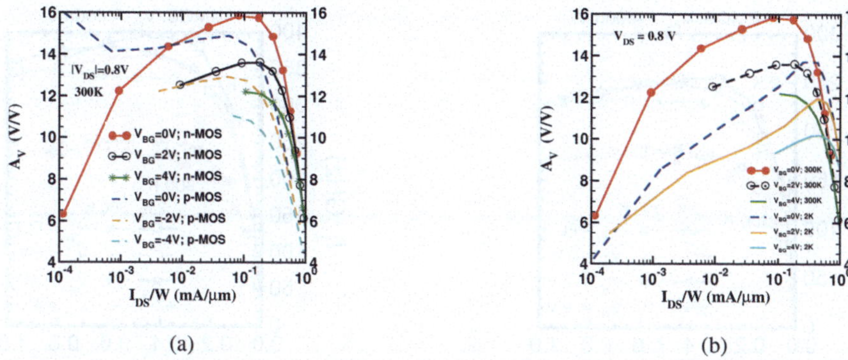

Fig. 5.16 (**a**) Measured A_V vs. I_{DS}/W at 300 K as a function of V_{BG} for 22-nm FDSOI (**a**) $40 \times 20\,\text{nm} \times 590\,\text{nm}$ p- and $40 \times 20\,\text{nm} \times 720\,\text{nm}$ n-MOSFETs with $2\times$ gate pitch and double gate contact. $|V_{DS}| = 0.8$ V (**b**) Measured A_V vs. I_{DS}/W at 2 K and 300 K as a function of V_{BG} for the n-MOSFET only

this short-channel n-MOSFET. This relative change is higher than that observed for transconductance, indicating that the output conductance also changes with V_{BG}. Figures 5.15b and 5.16a, highlight the fact that at 300 K the peak-A_V current density, $J_{p,Av}$, remains practically unchanged at about 0.1 mA/μm and 0.08 mA/μm, respectively, for n- and p-MOSFETs, respectively. Again, it confirms that circuit design based on constant current density ensures that the best voltage gain is achieved even in the presence of large threshold voltage variation. The ratio between $J_{p,Av}$ for p- and n MOSFETs seems to match that of $J_{p,gm}$ and also the mobility ratio of holes and electrons in these devices. The peak- A_v value for p-MOSFETs is also slightly lower than that of n-MOSFETs.

The $A_V - I_{DS}$/W characteristics of the n-MOSFET measured at 300 K and 2 K are compared in Fig. 5.16b. Unlike $J_{p,gm}$ and the other characteristic current densities $J_{p,fT}$ and $J_{p,fMAX}$, which will be discussed next and which remain constant with temperature, J_{pAv} increases significantly from ≈0.1 mA/μm to ≈0.3 mA/μm when the temperature is reduced from 300 K to 2 K. Also, unlike all the other figures of merit, the peak A_V value decreases by ≈10% at 2 K compared to 300 K. This experimental behavior indicates that opamp design must be optimized differently at 2 K than at 300 K, as will be discussed later in this chapter.

5.2.4 High-Frequency FoMs

The f_T, f_{MAX}, and MAG data measured at 300 K as a function of V_{BG} for the 28-nm FDSOI p- and n-MOSFETs in Figs. 5.5, 5.6, and 5.7 are complemented in Figs. 5.17 and 5.18 by the measured f_T and f_{MAX} characteristics vs. I_{DS}/W at 300 K and at 2 K for the 22-nm FDSOI devices whose measured DC and low-frequency

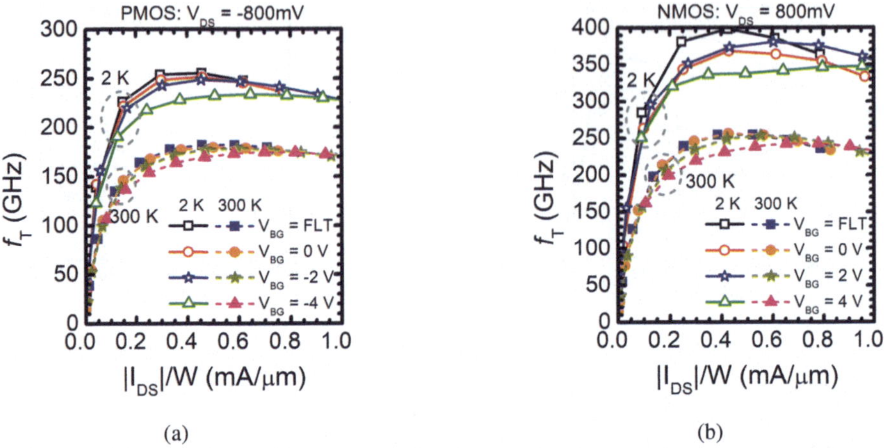

Fig. 5.17 Measured f_T as a function of I_{DS}/W and V_{BG} at 2 K and 300 K for 2× gate pitch 22-nm FDSOI (**a**) 40 × 20 nm × 590 nm *p*- and (**b**) 40 × 20 nm × 720 nm *n*-MOSFETs contacted on both sides of the gate

(LF) AC characteristics were presented earlier at different back-gate voltages. As first reported in [18], and as in planar bulk MOSFETs [10] and FinFETs [19], the peak-f_T, $J_{p,fT}$, and the peak-f_{MAX}, $J_{p,fMAX}$, current densities do not change with temperature. When the devices in Figs. 5.17 and 5.18 are biased in saturation at $|V_{DS}| = 0.8$ V and with floating back-gate, $J_{p,FMAX}$ is approximately 0.3 mA/μm and approximately 0.35 mA/μm for *p*- and *n*-MOSFETs, respectively, while $J_{p,fT}$ is approximately 0.5 mA/μm and approximately 0.45 mA/μm, respectively [19]. As can be seen in Fig. 5.17, the peak- f_T value improves from 180 GHz at 300 K to 255 GHz at 2 K for the *p*-MOSFET, while the *n*-MOSFET peak f_T increases from 255 GHz to 400 GHz. The increase in f_T is due to larger g_m and a reduction in C_{gs}, as shown in Figs. 5.20, 5.21, 5.22, and 5.23. The behavior is well-captured by the f_T formula (5.25). In (5.25), the impact of the source resistance, R_S, is already included in the g_m and C_{gs} values obtained either from simulations or measurements [10].

$$\frac{1}{2\pi f_T} = \frac{C_{gs} + C_{gd}}{g_m} + C_{gd} R_d \tag{5.25}$$

$$f_{MAX} \approx \frac{f_T}{2\sqrt{\left(R_i + R_s + R_g\right) g_o + 2\pi f_T R_g C_{gd}}} \tag{5.26}$$

As in the case of the 28-nm FDSOI technology, the 22-nm FDSOI MOSFET f_T and f_{MAX} vs. I_{DS}/W characteristics measured at different V_{BG} values overlap. They exhibit a wide range of current densities, from approximately 0.15 mA/μm to 1 mA/μm, where f_T and f_{MAX} values vary by less than − 10% from their peak values at a particular V_{BG}.

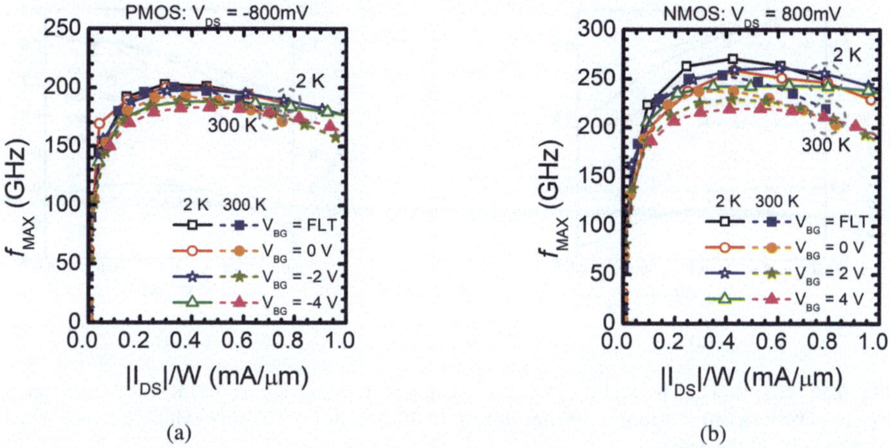

Fig. 5.18 Measured f_{MAX} as a function of I_{DS}/W and V_{BG} at 2 K and 300 K for 2× gate pitch 22-nm FDSOI (**a**) 40 × 20 nm × 590 nm$p$ - and (**b**) 40 × 20 nm × 720 nm-MOSFETs contacted on both sides of the gate

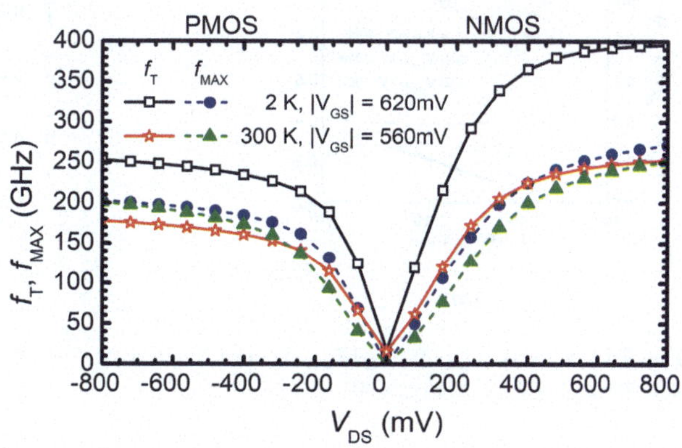

Fig. 5.19 Measured f_T and f_{MAX} as a function of V_{DS} at 2 K and 300 K for for 2× gate pitch 22-nm FDSOI 40 × 20 nm × 590 nm p- and 40 × 20 nm × 720 nm n-MOSFETs contacted on both sides of the gate

The measured dependence of f_T and f_{MAX} on V_{DS} is captured for the same 22-nm FDSOI devices at 2 K and 300 K in Fig. 5.19. At both temperatures, f_T and f_{MAX} increase rapidly with V_{DS} and tend to saturate at $V_{DS} = 0.4$ V. This means that there is little degradation of device speed in CMOS inverters or cascode topologies operating from the nominal supply voltage of $V_{DD} = 0.8$ V and $|V_{DS}| = 0.4$ V per transistor. Nevertheless, the highest f_T and f_{MAX}, just like the highest g_m, MAG, and linearity, are achieved at $|V_{DS}| = 0.8$ V at both temperatures.

The measured dependence of the transistor capacitances on V_{GS}, V_{BG}, and current density, shown in Figs. 5.20, 5.21, 5.22, and 5.23 at 2 K and 300 K and $|V_{DS}| = 0.8$ V, is relevant for understanding the f_T, f_{MAX}, and MAG dependence on bias current and terminal voltages, and provides important information for the hand design of mm-wave circuits. As can be observed in Figs. 5.20 and 5.22, the $C_{gs}(V_{GS})$ and $C_{ds}(V_{GS})$ characteristics shift with V_{BG} according to the V_t body

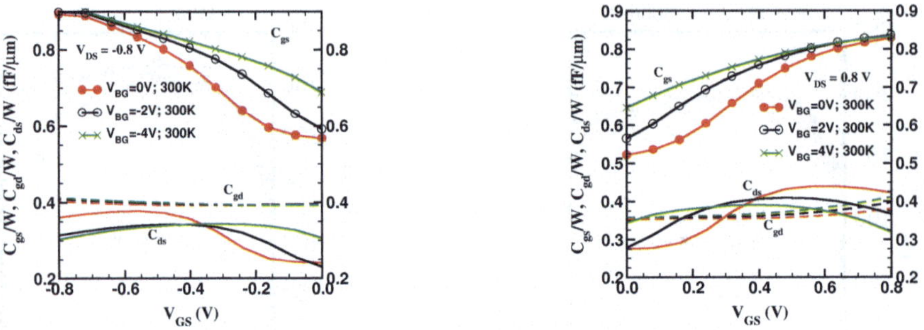

(a) (b)

Fig. 5.20 C_{gs}/W, C_{gd}/W, and C_{ds}/W as a function of V_{GS} and V_{BG} at 300 K for 2× gate pitch 22-nm FDSOI (**a**) 40 × 20 nm × 590 nm *p*- and (**b**) 40 × 20 nm × 720 nm *n*-MOSFETs contacted on both sides of the gate

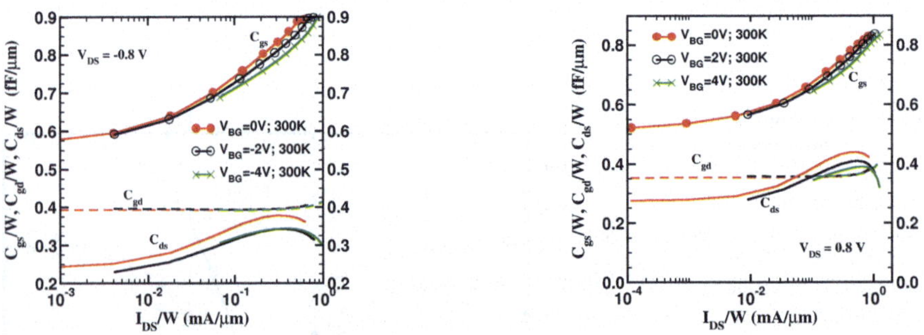

(a) (b)

Fig. 5.21 C_{gs}/W, C_{gd}/W, and C_{ds}/W as a function of I_{DS}/W and V_{BG} at 300 K for the (**a**) *p*- and (**b**) *n*-MOSFET in Fig. 5.20

factors α_n and α_p. C_{gd}, dominated in saturation by the gate-to-drain oxide spacer fringing capacitance, is practically independent of V_{GS} and V_{BG}. When plotted vs. drain current density, C_{gs} and C_{ds} follow a similar behavior as g_m, f_T, and f_{MAX}, with only a weak dependence on V_{BG} and remain insensitive to V_t variation. Unlike g_m, which peaks and then starts to decrease at large current densities, C_{gs} continues to increase with I_{DS}/W, explaining why $J_{p,fT}$ and $J_{p,fMAX}$ are smaller than $J_{p,gm}$. While C_{gd} remains constant over temperature, C_{gs} decreases at 2 K, especially at high current densities. This accounts for the significant improvement in f_T at 2 K compared to 300 K. The reasons for the decrease in C_{gs} at 2 K are not well-understood at this time. One explanation could be that t_{inv} is larger at 2 K than at 300 K. This reasoning is difficult to validate since commercial technology and device CAD software packages, like Sentaurus, do not converge at temperatures below 50 K. Another reason could be calibration errors due to large changes at 2 K in the material properties of the alumina standard substrate used for calibration. Unlike

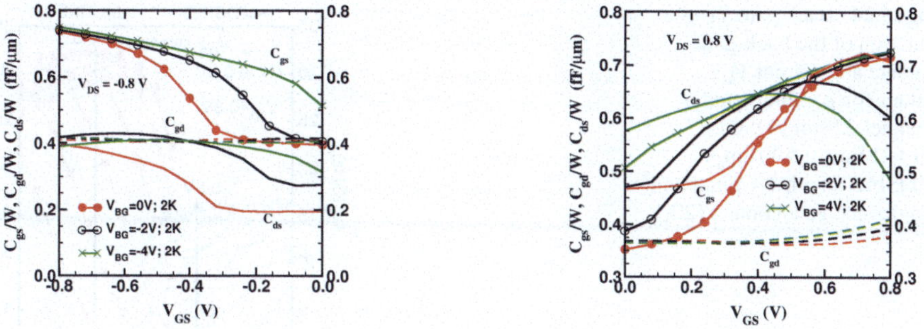

(a) (b)

Fig. 5.22 C_{gs}/W, C_{gd}/W, and C_{ds}/W as a function of I_{DS}/W and V_{BG} at 2 K and 300 K for the (**a**) *p*- and (**b**) *n* MOSFET in Fig.5.20

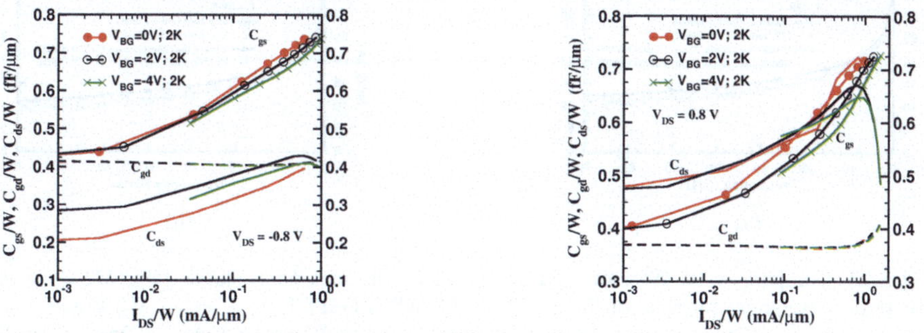

(a) (b)

Fig. 5.23 C_{gs}/W, C_{gd}/W, and C_{ds}/W as a function of I_{DS}/W and V_{BG} at 300 K for the (**a**) *p*- and (**b**) *n*-MOSFET in Fig. 5.20

C_{gd}, the measured C_{gs} and C_{ds} values are sensitive to the characteristics of the "open" standard. A similar behavior has been observed when measuring SiGe HBTs at 2 K.C_{bc} is practically constant with temperature, while C_{be} and C_{cs} decrease at 2 K compared to 300 K [13].

Unique among commercial CMOS technologies, FDSOI MOSFETs can be used as dual-gate transistors to implement a variety of high-frequency mixing and control functions in a very compact layout. In this case, high-frequency signals are applied to both the top and the back-gates. As predicted by (5.18), the back-gate transconductance, g_{mb}, is about 10 times smaller than the top gate transconductance. This impacts the cutoff frequency of the back gate, f_{Tbg}, measured in Fig. 5.24. The peak f_{Tbg} values are 16.3 GHz and 32.5 GHz for 22-nm FDSOI *p*- and *n*-MOSFETs, respectively. These values are high enough for signals in the GHz range to be adequately processed by the back gate. The f_{MAX} of the back gate is highly sensitive to the layout of the back-gate well, whose capacitance to the *p* substrate must be minimized.

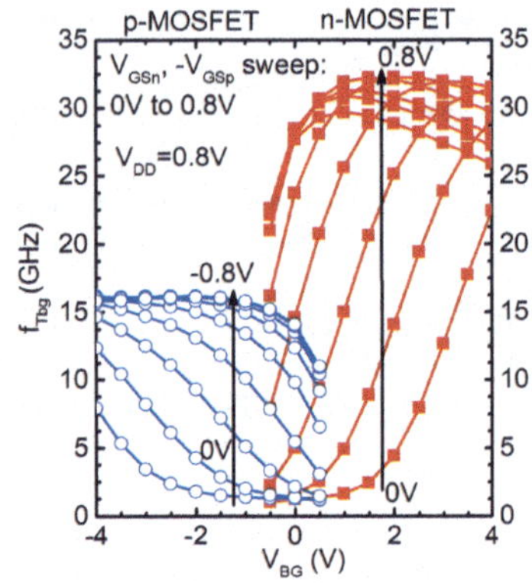

Fig. 5.24 Back gate f_T as a function of the back-gate voltage at different V_{GS} values for p - (blue) and n channel 22-nm FDSOI 40 × 20 nm × 590 nm MOSFETs with 1 × pitch and single-side gate contact [20]

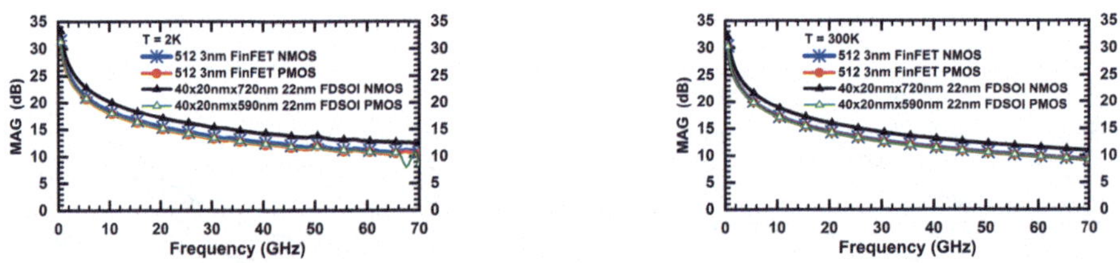

(a) (b)

Fig. 5.25 Comparison of the measured $MAG(f)$ (= the maximum stable gain, $MSG(f)$, in this frequency range) for state-of-the-art 22-nm FDSOI MOSFETs at $|V_{DS}| = 0.8$ V and 3-nm FinFETs at $|V_{DS}| = 0.75$ V at (**a**) 2 K and (**b**) 300 K. The FinFETs have $N_f N_{fin} = 256 \times 2 = 512$ unit MOSFETs connected in parallel

To finish this section on HF FoMs, Fig. 5.25 compares the measured MAG of fully wired 22-nm FDSOI MOSFETs with those of fully wired state-of-the-art commercial FinFETs as a function of frequency at both 300 K and 2 K. In this frequency range, $MSG(f)$, given by (5.27), is equal to $MAG(f)$ [10]. These measurements show that 22-nm FDSOI MOSFETs have comparable or higher MAG than FinFETs at all frequencies, making them most suitable for the applications discussed in this chapter.

$$\mathrm{MSG}(f) = \frac{g_m}{2\pi f C_{gd}} \tag{5.27}$$

5.3 Constant Current Density Design Method for FDSOI Circuits

The CMOS AMS, mm-wave, high-speed current-mode-logic (CML), and linear broadband fiberoptic front-end circuit topologies are very well-behaved and can be designed (sized and biased) based on deterministic algorithms that, in general, have unique optimal solutions [10]. As long as an optimal (in terms of minimizing parasitics) fully wired MOSFET layout with extracted RLC parasitics exists for each of the basic common-source (CS), common-gate (CG), and source-follower (SF) (i.e., common-drain) topologies, (i) the total gate width, W, (ii) drain bias current, I_{DS}, and (iii) V_{DS} of the MOSFETs in any of these topologies can be determined by hand design based on the particular circuit function specification. Moreover, given the previously reviewed and well-understood transistor AMS and HF FoMs as a function of bias current, gate width, and temperature, the performance of most of these circuit topologies is only weakly sensitive to bias current and process variation, leading to very robust, high-yield circuits. Therefore, applying computer optimization and, more recently, machine learning algorithms to the well-understood problem of transistor biasing and transistor sizing for these topologies is not energy-efficient and can lead to local optima, which may also be very sensitive to process and temperature variation. However, computer optimization and machine learning algorithms can be applied in designing complex tuned and broadband matching networks of arbitrary topology, for which there are no known equivalent circuits or parameterized models [21], and in local and global layout optimization to minimize interconnect, circuit footprint, and parasitics.

Next, we will review several algorithmic methodologies to optimally size and bias transistors in the most common AMS, mm-wave, and broadband circuit topologies to:

1. maximize speed, bandwidth, gain or minimize noise by setting the appropriate current density;
2. maximize linearity irrespective of bias current density by maximizing transistor V_{DS} using the back gate;
3. optimize performance irrespective of temperature using the back-gate voltage to maintain optimal current density and maximized transistor V_{DS}.

As can be observed in Fig. 5.26 and compiled in Table 5.2, although they do not depend on temperature, $J_{p,fT}$ and $J_{p,fMAX}$ increase with V_{DS}. The current range over which f_T and f_{MAX} remain within 10% of their peak value is maximized at the largest V_{DS} of 0.8 V per transistor. This means that, for circuit design, the current density at which a particular circuit topology should be biased must be decided only after the V_{DS} bias of the transistor(s) has been established.

The measurements in Fig. 5.27 at 300 K illustrate that the characteristic current density that maximizes MSG at 60 GHz, remains unchanged between 28-nm and 22-nm FDSOI n-MOSFETs, while the MAG improves in the more advanced technology node. If the layout of the MOSFET is optimally designed for high f_{MAX} and MAG, in this frequency range, MAG and MSG will be identical.

In the absence of measurements of minimum noise figure, NF_{min}, over temperature as a function of current density, V_{DS}, and V_{BG}, Fig. 5.28 compares the simulated room-temperature NF_{min} at 80 GHz as a function of current density, at different V_{BG} and V_{DS} values. J_{opt} is approximately 0.15 mA/μm, for both V_{DS} values, and increases slightly with V_{BG}. The lowest NF_{min} is obtained at $V_{DS} = 0.8$ V and at

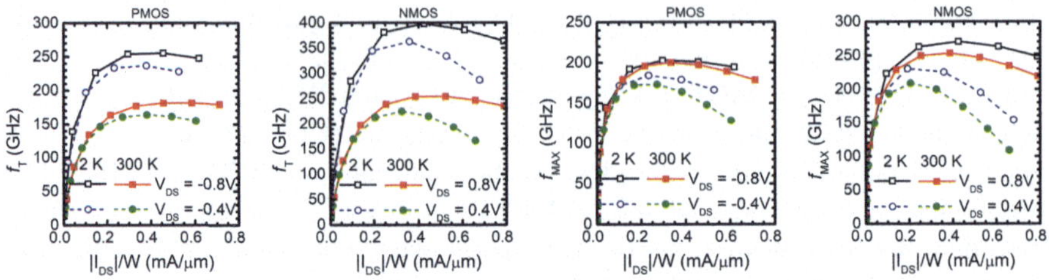

(a) (b) (c) (d)

Fig. 5.26 Measured f_T and f_{MAX} vs. I_{DS}/W for 2× gate pitch 22-nm FDSOI 40 × 20 nm × 590 nm p-and 40 × 20 nm × 720 nm-MOSFETs contacted on both sides at $|V_{DS}| = 0.8$ and 0.4 V at 2 K and 300 K

Table 5.2 $J_{p,fMAx}$ and $J_{p,fT}$ dependence on V_{DS} and temperature with floating back gate

	p-MOSFET				n-MOSFET			
	2 K		300 K		2 K		300 K	
V_{DS}(V)	−0.4	−0.8	−0.4	−0.8	0.4	0.8	0.4	0.8
$J_{p,fMAX}$ (mA/μm)	0.22	0.35	0.22	0.35	0.2	0.4	0.2	0.4
$J_{p,fT}$ (mA/μm)	0.38	0.45	0.38	0.46	0.35	0.45	0.35	0.45

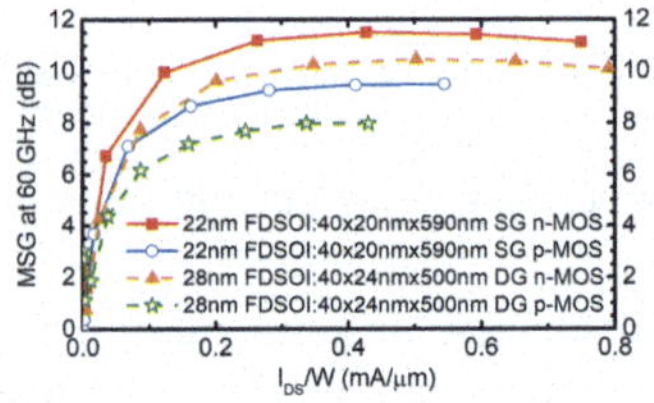

Fig. 5.27 Comparison of the measured MSG at 60 GHz (I_{DS}/W) for state-of-the-art 28-nm and 22-nm FDSOI at $|V_{DS}| = 1$ V and 0.8 V, respectively, with floating back gate at 300 K

$V_{BG} = -0.5$ V. Equation (5.28) [10] shows that NF_{min} depends on f_T and g_m. Like g_m, f_T, and f_{MAX}, it degrades at lower V_{DS} and as the back gate is more strongly forward-biased. Again, at a given V_{DS} and V_{BG}, NF_{min} remains practically flat for a relatively wide range of current densities, varying by less than 10% when I_{DS}/W changes from 0.1 to 0.3 mA/μm. This helps to design low-noise amplifiers (LNAs) that are robust to bias current variation.

$$F_{min} \approx 1 + \frac{2f}{f_T}\sqrt{k_1}\sqrt{g_m\left(R_s + R_g\right) + k_2\left(1 + \omega^2 R_i^2 C_{gs}^2\right)} \tag{5.28}$$

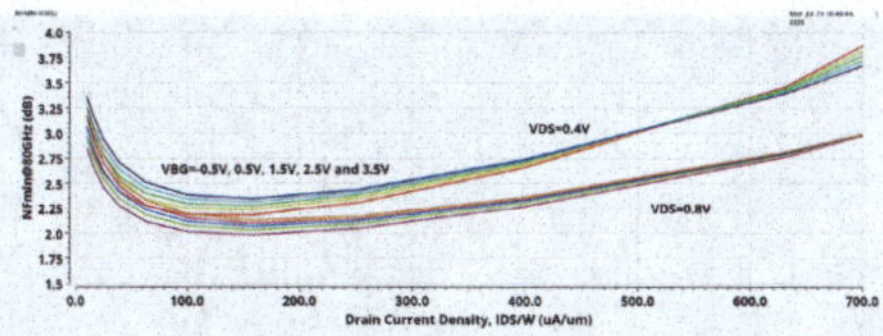

Fig. 5.28 Simulated $NF_{\min}$ at 80 GHz vs. I_{DS}/W for 40 × 20 nm × 720 nm 22-nm FDSOI n-MOSFET with 2× gate pitch and gate contacted on both sides at $V_{\mathrm{DS}} = 0.8$ and 0.4 V for V_{BG} swept from -0.5 V (lowest NF_{MIN}) to $+3.5$ V in 1 V steps. T = 65 °C

5.4 Unique Circuit Topologies that Benefit from the Back-Gate Control

1. Wide temperature range, low power (cryogenic) differential pairs with active loads;
2. CMOS logic circuits with switching speed control without changing V_{DD};
3. CMOS switches;
4. 0.8 V CML and quasi-CML logic circuits;
5. Low-voltage broadband linear transconductors and equalizer stages for analog-to-digital converter (ADC) front ends and wireline transceivers: CMOS source follower, super source follower, and g_m/g_m stages;
6. 4-terminal mm-wave varactor VCO with dual large/small KVCO slope;
7. Series-stacked doubler;
8. Cascode topologies that can be simultaneously optimized for minimum noise figure, maximum bandwidth, speed, output linear voltage swing, and linearity by maximizing V_{DS} independently of V_{GS} and drain current density;
9. Series-stacked, large-swing power amplifiers (PAs) and optical modulator drivers with individual output power or output voltage swing and gain control from the back gate.

5.5 AMS Circuit Design Examples

5.5.1 Differential Stage with Current-Mirror Load

The methodology for designing low-power and low-voltage CMOS opamps relies on the g_m/I_{DS} and A_V figures of merit [22]. Ideally, one would like to maximize the gain and gain-bandwidth product, GBW, of the differential pair with the lowest power consumption. Maximum gain is accomplished by using transistors with typically 10 times longer gate length than the minimum allowed by the technology node and biasing the differential pair at the maximum g_m/I_{DS}. Both design choices, however, compromise the bandwidth since longer gates and the small drain current

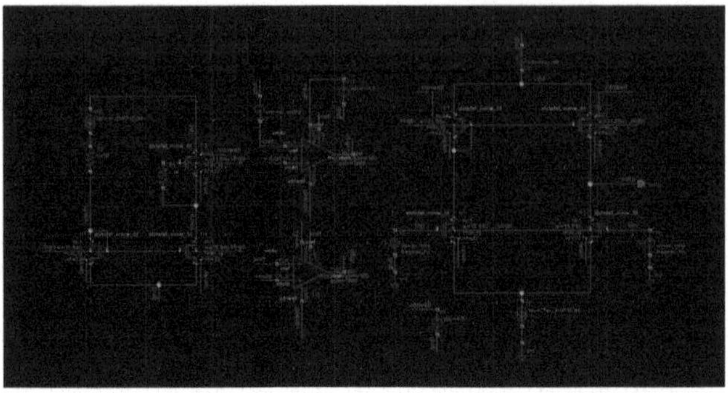

Fig. 5.29 Test bench schematic for the simulation of the differential stage with current-mirror load and back-gate-voltage-setting circuit for constant-current-density bias over temperature. The p- and n MOSFETs have $W_{fp} = 980$ nm and $W_{fn} = 720$ nm and identical L, which is varied from 18 nm to 180 nm. For simplicity, an ideal current source with 0 V DC drop is used in the differential pair

densities, needed for high g_m/I_{DS} ratios, reduce g_m and f_T [22]. Alternatively, a lower g_m/I_{DS} ratio is selected to increase the GBW at the expense of the low frequency gain and power consumption. The design then proceeds by adding a second stage and Miller compensation to guarantee the desired phase margin. Since the g_m/I_{DS} ratio depends only on the bias current density and remains constant as the gate widths of the transistors in the differential stage are scaled in tandem, the final gate width and bias current of the transistors in differential pair are determined based on the desired slew rate and current required to drive the specified load impedance.

The availability of the back gate allows for optimal differential-pair designs over a much wider temperature range than would otherwise be possible in bulk CMOS and FinFET CMOS technologies. If the g_m/I_{DS} dependence on current density is known from measurements, as in Figs. 5.10, 5.11, and 5.12, then it is also possible to design differential stages with optimal performance at any temperature, even in the absence of accurate design kit models valid in those temperature ranges. This is currently the situation faced by designers of cryogenic control and readout circuits for quantum computers.

The schematic of a basic differential pair with current-mirror load for differential-to-single-ended conversion is shown in Fig. 5.29. It includes relatively large transistors, suitable for typical AMS applications. The circuit was simulated in 22-nm FDSOI as a function of tail current density and gate length (L = 18 nm, 32 nm, 75 nm, 150, and 180 nm) using the mm-wave MOSFET models to accurately account for the back-gate well and substrate diodes. For each gate length value, the p-MOSFET and n-MOSFET back-gate voltages were optimized such that $V_{GS} = V_{DS} = 0.4$ V for maximum linearity. The gate finger width, W_{fp}, of the p-MOSFETs is 980 nm, while the W_{fn} of the n-MOSFETs is 720 nm. For the sake of simplicity, all transistors have the same gate length, but the method also applies if the p MOSFETs have different gate length than the n-MOSFETs. The test bench in Fig. 5.29 also includes a constant-current-density replica bias circuit with two ideal opamps, each with 60 dB of gain, placed in negative feedback loops. The outputs of the ideal opamps provide the corresponding p-MOSFET and n-MOSFET back-gate voltages, V_{BGP} and V_{BGN}, respectively, required to bias all the transistors in the differential pair with active load at the desired current density. The latter is set by the value of the reference current source and the gate width of the transistors in the replica bias circuit. This back-gate voltage-generating circuit works from 2 K to 400 K [7]. It ensures that all transistors in the differential stage have $V_{DS} = V_{DD}/2 = 0.4$ V at any current density in the 1μA/μm to 0.5 mA/μm range and at any temperature, thus allowing for the linearity of the differential stage to be maximized at each current density bias. It should be noted that by changing the value of V_{REF}, the $|V_{DS}|$ of the n- and p-MOSFETs could be set to a different value than $V_{DD}/2$. However, the latter is the most commonly encountered scenario for CMOS logic circuits, 0.8 V telescopic cascode LNAs, or in the readout TIAs of a quantum processor [12].

For the circuit to operate correctly with well-matched currents over a wide range of temperatures, the transistors in the replica bias circuit and those in the differential pair must have the same gate length and gate finger width and similar number of gate fingers and active area. If super low V_t (*slvt*) devices are used, as in this example, to cover the entire range of current densities and temperature ranges, the p substrate may need to be connected to a negative supply voltage at current densities below 0.1 mA/μm and cryogenic temperatures to avoid turning on the diode between the n-well and the p-substrate. The ideal opamps with 60 dB gain operate from $+/-$ 4.5 V supplies in order to provide the entire range of back-gate voltages required by both the n-MOSFETs and the p-MOSFETs in the differential-pair and in the replica bias circuit. They can be realized with high-voltage thick-oxide devices available in the same technology, but need not necessarily be integrated on the same die and operate over the same wide temperature range as the replica bias circuit and the differential pair.

As can be observed from the simulation results in Fig. 5.30 at 65 °C and $L = 180$ nm, the low-frequency voltage gain, A_v, is 42 dB at 1 MHz. The peak GBW is reached at a tail current density of 0.129 mA/μm. The GBW vs. I_{DS}/W characteristic has the same shape as the measured MOSFET g_m, f_T, and f_{MAX}vs.I_{DS}/W characteristics presented previously. As expected, the simulations in Fig. 5.31 versus gate length and current density show that at 65 °C, the highest gain is obtained for the longest gate length of 180 nm and at the lowest current density of 1 μA/μm. However, in the latter case the gain bandwidth product is the smallest, <2 GHz. It is also interesting to notice that the highest GBW of 360 GHz is obtained at $L = 38$ nm. A slight reduction of GBW to 350 GHz is observed when $L = 18$ nm, which also results in a low-frequency voltage gain of only 13 dB. The most interesting aspect of these simulations is the fact that the peak-GBW current density increases as the gate length decreases. Its value is 0.25 mA/μm for a gate length of 18 nm, similar to the measured $J_{p,fMAX}$ for a p- or n-MOSFET with 18 nm gate length biased at $|V_{DS}| = 0.4$ V included in Table 5.2.

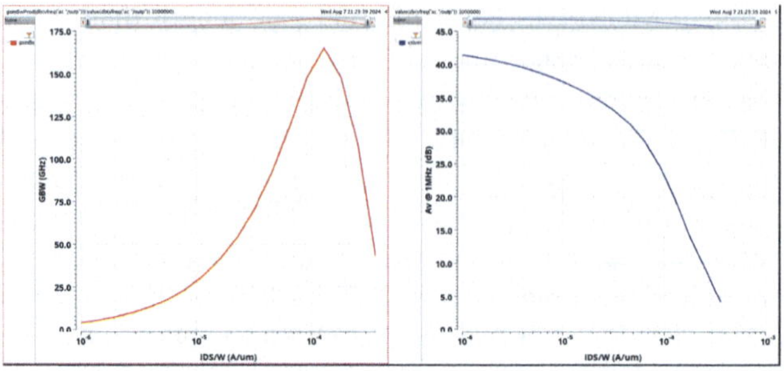

Fig. 5.30 Simulated a) GBW and b) low frequency gain, A_V, as a function of drain current density, $I_{DS}/W.L = 180$ nm, $T = 65\,°C$

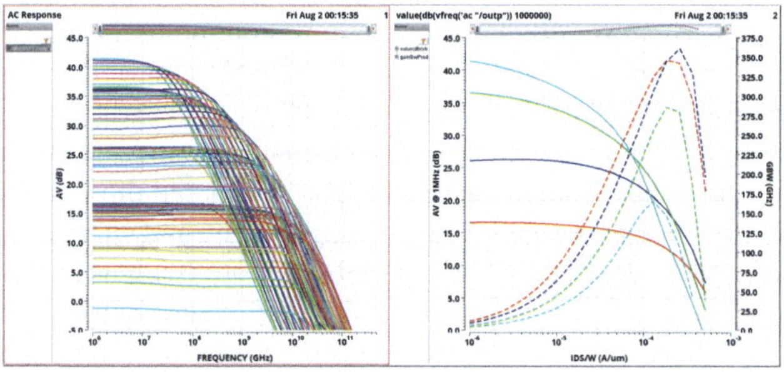

Fig. 5.31 Simulation at 65 °C of (**a**) voltage gain, A_v, vs. frequency for different I_{DS}/W and L, and (**b**) A_v @ 1 MHz (solid lines) and GBW (dashed lines) vs. I_{DS}/W, for L = 18 nm, 38 nm, 84 nm and 180 nm

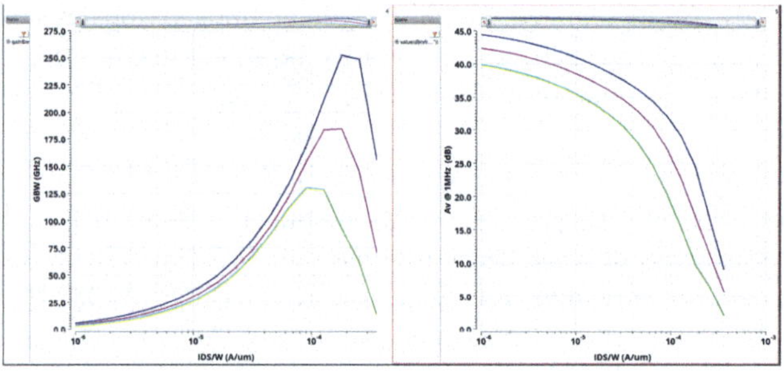

Fig. 5.32 Simulated (**a**) GBW and (**b**) A_v @ 1 MHz for $L = 180$ nm vs. I_{DS}/W, at −75 °C, 25 °C and 125 °C

Simulations of the small signal A_V and GBW were carried out in Fig. 5.32 for the highest-A_v differential-pair stage with a MOSFET gate length of 180 nm, from − 75 °C to + 125 °C as a function of current density. They show that A_V and the peak- GBW value increase as the temperature decreases and that the peak-GBW current density also increases as the temperature decreases, as observed for the n-MOSFET peak-A_v current density in Fig. 5.16b.

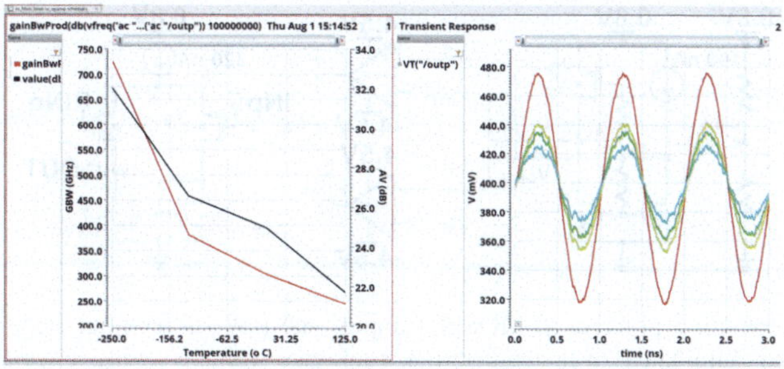

Fig. 5.33 Simulated (**a**) A_V@0.1 GHz (black) and GBW(red) and (**b**) transient output waveforms with noise, for $L = 50$ nm and $I_{DS}/W = 0.1$ mA/μm as a function of temperature from −250 °C to 125 °C

Small-signal A_V, GBW, and transient noise simulations with a 1 mV sinusoidal input at 1 GHz are presented in Fig. 5.33 over the extended temperature range from −250 °C to +125 °C for a differential pair with a MOSFET gate length of 50 nm, biased a fixed current density of 0.1 mA/μm. Although the foundry design kit models are not valid below −50 °C and predict higher V_t variation and improvement in g_m, f_T, and f_{MAX} at −250 °C than measured [7] and presented earlier, these simulations prove that the back-gate-voltage-generating circuit for constant-current-density bias of the differential pair performs its function correctly over the entire temperature range from −250 °C to 125 °C. The p- and n-MOSFET back-gate voltages adjust such that the V_{DS} of each transistor remains exactly 0.4 V at all temperatures and all current densities and, therefore, for any chosen g_m/I_{DS} ratio, with the output transient waveforms being nicely centered on 0.4 V in all cases. The transient noise simulations confirm that the amplitude of the output waveforms increases as the temperature decreases, proportional to A_V.

Cryogenic voltage amplifiers with minimum layout footprint are also needed in the readout path of monolithic quantum processors [7, 12]. To allow for the integration of millions of spin qubits with individual readout amplifiers on the same die for operation at 4 K within the thermal lift budget of liquid He cryostats, the power consumption of such readout amplifiers must be on the order of 1 μW (Fig. 5.34). The test bench with the DC operating point simulations in Figs. 5.34 and 5.35 at −250 °C and +125 °C, respectively, and the small signal and transient noise simulations in Figs. 5.36 and 5.37, respectively, demonstrate that such an amplifier, with less than 0.3 μW power dissipation, 20 dB gain, and low enough noise to handle 2 mV_{pp} input signals at 1 GHz, is realizable using MOSFETs with only 2 gate fingers, $L = 28$ nm, $W_{fn} = 160$ nm, and $W_{fp} = 200$ nm, operated in the subthreshold region at a current density of 1 μA/μm. Figure 5.37 predicts much lower noise, higher GBW, and similar gain and output swing at −250 °C compared to +125 °C.

The same back-gate-voltage-generating circuit for constant-current-density biasing over temperature can be used for all the other circuits discussed next.

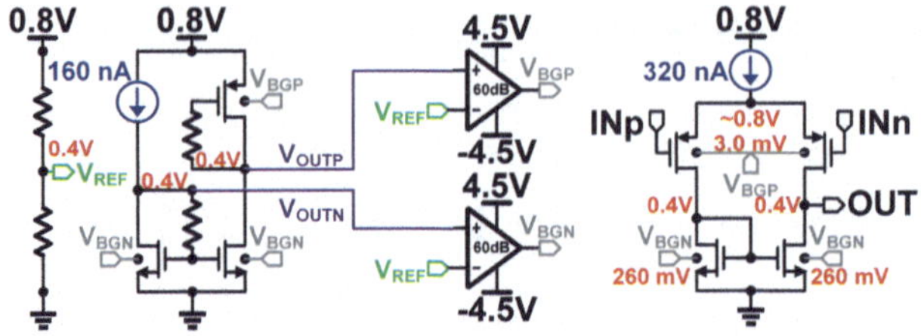

Fig. 5.34 DC-operating point of p-MOSFET input, sub-μW, differential stage with current-mirror load and automated over temperature back-gate voltage-setting circuit for quantum computing applications at $-250\,°C$. The stage draws 320 nA from 0.8 V supply. All transistors have 2 gate fingers, $W_{fp} = 200$ nm and $W_{fn} = 160$ nm, 28 nm gate length and are biased at a current density of 1 μA/μm with respect to the n-MOSFET gate width. For simplicity, an ideal current source with 0 V DC drop is used in the differential pair

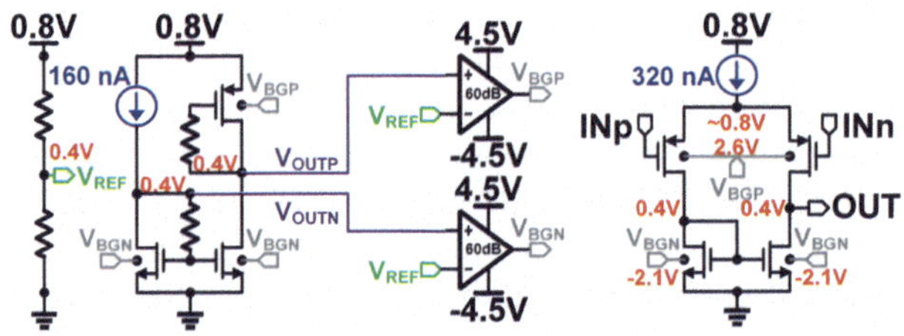

Fig. 5.35 DC-operating point of p-MOSFET input, sub-μW, differential stage with current-mirror load and automated over temperature back-gate voltage-setting circuit for quantum computing applications at $+125\,°C$. The stage draws 320 nA from 0.8 V supply. All transistors are biased at a current density of 1 μA/μm with respect to the n-MOSFET gate width

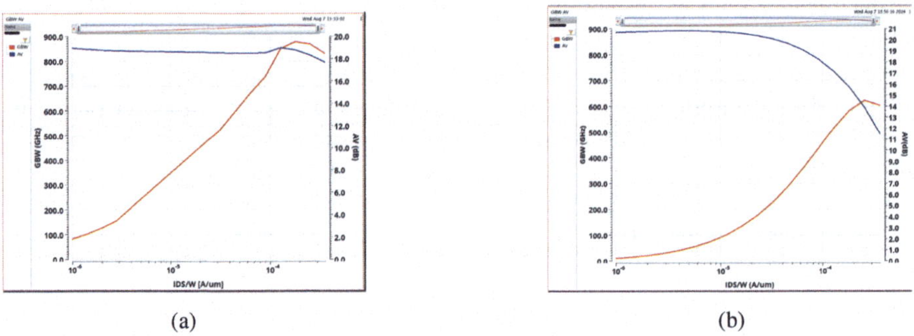

Fig. 5.36 Simulations of GBW (red) and A_V (blue) for the p-MOSFET input differential pair stage as the bias current density is swept from 1 μA/μm to 0.5 mA/μm at (**a**) $-250\,°C$ and (**b**) $125\,°C$, $L = 28$ nm

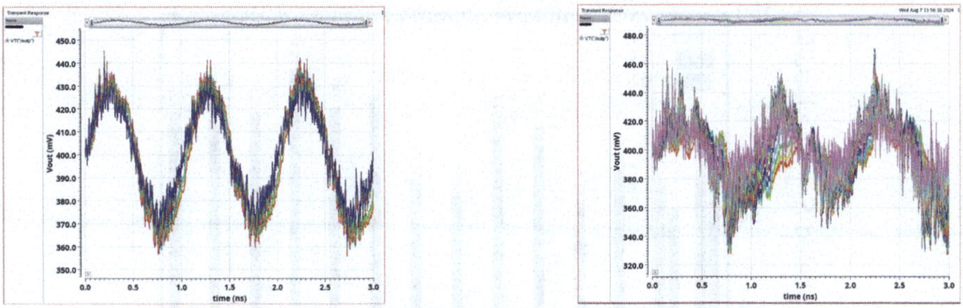

(a) (b)

Fig. 5.37 Transient noise simulation of the output waveforms with 2 mV, 1 GHz sinusoidal input as the bias current density is swept from 1 μA/μm to 0.5 mA/μm at (**a**) −250 °C and (**b**) 125 °C. This illustrates that the bias circuit works from −250 °C to 125 °C for any useful current density and that noise is considerably lower at 23 K than at 300 K. Note that the output waveforms remain centered on 400 mV DC level, irrespective of current density and temperature

5.5.2 *CMOS Logic*

The switching speed of the FDSOI CMOS logic gates is dictated by the MOSFET $g_{\rm m}$ and $I_{\rm ON}$, or equivalently, by the current density of the transistors in the logic gate when the logic gate is biased at the tripping point. As already discussed, $I_{\rm ON}$ and the bias current density can be controlled by changing the back-gate voltage without having to change $V_{\rm DD}$, as would be the case in planar bulk CMOS or FinFET technologies. To optimize the CMOS logic gate performance at the desired switching speed, at any temperature, the CMOS inverter back-gate voltages are set using the back-gate-voltage-generating circuit introduced in the differential stage section. This circuit ensures that the CMOS inverter $V_{\rm DSn} = V_{\rm SDp} = V_{\rm DD}/2$ at any bias current density set by the external reference current source when the CMOS inverter input voltage is $V_{\rm DD}/2$ [7]. In 22-nm FDSOI, the p-MOSFET $W_{\rm fp}$ or $N_{\rm fp}$ are sized so that the total gate width, $W_{\rm p}$, is 1.25× larger than that of the n-MOSFET, to match the mobility, $I_{\rm ON}/W$, and $g_{\rm m}/W$ ratios of the n- and p-type transistors. In 28-nm FDSOI, this ratio is closer to 2:1 [11]. The optimal transistor gate-finger-width and width ratios do remain largely invariant with $V_{\rm BG}$. However, since at 2 K the mobility, $I_{\rm ON}$, and transconductance of n-MOSFETs improve more than those of p-MOSFETs (30% versus 20%), the optimal p-MOSFET to n-MOSFET gate-width ratio increases for optimal operation at 2 K compared to that designed for best performance at 300 K. Therefore, when designing CMOS logic gates for operation at cryogenic temperatures, the width of the p-MOSFET should be increased compared to its value in CMOS gates designed for operation at 300 K.

Figure 5.38 compiles the CMOS inverter switching energy for FDSOI and FinFET technologies at room and cryogenic temperatures calculated based on measured transistor capacitances at $V_{\rm DS} = V_{\rm DD}/2$, using the expression $E_{inv} = C \cdot V_{DD}^2$, where $C = C_{\rm gs} + 2C_{\rm gd} + C_{\rm ds}$ for the smallest transistor available in the

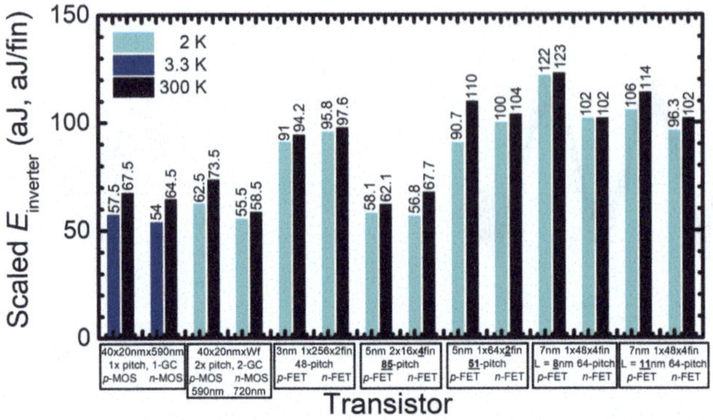

Fig. 5.38 CMOS-logic inverter switching energy in various CMOS technologies calculated from transistor I_{ON} and capacitance measurements at 2 K and 300 K

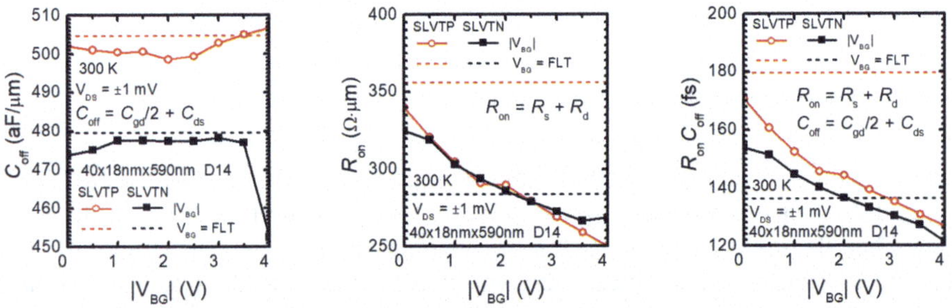

(a) (b) (c)

Fig. 5.39 Measured 22-nm FDSOI MOSFET switch parameters at 300 K (**a**) C_{off}, (**b**) R_{on}, and (**c**) $\tau_{\text{sw}} = R_{\text{on}} \times \text{C}_{\text{off}}$. FLT stands for floating back gate

technology [5]. The single-gate-finger 18 nm × 50 nm CMOS inverter was used in the case of 22-nm FDSOI, while in the case of FinFET technologies, the switching energy is specified per fin. At both 300 K and 2 K, the minimum size FDSOI CMOS inverter exhibits lower energy at nominal V_{DD} than all FinFET inverters, but it does occupy a larger die area. For all CMOS technologies, the switching energy improves (reduces) at cryogenic temperatures.

5.5.3 MOSFET Switch Performance as a Function of Back Gate

Figure 5.39 shows the definition and measured values as a function of the back-gate voltage of the on-resistance times gate width, R_{on}, off-capacitance per gate width, C_{off}, and the switch time constant, $\tau_{\text{sw}} = R_{\text{on}} \times C_{\text{off}}$, the main FoMs of the transistor

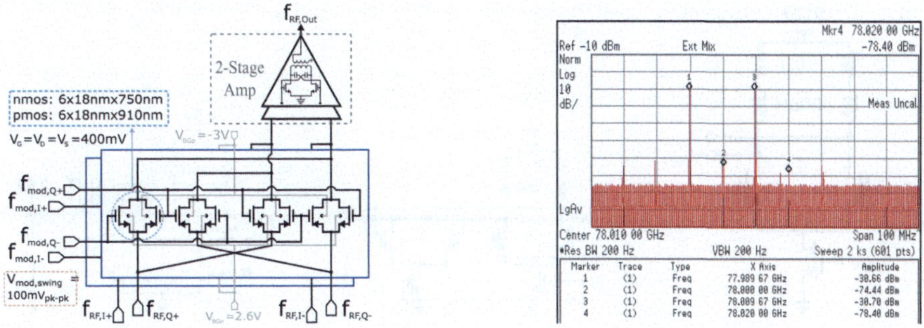

(a) (b)

Fig. 5.40 (**a**) Schematics of an SSB 80-GHz CMOS modulator. (**b**) Measured output spectrum showing >44 dB image and RF leakage suppression for two-tone input signals at 78 and 78.02 GHz, frequency shifted by IQ 10-MHz sinusoidal modulation signals [25]

when operated as a switch or passive mixer in the triode region, at $V_{DS} = \pm 1\,\text{mV}$. The parasitics due to the wiring of the transistor up to the top metal have been included in these measurements of 1× pitch 22-nm FDSOI transistors contacted on one side of the gate only. As can be seen, R_{on} and τ_{sw} improve as the body is more forward-biased, while the C_{off} remains unchanged, within the measurement error. When operated as a switch, the DC current through the transistor is small and, except in the case of fast samplers used in ADC front-ends [23], the gate resistance and f_{MAX} play a secondary role. Therefore, the gate-finger width can be increased to minimize the impact of the fringing capacitances on C_{off} and τ_{sw}. In 22-nm FDSOI technology, the *p*-MOSFET τ_{sw} is almost as good as that of the *n*-MOSFET, with the best values of $128 f_s$ and $120 f_s$ for *p*- and *n*-MOSFETs, respectively.

To demonstrate this unique feature of FDSOI technologies, the back-gate voltage of *n*- and *p* MOSFETs was optimized for (i) a fast CMOS series switch sampler in a 22-nm FDSOI 4× timeinterleaved ADC front end to maximize its bandwidth (>60 GHz) and isolation (>28 dB) [23], (ii) high isolation series-shunt and shunt switches in the qubit control path of a cryogenic quantum processor test vehicle [24], and (iii) in an 80-GHz single-sideband (SSB) CMOS passive modulator with >40 dB image and LO leakage suppression [25]. This type of passive IQ modulator is also popular in the control electronics of quantum processors based on superconducting qubits [26, 27]. For the 80-GHz CMOS IQ modulator in Fig. 5.40a, the SFDR, RF signal leakage, and image rejection, shown in Fig. 5.40b, were all improved to better than 40 dB. This was accomplished by setting the *p*- and *n*-MOSFET back-gate voltages in strong forward body bias to − 3 V and + 2.6 V, respectively, and adjusting the DC levels at the IQ baseband modulation inputs to compensate for transistor and layout parasitics mismatch.

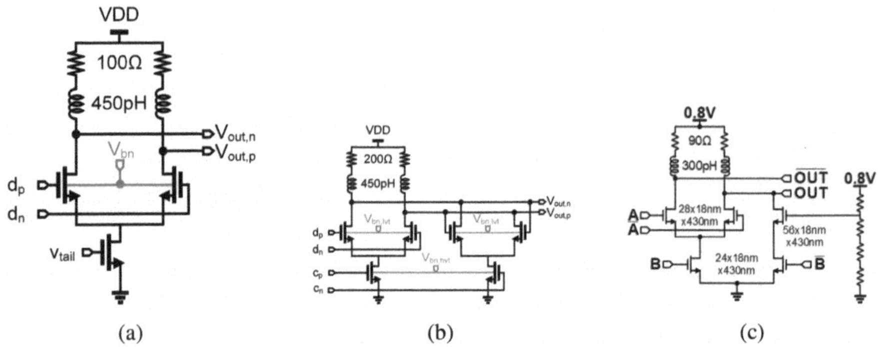

Fig. 5.41 (**a**) 4 mA CML inverter, (**b**) 2 mA quasi-CML latch, and (**c**) 4.4 mA quasi-CML AND gate schematics

5.5.4 Low-Voltage CML Logic

The schematic of the basic low-voltage CML and quasi-CML [10] gates is shown in Fig. 5.41. Normally, to obtain the maximum switching speed for a MOSFET CML gate, the MOSFETs should be switching between 0 and $J_{\text{p,fT}}$ current density, and the optimal logic swing should be set to [10]

$$\Delta V_{CML} = 3 \times \left[V_{GS}\left(J_{\text{p},f_T}\right) - V_{GS}\left(\frac{J_{\text{p,fT}}}{2}\right) \right] \tag{5.29}$$

This condition guarantees that the CML gate has adequate voltage gain over process and temperature variation to recover the full CML logic swing at its output. $V_{\text{GS}}\left(J_{\text{p,fT}}\right)$ is the MOSFET gate–source voltage when it is biased at the peak-f_{T} current density, and $V_{\text{GS}}\left(J_{\text{p,fT}}/2\right)$ is the gate–source voltage when the same MOSFET is biased at the half the peak$-f_{\text{T}}$ current density [10]. Because of the reduced supply voltage, it becomes increasingly difficult to satisfy this condition in the most advanced CMOS nodes while also maximizing V_{DS}.

In 22-nm FDSOI, the CML logic swing is set to 0.4 V $= V_{\text{DD}}/2$, which means that the DC voltage drop on the load resistance when the inputs are balanced is 0.2 V, leaving a V_{DS} headroom of 0.3 V for each of the two transistors in the stack. Since the gates of both top (data path) and bottom (clock path) transistors in the latch of Fig. 5.41b or AND gate of Fig. 5.41c are connected to the same DC level at the output of the preceding CML gate, nominally equal to 0.6 V, the top transistors must have a $V_{\text{GST}} = 0.3$ V, while the bottom transistors must have $V_{\text{GSB}} = 0.6$ V. At the same time, all transistors must switch from 0 to $J_{\text{p,fT}} \approx 0.4$ mA/μm. Therefore, the top MOSFETs must have a (super) low V_{t} while the bottom MOSFETs must have high V_{t}, 0.3 V higher than the V_{t} of the top MOSFETs. The back-gate voltages must be set accordingly and all transistors must have the same gate width and minimum gate length.

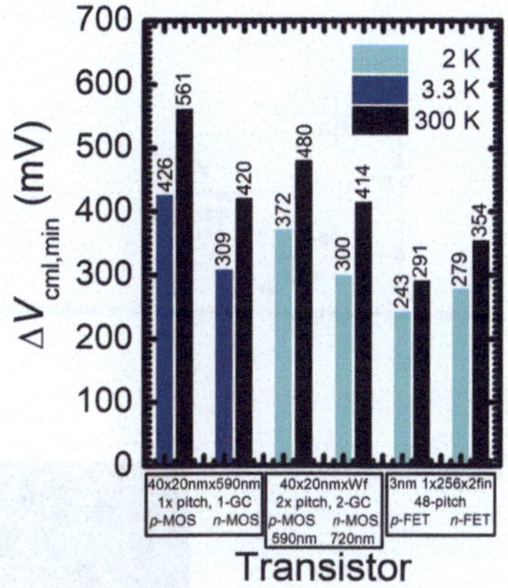

Fig. 5.42 Comparison of the measured CML logic swing, as defined in [10] as a function of temperature in 22-nm FDSOI and FinFET technologies

Figure 5.42 compares the measured optimal CML logic swing for FDSOI and FinFET technologies at room and cryogenic temperatures. The logic swing improves (i.e., decreases) at cryogenic temperatures. Because of the lower swing requirement and because the *p*- and *n*-MOSFETs are perfectly matched, the FinFET technology has an advantage over 22-nm FDSOI. In FDSOI, to achieve the highest speed with the lowest CML logic swing, only *n*-MOSFET topologies should be used. On the other hand, unlike in FinFET technologies, which must feature transistors with different, precisely controlled V_t flavors, by taking advantage of the back gate, in FDSOI, the V_t of the top and bottom transistors in a CML gate can be independently adjusted to optimize the logic swing and the V_{DS} of each transistor. This becomes particularly useful at cryogenic temperatures where the V_t of MOSFETs in all types of CMOS technologies increases by approximately 200 mV.

5.5.5 *Linear Broadband Transconductors*

Linear broadband transconductors are used in the front end of ADCs in wireline and fiberoptic receivers [28]. Although they do not require more than 0 dB gain, they must be designed for high linearity and very broad bandwidth (now >60 GHz) while driving the relatively large input capacitance, approximately 50 fF, of the ADC lane(s). Derived from the source follower (SF) stage, several low-voltage topologies featuring two vertically stacked transistors have become very popular in 5-nm FinFET ADC front ends [28]. The most common are: (i) the super source follower (SSF), (ii) CMOS SF, shown in Fig. 5.43a, and the *n*-MOSFET g_m/g_m stage, illustrated in its conceptual single-ended version in Fig. 5.43b, with the differential 22-nm FDSOI implementation in Fig. 5.43c.

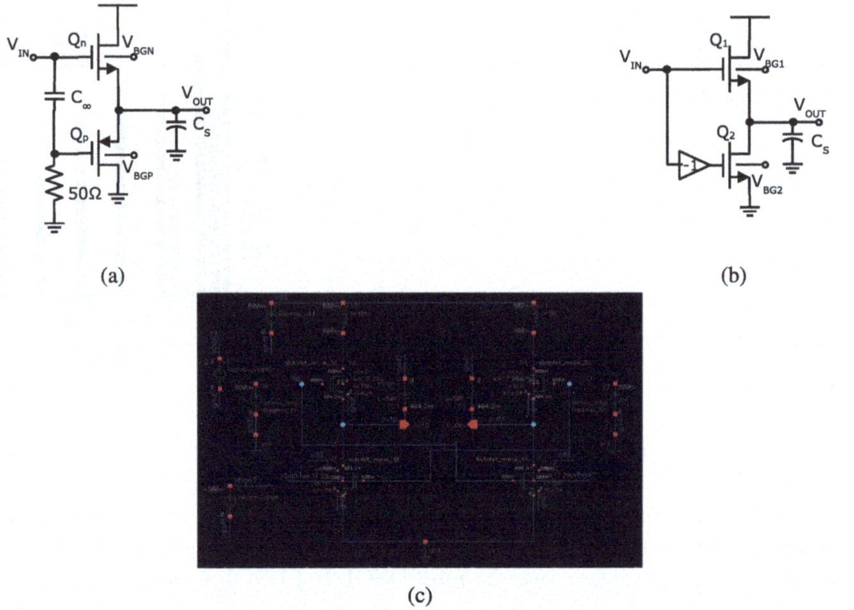

Fig. 5.43 Single-ended (**a**) CMOS source follower and (**b**) g_m/g_m stage concepts [28], and (**c**) differential n-MOSFET g_m/g_m transconductor schematics and DC operating point. The super low V_t, slvt, (top) high V_t hvt, (bottom) n-MOSFET size is $2 \times 24 \times 18\,\text{nm} \times 720\,\text{nm}$ with $V_{BG1} = -0.3$ V and $V_{BG2} = 2.9$ V

In FDSOI, it is possible to take advantage of the back-gate voltage to independently adjust the V_t of both transistors in the stack such that their $V_{DS} = V_{DD}/2$, and to simultaneously set the bias current density to 0.4 mA/µm, to maximize their bandwidth and linearity beyond what can be achieved in the finer lithography 5-nm and 3-nm FinFET technologies. To illustrate this point, test benches were created in 22-nm FDSOI for the single-ended CMOS SF stage, where, to first order, the nonlinearities of the p and n-MOSFETs cancel each other out, and for the differential g_m/g_m stage. For a fair comparison, the same transistor size and bias current were used in both cases. The simulation results depicted in Fig. 5.44 demonstrate that a bandwidth larger than 100 GHz and $SFDR$ of 40 dB can be achieved when driving a 50-fF capacitive load with a 100-GHz sinusoidal input at an amplitude of 200 mV $(400\,\text{mV}_{pp})$ per side. The bandwidth of the single-ended CMOS SF is the largest, 169 GHz, but with about 4 dB lower voltage gain (−1.7 dB) compared to that of the differential g_m/g_m n-MOSFET stage, which is + 2.2 dB.

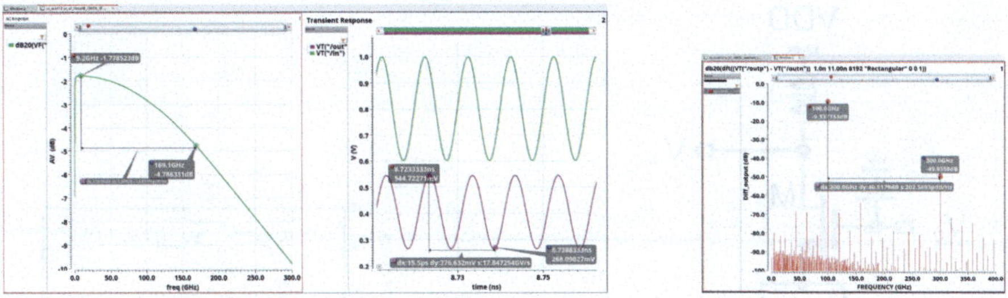

(a) (b)

Fig. 5.44 (**a**) Simulated frequency response and (**b**) input (green line) and output (pink line) waveforms of the CMOS SF, and (**c**) simulated output spectrum of the differential n-MOSFET g_m/g_m stage for 0.4 V_{pp} per side sinusoidal input at 100 GHz, demonstrating an $SFDR$ better than 40 dB

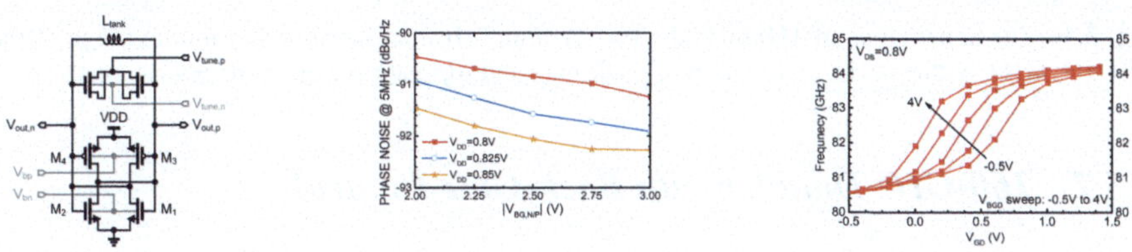

(a) (b) (c)

Fig. 5.45 (**a**) Schematics of an 80-GHz cross-coupled VCO. (**b**) Measured phase noise as a function of back-gate voltage and (**c**) tuning range as a function of varactor top gate and back-gate voltages [20]

5.5.6 80-GHz VCO with 4-Terminal Varactor Control

The presence of the independently controlled back gate can be taken advantage of to realize a dual-slope varactor, which can be used in a variety of voltage-controlled circuit topologies. Such an example is the 22-nm FDSOI 80-GHz cross-coupled CMOS-inverter LC VCO shown in Fig. 5.45a. Its measured phase noise is plotted in Fig. 5.45b as a function of the back-gate voltages of the transistors in the cross-coupled negative resistance generator at three different supply voltage values. The VCO phase noise improves as the back gate becomes more forward-biased, increasing the current density in the negative resistance generator and moving the charge centroid away from the top gate-to-channel interface. Figure 5.45c illustrates the measured tuning range as a function of the control voltages applied to the top and back gates of the varactor [20]. The typical tuning characteristic versus the top gate voltage can be shifted and centered at the desired top gate voltage from the back-gate control voltage. This feature is particularly handy when the VCO is inserted in a phase lock loop (PLL).

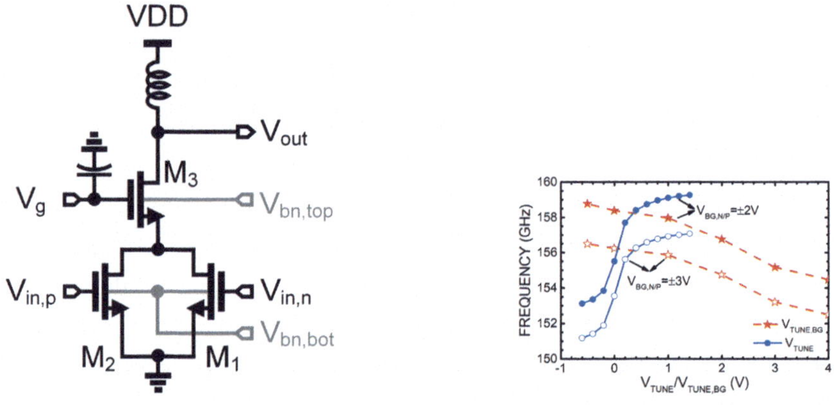

(a) (b)

Fig. 5.46 (**a**) Schematics of 160-GHz doubler [20] and (**b**) measured output tuning range of the 150–160 GHz synthesizer as a function of back-gate and top-gate control voltages

5.5.7 160-GHz Doubler with Back-Gate Control

Different back-gate voltages have also been used to optimize the performance of a 22-nm FDSOI doubler from 80 GHz to 160 GHz, whose schematic is depicted in Fig. 5.46a [20]. The back-gate voltage of the bottom differential *n*-MOSFET pair, $V_{\text{bn,bot}}$ is set for class-AB operation, at the optimal conduction angle, which maximizes the output power and power added efficiency, PAE, at the second harmonic [10], while the $V_{\text{bn,top}}$ of the common-gate *n*-MOSFET is set for maximum linear gain at 160 GHz. This doubler was paired with the VCO in Fig. 5.45a to realize a 160-GHz dual-slope signal source whose measured tuning characteristics are shown in Fig. 5.46b. As can be seen, the top gate KVCO_{TG} is approximately 4 GHz/V while the KVCO_{BG} associated with the back gate has negative sign and is approximately 10 times smaller. Such dual-slope VCOs are useful in the design of low-noise stable PLLs when the ratio of the output and reference frequencies is very large.

5.5.8 Cascodes

The nonlinearity of MOSFET CS and cascode amplifier stages is dominated by (i) the nonlinearities of the MOSFET $C_{gs}(V_{GS})$ and $g_m(V_{GS})$ characteristics, both of which can be reduced by biasing the stage at high current densities in the 0.3–0.5 mA/μm range, and (ii) the nonlinear dependence of all small signal equivalent circuit parameters $\left(g_m, g_o, C_{db}, C_{gd}\right)$ on V_{DS}. The latter can be minimized by setting V_{DS} in deep saturation, away from the triode and drain-gate

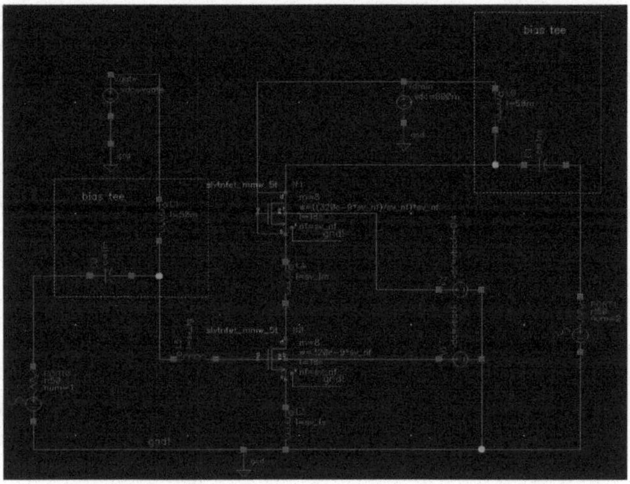

Fig. 5.47 Schematics of 28-GHz LNA design test bench

breakdown regions, in an identical approach to the DC load-line technique used to maximize the linear output voltage swing in an amplifier stage. In a regular telescopic cascode stage like the one in Fig. 5.47, the V_{DS} of both transistors can be set to $V_{DD}/2$. Alternatively, since the voltage swing at the middle drain/source node is small, the V_{DS} of the CG device is set to $0.6V_{DD}$, while that of the CS MOSFET to $0.4V_{DD}$. The second V_{DS} allocation scheme should lead to larger linear output voltage swing. Assuming that both MOSFETs are biased at the same current density, then V_{BG2} must satisfy the equation $V_{BG2} = V_{BG1}+V_{DS1} = V_{BG1}+0.4V_{DD}$. For cascode LNA stages, the drain bias current density is set to $I_{DS}/W = 0.2$ to 0.25 mA/μm to minimize NF_{MIN} while still maintaining reasonable linearity of the $C_{gs}(V_{GS})$ and $g_m(V_{GS})$ characteristics.

How do we choose V_{GS1} and V_{BG1}? For the first case aforementioned, V_{BG1} and $V_{BG2} = V_{BG1} + V_{DD}/2$ can be tuned in tandem, while keeping the bias current density fixed at the desired value, such that $V_{GS1} = V_{GS2} = V_{DS1} = V_{DS2} = V_{DD}/2$, with the DC voltages at gates 1 and 2, V_{G1} and V_{G2} set to $V_{DD}/2$ and V_{DD}, respectively.

For the second case, a good choice is to set $V_{GS1} = V_{GS2} = 0.6V_{DD}$, which also allows simple biasing of the gate of M_2 from V_{DD}.

We should note that in the case of FinFET and planar bulk MOSFET cascodes, there is no flexibility in independently controlling the current density and V_{GS} since, by setting the current density, V_{GS1} and V_{GS2} are also fixed because the threshold voltages of both MOSFETs in the cascode are fixed. In FDSOI, once V_{G1}, V_{G2}, and the transistor sizes are fixed, the gain of the LNA can be changed from the back-gate voltages of M_1 and M_2 without affecting the loading on the signal path.

We next provide an example of how the traditional, computer-driven, algorithmic methodology for designing cascode LNAs [10] can be adapted to take advantage of the back-gate voltage to maximize the linear output voltage swing of the 22-nm FDSOI LNA stage, without compromising its noise figure, gain, and range of frequencies over which it operates.

5.6 Design Example: Biasing and Sizing a 28-GHz Single-Stage Cascode LNA to be Noise Matched to 50 Ω with Minimum Noise Figure

The single-stage cascode LNA test bench is depicted in Fig. 5.46 along with the component size scaling equations, which allow us to carry out the design in an algorithmic manner. The foundryprovided mm-wave design kit models of the MOSFET wired up to the 4th. metal are employed, with the associated RC layout parasitics captured by the model. In this case, the V_{DS} of both transistors is set to $V_{DD}/2$ and $V_{DD} = 0.8$ V.

1. Start by setting $W_f = 320$ nm to minimize NF_{min}. Use n-MOSFETs with double gate contact and 2× gate pitch to maximize f_T and f_{MAX} and minimize NF_{min}. Set L_S, L_M, $L_G = 0.1$ pH (small enough to be negligible beyond 200 GHz). Set the number of gate fingers, $N_f = 25$, and the number of MOSFETs connected in parallel, $m = 4$ to 8, to allow for scaling the total gate width, W, up or down from N_f and m during the computer-run design flow.
2. Set $V_{GS1} = 0.4$ V. Bias at $J_{opt} = 0.2$ mA/μm by adjusting V_{BG1} and $V_{BG2} = V_{BG1} + 0.4$ V. This is only possible in FDSOI CMOS. Since the two transistors are identical and $V_{G2} = V_{DD}$, it also results in $V_{DS1} = V_{DS2} = V_{GS1} = V_{GS2} = V_{DD}/2$ which maximizes the linear output voltage swing. Note that it is also possible to set $V_{DS1} = 0.3$ V and $V_{DS2} = 0.5$ V by adjusting V_{G2}, V_{BG1}, and V_{BG2} such that J_{opt} remains unchanged at 0.2 mA/μm.
3. Sweep L_M to maximize f_T, as illustrated in Fig. 5.48 while keeping the transistor size, V_{BG1}, V_{BG2} and the current density, I_{DS}/W, unchanged. This step can be skipped when designing LNAs at frequencies below $f_T/10$ of the MOSFET. However, at frequencies beyond $f_T/5$, adding L_M can significantly improve the LNA noise figure.
4. Starting with the L_M value from step 3 aforementioned and $N_f = 25$, sweep N_f (also m if necessary) and $L_M \sim 1/N_f$ in tandem until $R_{SOPT} = 50\Omega$, as illustrated in Fig. 5.49.
5. With N_f, m, and L_M fixed at the values determined earlier, sweep L_S until $R_{in} = 50\Omega$, as depicted in Fig. 5.50. It is sufficient to reach within 10% of 50Ω, otherwise L_S becomes too large, increasing R_{SOPT} above 50Ω and limiting the maximum achievable gain.
6. Sweep L_G until the imaginary parts of Z_{in} and $Z_{SOPT} = 0\Omega$. In reality, they differ by ≈10%; therefore, a compromise L_G value must be chosen to minimize both X_{in} and X_{SOPT}, as shown in the final circuit simula-

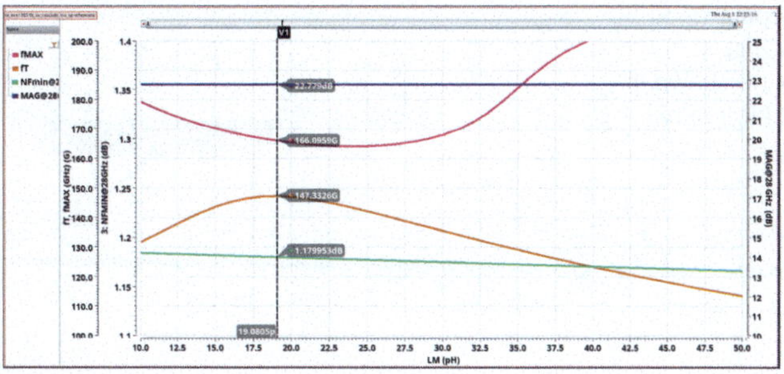

Fig. 5.48 Sweeping L_M to maximize cascode f_T

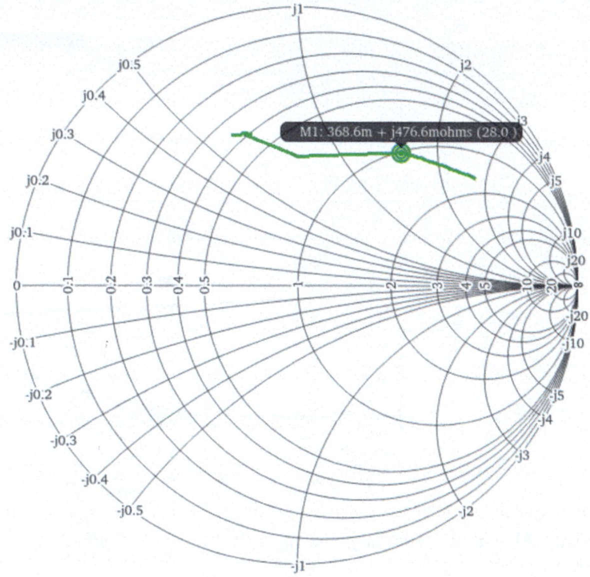

Fig. 5.49 Sweeping N_f to make $R_{SOPT} = 50\Omega$.Z_{SOPT} is plotted at 28 GHz as a function of N_f

tions of Γ_{OPT} (G_{min}), in Fig. 5.51S_{11}, NF, NF_{min}, MAG, and S_{21} compiled in Fig. 5.51.

Figure 5.52 reproduces the DC operating point schematics of the designed LNA cascode stage. The final component values are $L_S = 70$ pH, $L_G = 400$ pH, $L_M = 18$ pH. Both MOSFETs have a total gate width of $8 \times 28 \times 0.32$ nm $=$ 71.68 μm, $V_{DS} = V_{GS} = 0.4$ V, $I_{DS} = 13.3$ mA, $I_{DS}/W = 0.19$ mA/μm. At 65 °C, $V_{BG1} = 0.7$ V.

As can be seen from the simulations results in Fig. 5.51, at 28 GHz, both S_{11} and Γ_{Sopt} (G_{min}) are better than $-$ 14 dB at 28 GHz, and the 50$-\Omega$ NF is only 0.075 dB higher than NF_{MIN}. These values prove that simultaneous noise and input impedance matching has been accomplished. An output matching network still needs to be added to maximize S_{21}, ideally to match the MAG value at 28 GHz.

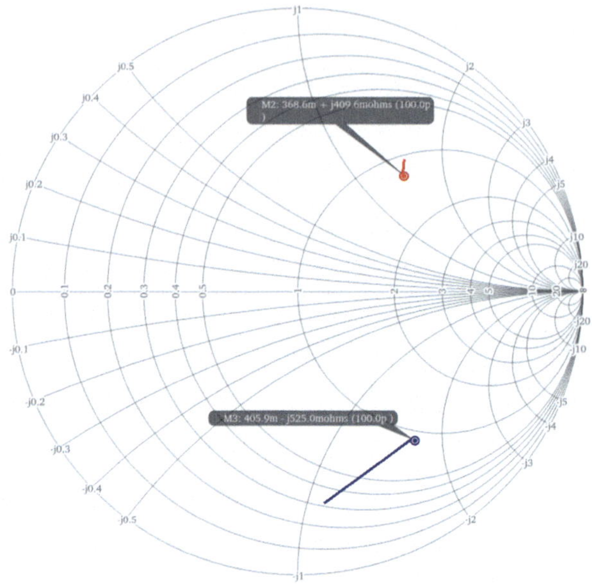

Fig. 5.50 Sweeping L_S to make $R_{in} = 50\Omega$ at 28 GHz. S_{11} (L_S) is shown in blue and Z_{SOPT} (L_S) in red

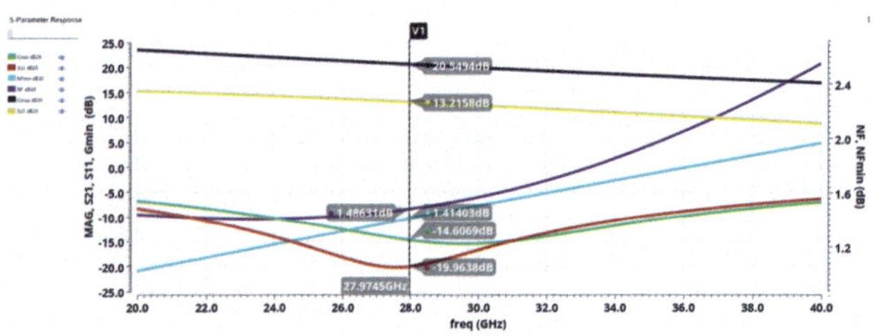

Fig. 5.51 Final frequency sweep with $L_G = 400\,\text{pH}, L_S = 70\,\text{pH}, L_M = 17.85\,\text{pH}, N_f = 28, m = 8$

Finally, it should be mentioned that the back-gate voltage can be adjusted by the automatic backgate-voltage-generating circuit discussed in the differential-stage section to maintain the same bias conditions of $V_{GS1,2} = V_{DS1,2} = V_{DD/2}$ at the same (or other desired) bias current density over the entire temperature range from 2 K to 400 K. This might be useful ins LNAs designed for space applications.

5.6.1 Series-Stacked Cascodes in PAs and Large-Swing Optical Modulator Drivers

Just like placing an inductor between the source node of a MOSFET and ground creates a positive resistance at its gate terminal and is commonly employed in LNAs,

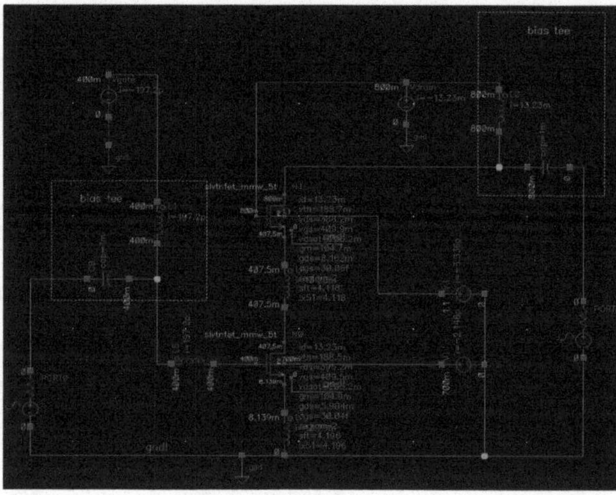

Fig. 5.52 Final component values and DC operating point of the 28-GHz LNA cascode stage

an appropriately sized capacitor connected between the gate and ground creates a positive resistance at the MOSFET source and drain terminals. Capacitively loading the base or gate of a transistor has been used in mm-wave common-base InP HBT LNAs to match the real part of the input impedance, and in series-stacked CMOS PAs to match the real part of the output impedance to 50Ω, without having to add a (lossy) matching network between the emitter and the signal source, or between the drain and the load, respectively. If the capacitor is replaced by an FDSOI varactor, C_{var}, as in [29], then the output and input admittances of a CG stage become tuneable from the top-gate or back-gate voltage of the varactor according to the formulae:

$$Y_{\text{in}} \approx \frac{g_m + j\omega C_{gs}}{1 + \frac{C_{gs}}{C_{var}}} \tag{5.30}$$

$$Y_{out} \approx \frac{g_o}{1 + \frac{C_{gd}}{C_{var}}} + j\omega \left(C_{gd} + C_{db}\right) \tag{5.31}$$

This scheme of capacitively loading the gates of series-stacked cascodes has been exploited in the design of large voltage swing 28-nm FDSOI optical modulator drivers [30] and 22-nm FDSOI power DACs [31], and SOI [32] and FDSOI CMOS power amplifiers up to the *D*-Band [33]. It creates an artificial large-voltage composite MOSFET using only the highest f_{T} and f_{MAX} low-voltage, thin-oxide devices [32], as illustrated in Fig. 5.53 for 3-stack and 4-stack *n*-MOSFET, and 3-stacked CMOS-inverter cascodes. In an FDSOI CMOS technology, the back-gate voltage of each transistor in the stack can be optimized for (i) linearity, as in the measurements presented in Fig. 5.54 of a 28-nm FDSOI series-stacked

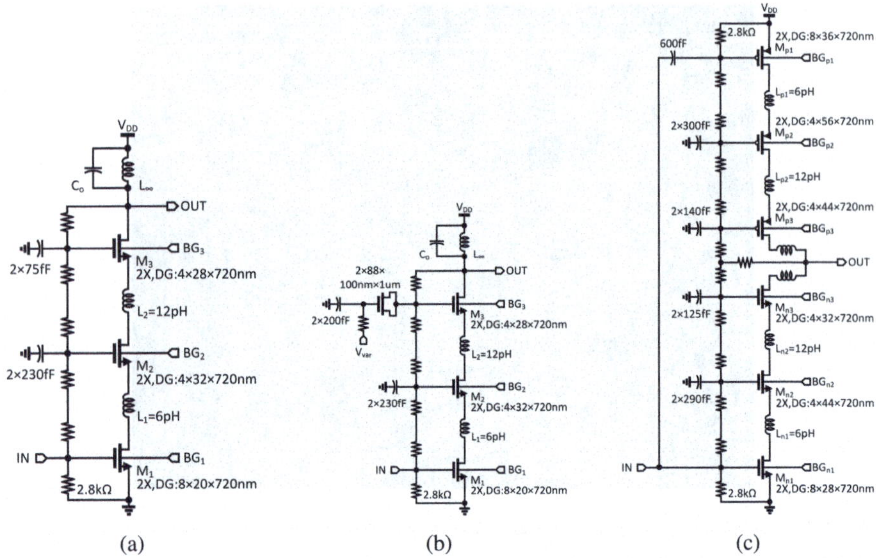

Fig. 5.53 Schematics of series-stacked, large-swing cascodes: (**a**) *n*-MOSFET triple cascode, (**b**) *n*-MOSFET triple cascode with varactor output impedance tuning, and (**c**) CMOS triple cascode inverter [29]

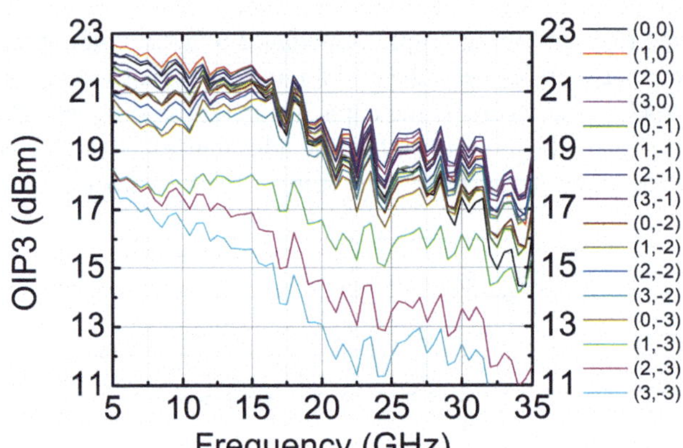

Fig. 5.54 Measured third-order output intercept point, OIP3, of a 28-nm FDSOI 3-stack CMOS inverter cascode vs. the back-gate voltages of the *n*- and *p*-MOSFETs in the stack, shown in brackets (,) [30]

cascoded CMOS inverter [30], (ii) output power/swing, and (iii) PAE by setting the bias current density of the MOSFETs for class- A ($I_{DS}/W = 0.4$ mA/μm), class-AB ($I_{DS}/W = 0.2$–0.3 mA/μm), class-B ($I_{DS}/W = 0.05$–0.1 mA/μm) or class-F operation from V_{BGI}. Typically, one sets $V_{DSi} = 0.8$ V (the nominal maximum

V_{DS} allowed per transistor junction in 22-nm FDSOI) for maximum linear output voltage swing, and with $V_{GSi} = 0.4$ V and $V_{BG1} = V_{BGi} + (i - 1)*0.8$ V, where the subscript i defines the ith MOSFET in the stack, starting with $i = 1$ for the CS MOSFET.

It can be demonstrated using charge conservation analysis [34] that, for the output voltage swing of the transistors in the stack to add up in series at the output node, the value of the capacitors $C_{Gn,i}$ connected between the gate of n-MOSFET M_i and ground must satisfy the following equation

$$C_{Gn,i} = \frac{C_{gsn,i} + 3C_{gdn,i}}{2i - 3} \quad (5.32)$$

where C_{gsni} and C_{gdi} are the gate-to-source and gate-to-drain capacitances, respectively, of the ith n-MOSFET in the stack in Fig. 5.53.

Two examples are provided next on how to use the formulae above and the measured transistor capacitances in Figs. 5.21, 5.22, 5.23, and 5.24 to optimally size series-stacked 22-nm FDSOI cascodes for operation as class- F and class- AB PA output stages at 28 GHz and 60 GHz, respectively.

5.7 Design Example 1: Class-F Differential Stacked PA Output Stage with P_{out} = 100 mW at 28 GHz

The number of n-MOSFETs in the stack, typically 3–4, is chosen so that the saturated output power of 20 dBm can be achieved at 28 GHz with a reasonably high R_{LOPT}, approximately 50Ω per side, with moderate transistor size ($W < 200\,\mu$m). This obviates the need for a rather lossy output matching network with a large impedance transformation ratio. Assume that the DC V_{DS} drop per transistor is 0.8 V and that the maximum allowed peak-to-peak voltage swing per transistor is 1.6 V_{pp}. Also, assume that up to three harmonics are properly terminated by the output matching network, which limits the maximum achievable efficiency to 80% [10].

Using the measured transistor characteristics presented earlier, fully wired n-MOSFETs have $L = 18$ nm, $J_{MAX} = 1$ mA/μm, $J_{pfTn} = 0.4$ mA/μm, $g'_{mn} = 1.95$ mS/μm; $C'_{gs} = 0.75$ fF/μm; $C_{ngd} = 0.35$ fF/μm; $C'_{sb} = C_{db} = 0.4$ fF/μm; $R'_s = R'_d = 200$ Ω*μm; $R'_{on} = 400$ Ω*μm.

5.7.1 Solution

Per side we need a saturated output power of 17 dBm = 50 mW. Since there are m transistors in the stack, $V_{MAX} = m*1.6$ V = 4.8 V if $m = 3$ and 6.4 V if $m = 4$.

In a class-F PA stage, $P_L = \frac{V_1 I_1}{2} = \frac{V_{MAX} I_{MAX}}{2\pi}$ and $R_{\mathrm{LOPT}} = 4V_{\mathrm{MAX}}/\pi \mathrm{I_{MAX}}$ [10]. This means that $R_{LOPT} = 2\frac{V_{MAX}^2}{\pi^2 P_L} = m^2 \frac{1.6^2 V^2}{9.869 \cdot 0.05} = 10.38 \cdot m^2 \Omega$. In this case, we set $m = 3$, since $m = 4$ leads to R_{LOPT} per side much larger than 50Ω. Smaller m leads to higher maximum PAE because there are fewer transistors in the stack. $P_L = \frac{V_1 I_1}{2} = \frac{V_{MAX} I_{MAX}}{2\pi} = 0.05$ W => $I_{\mathrm{MAX}} = 0.05$ W*2 * 3.14/4.8 V = 65.45 mA per side.

We must now calculate the optimal transistor ($\mathrm{M_1} - \mathrm{M_3}$) and gate capacitor sizes $\left(C_{g2}, C_{g3}\right)$, the optimal differential load resistance and the power consumption when accounting for the MOSFET R_{on}. If, because only the first three harmonics can be optimally terminated, PAE is at best 80%, I_{MAX} must be increased by $1/0.8 = 1.25$ to compensate for the reduced efficiency and still achieve $P_{\mathrm{out}} = 100$ mW. This implies that we should design for an $I_{\mathrm{MAX}} = 82$ mA per side instead of 65.45 mA per side.

To find the transistor size: $W = I_{\mathrm{MAX}}/J_{\mathrm{MAX}} = I_{\mathrm{MAX}}/(1\ \mathrm{mA}/\mu\mathrm{m}) = 82\,\mu\mathrm{m}$ for all MOSFETS in stack. $R_{\mathrm{LOPT}} = 4V_{\mathrm{MAX}}/\pi I_{\mathrm{MAX}} = 4 * 4.8/(3.14 * 0.082) = 74.53\Omega$. The bias current density is $J_{\mathrm{MAX}}/\pi = 0.318\ \mathrm{mA}/\mu\mathrm{m} \Rightarrow I_{\mathrm{DC}} = 2 \times 26$ mA, (accounting for the differential design). $V_{\mathrm{DD}} = 2.4\ \mathrm{V} = V_{\mathrm{MAX}}/2. P_{\mathrm{DC}} = V_{\mathrm{DD}} I_{\mathrm{DC}} = 124.8$ mW.

From the measurements presented in the switch section, $R_{\mathrm{on}} \approx 300/82 = 3.65\Omega$. For three transistors in the stack, $V_{\mathrm{MIN}} = 3R_{\mathrm{on}} I_{\mathrm{MAX}} = 0.9$ V. This again degrades the PAE by $\approx (1 - V_{\mathrm{MIN}}/V_{\mathrm{MAX}}) = 0.8125$ [10], which means that the final efficiency is degraded to at most 0.812*0.8 = 65%. It also implies that P_{SAT} is only 81 mW and that we need to again increase the transistor size and bias current by $1/0.81 = 1.24$ times to 101.2μm and $I_{\mathrm{MAX}} = 101.2$ mA. Accounting for R_{on} also changes the peak-to-peak linear output voltage swing to 2*(2.4 V − 0.9 V) = 3 $\mathrm{V_{pp}}$ and $R_{\mathrm{LOPT}} = 3/0.101 = 30.4\Omega$ per side. $P_{\mathrm{DC}} = 2 \times 32.2$ mA*2.4 V = 154.6 mW. Now we can proceed to calculate the values of C_{g2} and C_{g3}.

$$C_{gs} = W^* 0.75\ \mathrm{fF}/\mu\mathrm{m} = 101.2 * 0.75\ \mathrm{fF} = 75.9\ \mathrm{fF},$$

$$C_{gd} = W^* 0.35\ \mathrm{fF}/\mu\mathrm{m} = 101.2 * 0.35\ \mathrm{fF} = 35.42\ \mathrm{fF},$$

$$C_{db} = C_{sb} = W^* 0.4\ \mathrm{fF}/\mu\mathrm{m} = 101.2 * 0.4\ \mathrm{fF} = 40.5\ \mathrm{fF},$$

$$C_{g2} = C_{gs} + 3C_{gd} = 75.9\ \mathrm{fF} + 3 * 35.42\ \mathrm{fF} = 182.16\ \mathrm{fF}.$$

$$\mathrm{C_{g3}} = C_{g2}/3 = 182.16\ \mathrm{fF}/3 = 60.72\ \mathrm{fF}.$$

A matching network will also have to be designed to properly terminate the output for class-F operation on R_{LOPT} at 28 GHz, and to provide a short circuit at the second harmonic (56 GHz), and an open circuit at the third harmonic, 84 GHz.

5.8 Design Example 2: Class- AB Differential Stacked PA Output Stage with $P_{SAT} = 100$ mW at 60 GHz.

$J_{p,MAG} = 0.5$ mA/µm, $J_{MAX} = 1$ mA/µm, $J_{DC} = 0.3$ mA/µm, $V_{DS} = 0.8$ V. This means that for a 3-stack stage, $V_{DD} = 2.4$ V, $P_{SAT} = 100$ mW $= V_{MAX} I_{MAX}/8$. By accounting for the impact of R_{on} at I_{MAX}, V_{MAX} becomes: $2\,(V_{DD} - 3R_{ON} * I_{MAX}) = 2 * (2.4\text{–}0.9)\text{V} = 3$ V.

$\Rightarrow I_{MAX} = 8 * P_{SAT}/V_{MAX} = 266.6$ mA, i.e. 133.3 mA/side. $W = I_{MAX}/J_{MAX} = I_{MAX}/(1\ \text{mA}/\mu\text{m}) = 133.3\,\mu$m. Choose $W = 4 \times 56 \times 18\,\text{nm} \times 590$ nm $= 132.16\,\mu$m => $I_{DC} = J_{DC} W = (0.3\ \text{mA}/\mu\text{m})^*(132.16\,\mu\text{m}) = 39.64$ mA per side. As a result, $P_{DC} = I_{DC} V_{DD} = 190.3$ mW and $R_{LOPT} = V_{MAX}/I_{MAX} = 22.7\Omega$ per side.

Since, as in the previous example, $C_{gd} = 0.35$ fF/µm, $C'_{gs} = 0.75$ fF/µm, we calculate $C_{g2} = 237.9$ fF, $C_{g3} = 79.3$ fF from (5.32). The final simulated design in Fig. 5.54 has transistors with $W = 4 \times 50 \times 20\,\text{nm} \times 590$ nm $= 118\,\mu$m, $C_{g2} = 210$ fF, and $C_{g3} = 70$ fF, within less than 10% from the hand design values, $L_D = 170$ pH, $L_M = 20$ pH, $V_{GS1,2,3} = 0.4$ V, $V_{BG1} = 1.17$ V, $V_{BG2} = 1.97$ V, $V_{BG2} = 2.77$ V.

The simulated performance with 50Ω loads per side and no output matching network, other than $L_D = 170$ pH connected between the output node and V_{DD}, gives $G = 18$ dB, PAE = 30.7%, and $P_{SAT} = 16.5$ dBm per side, 19.5 dBm differential. Simulations of the circuit using the hand-designed component values show a slightly lower PAE of 30% and similar power gain and P_{SAT} (Fig. 5.55).

Figure 5.56 illustrates the large-swing series-stacked cascode stage concept applied to the 160-GHz doubler and the 160-GHz LO-path buffers of the 22-nm FDSOI dual-polarization 160-GHz radar sensor transceiver in [33]. In this case, the bottom differential pair is biased in class B with $V_{BG1} = 0$ V. The CG devices in the doubler and the cascode MOSFETs in the LO buffer are biased at 0.3 mA/µm with backgate voltages of 1.9 V and 2.7 V, respectively. All transistors in both circuits have $V_{DS} = 0.8$ V.

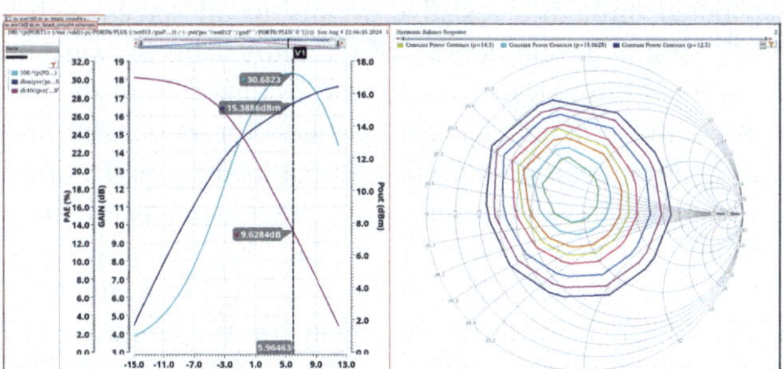

Fig. 5.55 Simulated (**a**) power gain (pink line), P_{out} (blue line) and PAE (cyan line) as a function of input power and (**b**) load-pull contours for the 60-GHz class $- AB$ PA design example

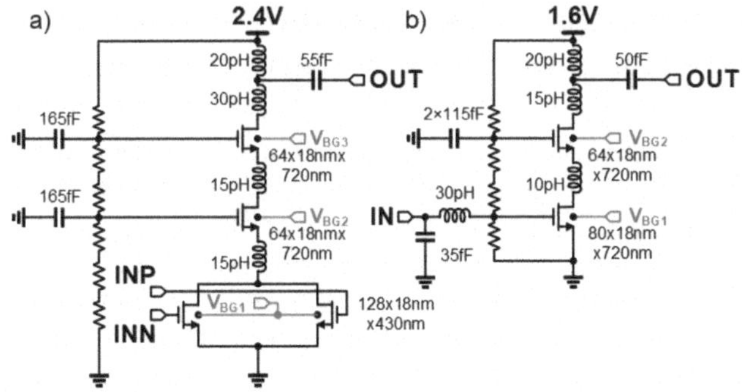

Fig. 5.56 (**a**) 160-GHz high-power, series-stacked cascode doubler and (**b**) 160-GHz series-stacked cascode LO buffer [33]

5.9 160-GHz LNA with Back-Gate Gain Control

The schematic of a 160-GHz 3-stage cascode LNA with gain control is shown in Fig. 5.57. It was used at the input of the two receivers in the dual-polarization radar sensor transceiver reported in [33]. The LNA stages employ inductive broadbanding to improve the f_T and the NF_{MIN} of the cascode [35], and consist of only minimum gate length (18 nm) slvt n-MOSFETs biased at 0.25 mA/μm to minimize the noise figure while maintaining high gain. The LNA design follows the traditional mm-wave methodology discussed earlier, with inductive degeneration in the input stage to set the real part of the input impedance to 50Ω and the size of the first stage MOSFETs optimized to match the real part of the optimum noise impedance to 50Ω. A transformer is used at the input to tune out the pad capacitance over a broad bandwidth and to provide some ESD protection. As discussed, to increase gain and linearity, the V_{DS} of each transistor is set to 0.8 V, the maximum safe nominal DC voltage.

The supply voltage of the LNA is $V_{DD} = 1.6$ V. Gain control is applied in each stage via the back-gate voltages V_{BG1} and V_{BG2} of the CS and CG transistors, which are distributed through a resistive divider to all stages such that $V_{BG1} \approx V_{BG2} - 0.8$ V at a drain current density of 0.25 mA/μm. Because the V_{GS} of the CS MOSFETs is fixed through resistive dividers connected to the top gates of the MOSFETs, this control scheme changes the bias current density in each cascode stage, therefore changing their power gain without loading the 160-GHz signal path and without affecting the input impedance. The latter is practically insensitive to the gain control setting. As can be seen in Fig. 5.58a, when V_{BG2} is swept from $+4$ V down to -0.5 V, the LNA gain changes from 19.8 dB to 8 dB, without affecting the input match and RF bandwidth. This gain-control range and RF-bandwidth invariance are also reflected in the measured conversion gain of the entire 160 GHz receiver in Fig.5.58b. Figure 5.59 shows the measured receiver noise figure as a function of IF frequency at 159 GHz and 151 GHz. As the gain of the LNA is reduced, the receiver NF increases by a maximum of 7 dB at the lowest gain setting.

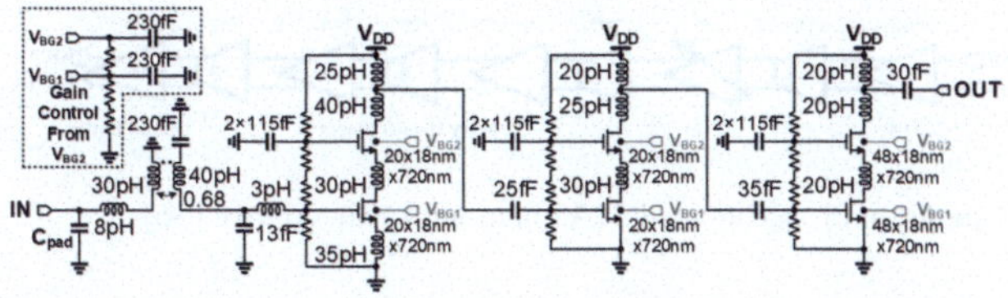

Fig. 5.57 160-GHz LNA schematics [33]

(a) (b)

Fig. 5.58 Measured (**a**) *S*-parameters of the LNA and (**b**) receiver down-conversion gain with control from the back-gate voltage [33]

(a) (b)

Fig. 5.59 Measured receiver noise figure with back-gate gain control as a function of IF frequency at the LO frequency of (**a**) 151 GHz and (**b**) 159 GHz

In agreement with the 22-nm FDSOI n-MOSFET noise figure simulations presented in Fig. 5.28, the receiver noise figure is minimized and changes little for V_{BG2} values between 2.5 V and 4 V, roughly corresponding to drain current densities between 0.1 mA/μm and 0.4 mA/μm[32]. The maximum receiver gain and the minimum value of the noise figure are obtained at $V_{BG2} = 2.5$ V. For V_{BG2} values between + 0.5 V and − 0.5 V the current density drops below 0.15 mA/μm causing the receiver noise figure to degrade rapidly. This is due to the increased noise figure and lower power gain in each stage of the LNA, increasing the contribution of the downconvert mixer to the overall receiver noise figure.

5.9.1 80-GBaud Fiberoptic TIA with Stacked Cascodes

Transimpedance amplifiers (TIAs) are used as ultra wide band low noise linear amplifiers at the input of the most advanced 64-GBaud and 128-GBaud fiberoptic

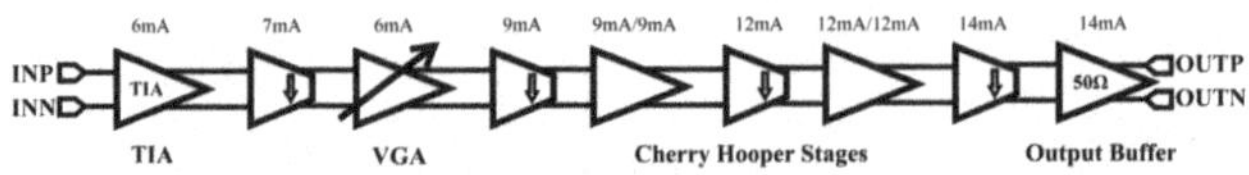

Fig. 5.60 Block diagram of g_m-boosted TIA chain with gain control [36]

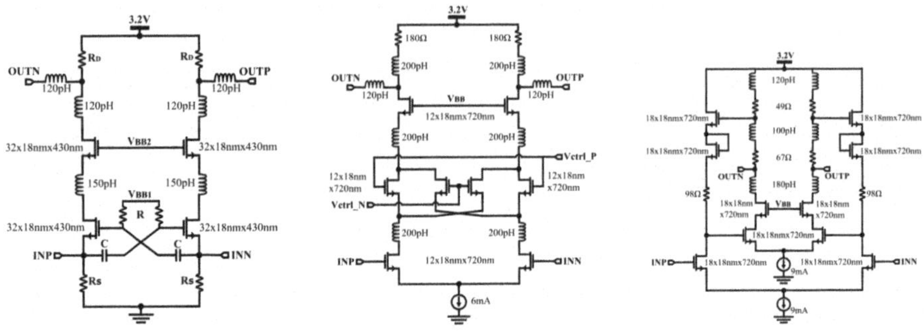

(a) (b) (c)

Fig. 5.61 (**a**) TIA stage, (**b**) VGA stage and (**c**) Cherry-Hooper gain stage schematics [36]

receivers operating with 4-PAM and 16-QAM modulation format. As in the tuned 160-GHz LNA presented earlier, the input stage of the TIA must operate with the minimum possible noise when driven by the photodiode signal. Unlike the tuned LNA, which must be matched to 50Ω, in fiberoptic receivers, the input impedance of the TIA must be as low as possible in order for its bandwidth to be maximized and to be relatively insensitive to the photodiode capacitance. The block diagram of a multi-stage 80-GBaud TIA with over 80 dBΩ differential transimpedance gain and more that 40 dB of gain control is shown in Fig. 5.60 [36].

The design challenge in this case is that the linearity, gain, and bandwidth requirements for operation at over 100 GBaud with up to 4 mA_{pp} linear differential input current and a photodiode capacitance of 35 fF cannot be satisfied with a traditional n-MOSFET or CMOS inverter input stage with transimpedance feedback. Instead, a cascoded, g_m-boosted, CG topology with cross-coupled input capacitors, shown in Fig. 5.61a, is employed to provide the lowest possible differential input noise current and input impedance with the maximum gain-bandwidth product. To minimize the noise of the input stage, the transistors are biased at $I_{DS}/W = 0.21$ mA/μm. Unlike in the 160-GHz LNA, where the back-gate voltages were used for gain control, the back-gate voltages of each transistor in the input stage and in the following SF (blocks with downward pointing arrow in Fig. 5.60), variable gain amplifier (VGA), Fig. 5.61b, and fixed-gain Cherry Hooper (CH), Fig. 5.61c, stages (needed to achieve the gain specification), are optimized to maximize the linearity and bandwidth of the entire chain. The bias conditions of each stage are summarized as follows:

(a) (b)

Fig. 5.62 (**a**) Simulated TIA chain differential gain and equivalent input noise current, I_{neq}, with gain control and (**b**) measured (solid lines with symbols) single-ended Z_{21}, S_{22}, and input resistance, R_{in} for three gain settings compared to simulation (dashed lines) at maximum gain setting [36]

(a) (b)

Fig. 5.63 (**a**) Simulated 150-GBaud PAM-4 eye diagrams at the output of the g_m-boosted TIA stage for a (**a**) 0.1 mA_{pp} per side and (**b**) 3 mA_{pp} per side PRBS11 input current

TIA stage: $V_{BG1} = 1.2$ V, $V_{BG2} = 2$ V, $V_{DS1,2} = 0.8$ V, $I_{DS} = 3$ mA/side, $V_{RD} = 750$ mV, $R_D = 250\Omega$.

VGA stage: $V_{BG1} = 1.6$ V, $V_{BG2} = 2.4$ V, $V_{BG3} = 3.2$ V, $V_{DS1,2,3} = 0.8$ V, $I_{DS} = 3$ mA/side, $I_{DS}/W = 0.37$ mA/μm, $V_{RD} = 570$ mV, $R_D = 180\Omega$.

CH stage: $V_{BG1} = 1.6$ V, $V_{BG2} = 2.4$ V, $V_{BG3} = 3.0$ V, $V_{DS1,2} = 0.8$ V, $I_{DS} = 3$ mA/side, $I_{DS}/W = 0.37$ mA/μm; $V_{RD} = 570$ mV, $R_D = 180\Omega$.

The simulated differential transimpedance gain, Z_{21}, (blue lines) and differential equivalent input noise current (red lines) of the entire chain for nine variable gain settings are shown in Fig. 5.62a along with the Z_{21} (green line) of the input stage. The input stage has over 70 GHz bandwidth, while the bandwidth of the entire chain is reduced to 45 GHz. The simulated single-ended Z_{21}, S_{22}, and R_{in} are compared with measurements up to 67 GHz in Fig. 5.62b. A feature of this g_m-boosted topology is that its bandwidth and noise improvement only applies in differential mode. The single-ended bandwidth is smaller, ≈35 GHz, and the single-ended input resistance is equal to the differential input resistance.

The over 70 GHz small-signal bandwidth of the input stage and its 2.2 V_{pp} linear differential output swing are confirmed by the 150-GBaud 4-PAM (i.e. 300 Gb/s) eye diagrams shown in Fig. 5.63, obtained using transient noise simulations with noise integrated up to 200 GHz. The eye diagram measurements in Fig. 5.64, presented without de-embedding setup, cable, and probe losses, are limited to 80 GBaud by the maximum symbol rate of the arbitrary waveform generator used to provide the input signal and by the bandwidth of the CH stages.

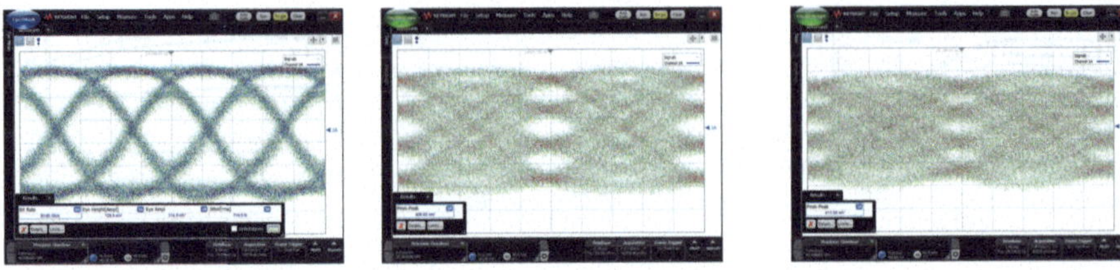

(a) (b) (c)

Fig. 5.64 Measured eye diagrams at the output of the TIA chain (**a**) 80 Gb/s NRZ, (**b**) 70 GBaud PAM-4, and (**c**) 80 GBaud PAM-4 [36]

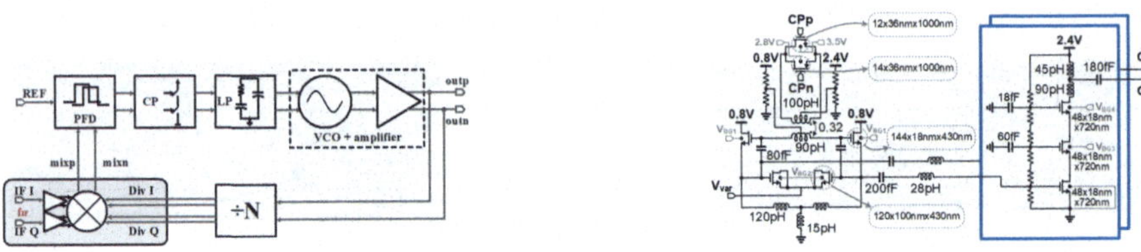

(a) (b)

Fig. 5.65 (**a**) PLL block diagram. (**b**) 80-GHz VCO and output buffer schematic [33]

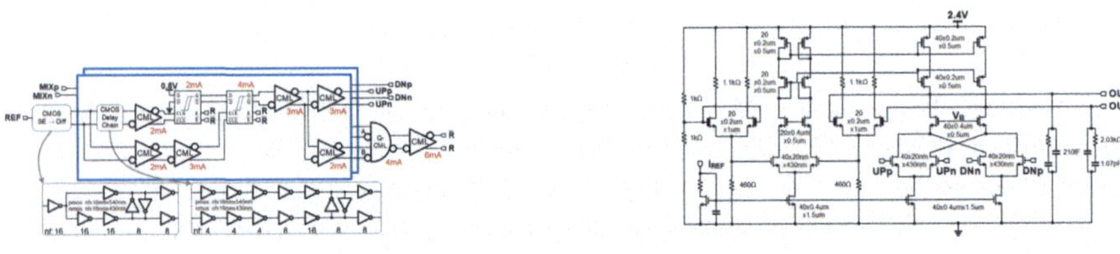

(a) (b)

Fig. 5.66 (**a**) 10-GHz PFD and (**b**) 10-GHz charge pump with low-pass filter schematics [33]

5.9.2 80-GHz Cryogenic PLL

The 22-nm FDSOI PLL in Fig. 5.65 was designed for fast settling, room-temperature 80-GHz OFDM and stepped-frequency automotive radar and 160-GHz radar sensor applications [33]. It features adjustable PLL bandwidth from 50 MHz to 200 MHz and consumes 300 mW. As illustrated in Figs. 5.66 and 5.67, it includes all the AMS, CMOS logic, high-speed CML logic, analog-mixed-signal, and mm-wave circuits discussed so far.

By tuning the back-gate voltages in all circuits to compensate for the MOSFET threshold voltage increase at cryogenic temperatures and bias the circuits at the same current densities as at room temperature, its performance was optimized for operation at 2–70 K without any redesign and without changing the power consumption. This is useful for operation in a would-be monolithic quantum

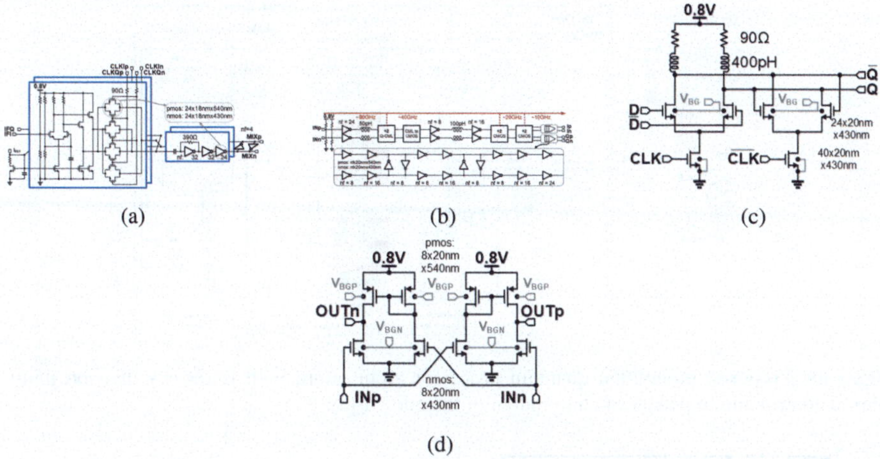

Fig. 5.67 (**a**) 10-GHz IQ upconverter and (**b**) 80-GHz static divider chain (top) with schematics of the 80 GHz CML static-divide-by-2 stage (bottom center) and CML-to-CMOS converter stage (bottom right) [33]

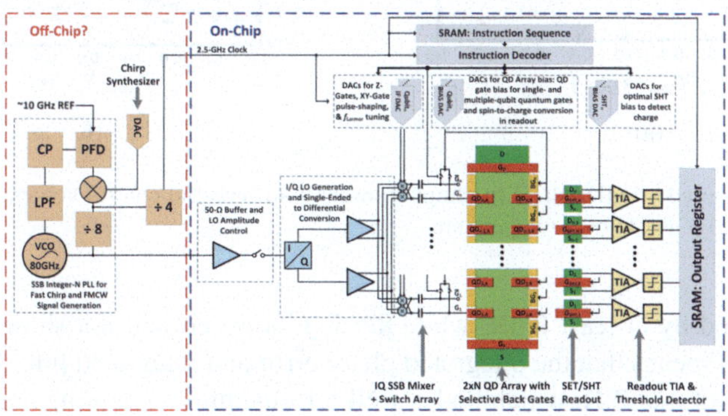

Fig. 5.68 Proposed monolithic quantum processor architecture with individual control of each qubit gate and direct detection single-electron (SET) or single-hole (SHT) transistor readout array

processor where the classical control and readout electronics are integrated together on the same die with thousands and potentially millions of semiconductor spin qubits formed with large arrays of coupled quantum dots [37]. Two potential monolithic quantum processor architectures, realizable in FDSOI CMOS, with different qubit readout approaches, both involving an 80-GHz PLL and arrays of 80-GHz passive IQ upconverters to generate the spin manipulation signals, are depicted in Figs. 5.68 and 5.69. In the 2D array with 2048 coupled QDs from [37], selective back gates are used to tune the tunnel coupling between adjacent QDs. As illustrated in Fig. 5.70, the measured tuning range and the K_{VCO} of the free-running 22-nm FDSOI 80-GHz Colpitts VCO with dual voltage control remains practically unchanged between 300 K and 2 K. At 2 K, the output power increased and the noise

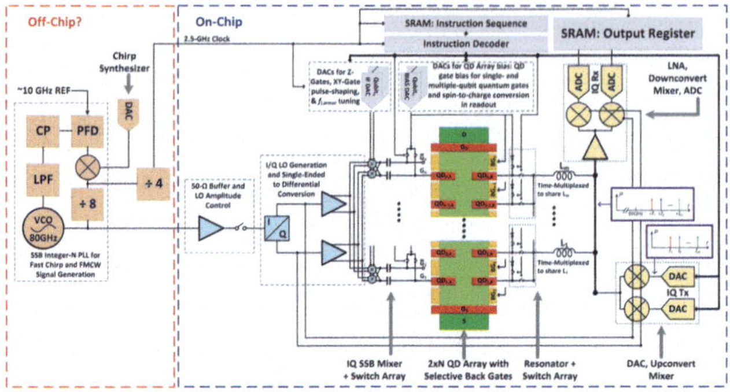

Fig. 5.69 Proposed monolithic quantum processor architecture with frequency division multiplexed control and impedancemetry readout of the qubit array

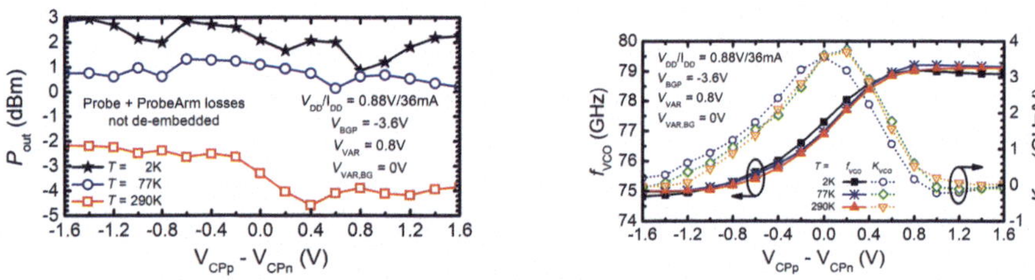

(a) (b)

Fig. 5.70 Measured 80-GHz VCO (**a**) output power vs. differential control voltage range and (**b**) tuning range and $KVCO$ over temperature

floor improved by at least 6 dB, while the $1/f$ noise corner, not shown, remained unchanged. It means that the integrated phase error and integrated jitter reduce by at least a factor of 4 at 2 K, which bodes well for quantum computing and cold space applications.

Figure 5.71 compares the measured phase noise performance at 300 K and 50 K as a function of offset frequency at the 80-GHz output of the entire PLL, with the same 10-GHz PLL reference source placed outside the cryostat and operating at room temperature. Therefore, the phase noise contribution of the reference source within the bandwidth of the PLL is the same at both temperatures. The phase noise floor of the PLL improved at 50 K by approximately 4 dB, even at offset frequencies within the bandwidth of the PLL, a proof that the phase noise contribution of the phase-frequency-detector (PFD) also decreases at cryogenic temperature. The improvement in phase noise and integrated phase error at 50 K is even greater when the PLL bandwidth is reduced to 50 MHz, as illustrated in Fig. 5.71b. Nevertheless, at 50 K, the phase noise contributions of the PLL circuits within the bandwidth of the PLL are still about 10 dB above the setup noise floor, indicating that there still is room for lower-noise circuit design of this record low phase noise 80-GHz cryogenic PLL.

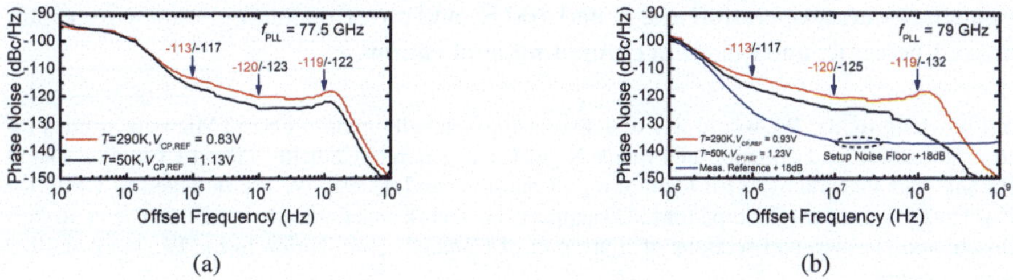

Fig. 5.71 (**a**) Measured 80-GHz PLL phase noise at 300 K and 50 K at best charge pump settings. (**b**) Comparison of phase noise contributions when the PLL bandwidth at 50 K is reduced to 50 MHz

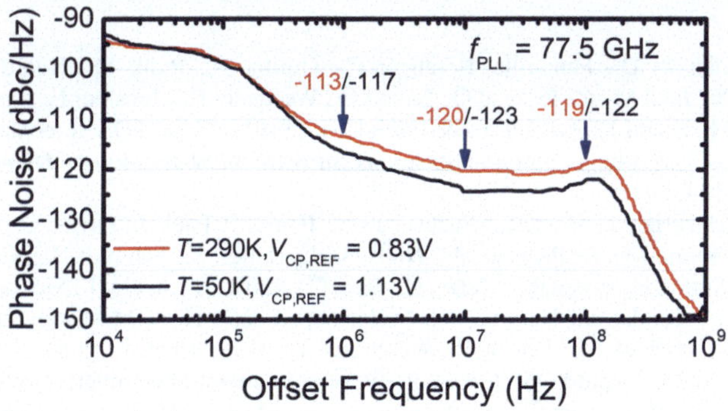

Fig. 5.72 Calculated single-spin gate XY rotation error limitation based on the measured PLL phase noise

Finally, using the PLL phase noise measurements from Fig. 5.71, the spin rotation qubit gate fidelity limit due solely to the classical control electronics is plotted in Fig. 5.72. It indicates that the error rate due to the control electronics operating at 80 GHz and 2–4 K, will be better than 10^{-4} as long as the qubit Rabi frequency, f_{Rabi}, is larger than 0.5 GHz. It becomes obvious that for the best quantum processor performance, the PLL and its reference should be operated at cryogenic temperature to achieve even lower jitter and error rates.

5.10 Conclusions

This chapter has reviewed the main AMS, digital, and HF figures of merit of FDSOI MOSFETs as a function of the back-gate voltage and discussed how algorithmic design methodologies can be adapted to take advantage of the back-gate control voltage to optimize circuit performance over a wide temperature range from 2 K to 400 K, beyond that achievable with planar bulk and FinFET CMOS technologies. Examples were provided of experimental circuits spanning the DC to 200 GHz frequency range, operation at 2 K and 300 K, and covering analog, mm-wave radar, radio, fiberoptic, and quantum computing applications.

Acknowledgments We would like to acknowledge GlobalFoundries and STMicroelectronics for chip fabrication and donation and thank Nigel Cave, Andreia Cathelin, Pascal Chevalier, David Harame and Peter Schvan for technology discussions and access. We are indebted to CMC for CAD tools and to Jaro Pristupa for CAD support. This work was funded by NSERC, Ciena, Robert Bosch, and fJscaler, and sections of it are part of Canada's contribution to the EU Horizon2020 IQubits project.

References

1. Petit D., Golanski D., Fenouillet-Béranger C., Guillot N., Rafik M., Huard V., Puget S., Montagner X., Jaud M.-A., Rozeau O., Saxod O., Wacquant F., Monsieur F., Barge D., Pinzelli L., Mellier M., Boeuf F., Arnaud, F., & Haond, M. (2012). 28-nm FDSOI technology platform for highspeed low-voltage digital applications. In *IEEE VLSI Technology Digest of Technical Papers* (pp. 133–134).
2. Carter, R., Mazurier, J., Pirro, L., Sachse, J.-U., Baars, P. Faul, J., Grass, C., Grasshoff, G., Javorka, P., Kammler, T., Preusse, A., Nielsen, S., Heller, T., Schmidt, J., Niebojewski, H., Chou, P.-Y. Smith, E., Erben, E., Metze, C., Bao, C., Andee, Y., Aydin, I., Morvan, S., Bernard, J., Bourjot, E., Feudel, T., Harame, D., Nelluri, R., Thees, H.-J., Meskamp, L.M., Kluth, J., Mulfinger, R., Rashed, M., Taylor, R., Weintraub, C., Hoentschel, J., Vinet, M., Schaeffer, J., & Rice, B. (2016). 22-nm FDSOI technology for emerging mobile, internet-of-things, and RF applications. In *IEEE International Electron Devices Meeting (IEDM)*, San Francisco (pp. 27–30).
3. Shopov, S., & Voinigescu, S. P. (2014). Characterization of the high frequency performance of 28-nm UTBB FDSOI MOSFETs as a function of backgate bias. In *IEEE CSICS Digest*, San Diego.
4. Pati Tripathi, S., Bonen, S., Bharadwaj, A., Jager, T., Nastase, C., Iordănescu, S., Boldeiu, G., Nicoloiu, A., Păşteanu, M., Müller, A., & Voinigescu, S. P. (2022). Characterization and modelling of quantum dot behaviour in FDSOI devices. *IEEE Journal of the Electron Devices Society, 10*, 600–610. https://doi.org/10.1109/JEDS.2022.3176205
5. Taur, Y., & Ning T. H. (2009). *Fundamentals of modern VLSI devices* (2nd ed., Ch. 3). Cambridge University Press.
6. Cardoso Paz, B., Cassé, M., Theodorou, C., Ghibaudo, G., Kammler, T., Pirro, L., Vinet, M., de Franceschi, S., Meunier, T., & Gaillard, F. (2020). Performance and low-frequency noise of 22-nm FDSOI down to 4.2 K for cryogenic applications. *IEEE Transactions on Electron Devices, 67*(11), 4562–4567.
7. Bonen, S., Dadash, M. S., Zandieh, A., Alakuşu, U., Gong, M. J., Rafique, J., Wu, L., Checca, E., Hsu, H. Y., Tripathi, S. P., Cooke, G., & Voinigescu, S. P. (2022). Harnessing the unique features of FDSOI CMOS technology in fibreoptic, millimetre-wave, and quantum computing circuits from 2 K to 400 K. *Solid State Electronics, 194*, 108343.
8. Han, H.-C., Zhao, Z., Lehmann, S., Charbon, E., & Enz, C. (2023). Novel approach to FDSOI threshold voltage model validated at cryogenic temperatures. *IEEE Access, 11*, 56951–56957.
9. Dickson, T. O., Yau, K. H. K., Chalvatzis, T., Mangan, A., Beerkens, R., Westergaard, P., Tazlauanu, M., Yang, M.T., & Voinigescu, S. P. (2006). The invariance of characteristic current densities in nanoscale MOSFETs and its impact on algorithmic design methodologies and design porting of Si(Ge)(Bi) CMOS high-speed building blocks. *IEEE Journal of Solid-State Circuits, 41*(8), 1830–1845.

10. Voinigescu, S. (2013). *High frequency integrated circuits* (Chapter 4). Cambridge University Press.
11. Kao, K.-H., Wu, T.R., Chen, H.-L., Lee, W.-J., Chen, N.-Y., Ma, W. C.-Y., Su, C.-J., & Lee, Y.-J. (2020). swing saturation of nanoscale MOSFETs due to source-to-drain tunneling at cryogenic temperatures. *IEEE Electron Device Letters, 41*(9), 1296–1299.
12. Bonen, S., Alakusu, U., Duan, Y., Gong, M. J., Dadash, M. S., Lucci, L., Daughton, D. R., Adam, G. C., Iordănescu, S., Păşteanu, M., Giangu, I., Jia, H., Gutierrez, L. E., Chen, W. T., Messaoudi, N., Harame, D., Müller, A., Mansour, R. R., Asbeck, P., & Voinigescu, S. P. (2019). Cryogenic characterization of 22-nm FDSOI CMOS technology for quantum computing ICs. *IEEE Electron Device Letters, 40*(1), 127–130.
13. Bonen, S., Cooke, G., Jager, T., Bharadwaj, A., Pati Tripathi, S., Céli, D., Chevalier, P., Schvan, P., & Voinigescu, S. P. (2022). Cryogenic characterization of the high-frequency and noise performance of SiGe HBTs from DC to 70 GHz and down to 2 K. *IEEE Microwave and Wireless Components Letters, 32*(6), 696–699.
14. Bohuslavskyi, H., Jansen, A. G. M., Barraud, S., Barral, V., Cassé, M., Le Guevel, L., Jehl, X., Hutin, L., Bertrand, B., Billiot, G., Pillonnet, G., Arnaud, F., Galy, P., De Franceschi, S., Vinet, M., & Sanquer, M. (2019). Cryogenic subthreshold swing saturation in FD-SOI MOSFETs described with band broadening. *IEEE Electron Device Letters, 40*(5), 784–787.
15. Jin, X., Kumar, P., & Schröter, M. (2024, October). On the emitter back-injection current in advanced SiGe HBTs at cryogenic temperatures. In *IEEE BCICTS Proceedings* (pp. 274–277).
16. Han, H.-C., Jazaeri, F., D'Amico, A., Baschirotto, A., Charbon, E., & Enz, C. (2021). Cryogenic characterization of 16 nm FinFET technology for quantum computing. In *IEEE ESSCIRC Proceedings*.
17. Pati Tripathi, S. (2022). Cryogenic characterization and modelling of FinFET technology. Master of Applied Sciences Thesis, University of Toronto.
18. Gong, M. J., Alakusu, U., Bonen, S., Dadash, M. S., Lucci, L., Jia, H., Gutierrez, L. E., Chen, W. T., Daughton, D. R., Adam, G. C., Iordănescu, S., Păşteanu, M., Messaoudi, N., Harame, D., Müller, A., Mansour, R. R., & Voinigescu, S. P. (2019). Design considerations for spin readout amplifiers in monolithically integrated semiconductor quantum processors. In *IEEE RFIC*.
19. Voinigescu, S. P., Bonen, S., Pati Tripathi, S., Cooke, G., Jager, T., & Bharadwaj, A. (2023). A circuit designer's perspective on transistor modelling challenges for 6G, fiberoptics, and quantum computing ICs. In *IEEE Bipolar, BiCMOS and Compound Semiconductor IC and Technology Symposium Digest* (pp. 36–43). Monterey.
20. Dadash, M. S., Bonen, S., Alakusu, U., Harame, D., & Voinigescu, S. P. (2018). DC-170 GHz characterization of 22-nm FDSOI technology for radar sensor applications. In *IEEE European Microwave Integrated Circuits Conference*, Madrid.
21. Roy, C., & Wu, K. (2024). A review of electromagnetics-based microwave circuit design optimization. *IEEE Microwave Magazine, 25*(7), 16–40.
22. Silveira, F., Flandre, D., & Jespers, P. G. A. (1996). A g_m/I_D based methodology for the design of CMOS analog circuits and its application to the synthesis of a silicon-on-insulator micropower OTA. *IEEE Journal of Solid State Circuits, 31*(9), 1314–1319.
23. Zandieh, A., Nguyen, T.-L., Harame, D., & Voinigescu, S. P. (2018). 128-GS/s ADC front-end with over 60-GHz input bandwidth in 22-nm Si/SiGe FDSOI CMOS. In *Digest. IEEE Bipolar, BiCMOS and Compound Semiconductor IC and Technology Symposium*, San Diego.
24. Wu, L., Hsu, H. Y., & Voinigescu, S. P. (2021). A DC to 220 GHz high-isolation SPST switch in 22-nm FDSOI CMOS. *IEEE Microwave and Wireless Components Letters, 31*(6), 775778.
25. Liu, T., Hsu, H. Y., Hash, J., & Voinigescu, S. P. (2022). W-band active repeater arrays and cognitive receivers for OFDM radar networks. In *IEEE IMS*.
26. Bardin, J. C., et al. (2019). Design and characterization of a 28-nm bulk-CMOS cryogenic quantum controller dissipating less than 2 mW at 3 K. *IEEE J. Solid-State Circuits, 54*(11), 3043–3060.

27. Chakraborty, S., et al. (2022). A cryo-CMOS low-power semi-autonomous transmon qubit state controller in 14-nm FinFET technology. *IEEE Journal of Solid-State Circuits, 57*(11), 3258–3273.
28. Nguyen, R. L., et al. (2024). A 200GS/s 8b 20fJ/c-s receiver with >60 GHz AFE bandwidth for 800Gb/s optical coherent communications in 5-nm FinFET. In *IEEE International Solid-State Circuits Conference Digest of Papers* (pp. 344–345).
29. Dadash, M. S., Harame, D., & Voinigescu, S. P. (2018). Large-swing 22 nm Si/SiGe FDSOI stacked cascodes for 56G baud drivers and 5G PAs. In *Digest. IEEE Bipolar, BiCMOS and Compound Semiconductor IC and Technology Symposium*, San Diego.
30. Shopov, S., & Voinigescu, S. P. (2016). A 3 × 60 Gb/s transmitter/repeater front-end with 4.3 Vpp single-ended output swing in a 28 nm UTBB FD-SOI CMOS technology. *IEEE JSSC, 51*, 1151–1662.
31. Rafique, J., Nguyen, T., & Voinigescu, S. P. (2019). A 4.6V, 6-bit, 64GS/s transmitter in 22-nm FDSOI CMOS. In *IEEE Bipolar, BiCMOS and Compound Semiconductor IC and Technology Symposium*, Nashville.
32. Pornpromlikit, S., Jeong, J., Presti, C. D., Scuderi, A., & Asbeck, P. M. (2009). A Watt-level stacked-FET linear power amplifier in silicon-on-insulator CMOS. *IEEE Transactions on Microwave Theory and Techniques, 58*(1), 57–64.
33. Zandieh, A., Bonen, S., Dadash, M. S., Gong, M. J., Hasch, J., & Voinigescu, S. P. (2021). "155-GHz FMCW and stepped-frequency carrier OFDM radar sensor transceiver IC featuring a PLL with < 30-ns settling time and 40-fs RMS jitter. *IEEE Transactions on Microwave Theory and Techniques, 69*(11), 4908–4924.
34. Sarkas, I. (2012). PhD Thesis, University of Toronto.
35. Yao, T., Gordon, M. Q., Tang, K. K. W., Yau, K. H. K., Yang, M.-T., Schvan, P., & Voinigescu, S. P. (2007). Algorithmic design of CMOS LNAs and PAs for 60-GHz radio. *IEEE Journal of Solid State Circuits. 42*(5), 1044–1057.
36. Zhao, J., & Voinigescu, S. P. (2023). An 80-GBaud PAM-4 Gm-boosted variable-gain TIA in 22-nm FDSOI. In *IEEE Bipolar, BiCMOS and Compound Semiconductor IC and Technology Symposium Digest* (pp. 199–202), Monterey.
37. Bonen, S., Pati Tripathi, S., McIntosh, J., Jager, T., & Voinigescu, S. P. (2024). Investigation of p- and nType quantum dot arrays manufactured in 22-nm FDSOI CMOS at 2–4 K and 300 K. In *IEEE Electron Device Letters, Early Access*. https://doi.org/10.1109/LED.2024.3435380

Part III
Data Converter Techniques

Chapter 6
A Back-Gate Linearization Technique for Current-Steering DACs

Marcel Runge and Friedel Gerfers

6.1 Multi-bit DAC Nonlinearity

Due to local process variations, the DAC weights (unit currents) in multi-bit DACs can vary, leading to errors of both stochastic (random) and gradient (systematic) nature. These variations are significant because they result in a nonlinear transfer function of the DAC, which can adversely affect the overall system's operation by reducing the accuracy and reliability of signal processing. The technique presented here is demonstrated using a DAC operating within a CT Sigma Delta modulator, as highlighted in Fig. 6.1. The modulator comprises an analog loop filter $H(s)$, a multi-bit quantizer, and DAC. The DAC output $y_{\mathrm{DAC}}(t)$ is subtracted from the modulator input $u(t)$ prior to entering the loop filter $H(s)$. In this study, a thermometer-coded DAC with N unit weights w_{i} is chosen. Thermometer coding is preferred over binary coding because it effectively minimizes output glitches, particularly during transitions near the midpoint of the code range. Thanks to the multi-bit architecture and oversampling working principle, the DAC closely follows the modulator input signal. Hence, only a few least significant bits (LSBs) switch among consecutive samples, ensuring smooth and accurate signal transitions. A particular challenge arises from the fact that the DAC outputs are interpreted as part of the modulator's input signal, as illustrated in Fig. 6.1 [1]. This means that any nonlinearity in the DAC is directly transferred to the input signal, compromising the accuracy of the entire system. To ensure reliable system performance, the DAC's linearity must at least match, or ideally exceed, the overall linearity specification of the modulator. The internal resolution of the quantizer and the DAC in multi-bit systems is typically between 3 and 5 bit. Figure 6.2 illustrates this issue by showing two exemplary mean-free unit weights with slightly different values. Even when all DAC weights

M. Runge (✉) · F. Gerfers
Technische Universität Berlin, Berlin, Germany
e-mail: marcel.runge@tu-berlin.de; friedel.gerfers@tu-berlin.de

F. Gerfers (ed.), *Bulk-Driven Circuit Techniques for CMOS FDSOI Processes*,
https://doi.org/10.1007/978-3-031-85114-8_6

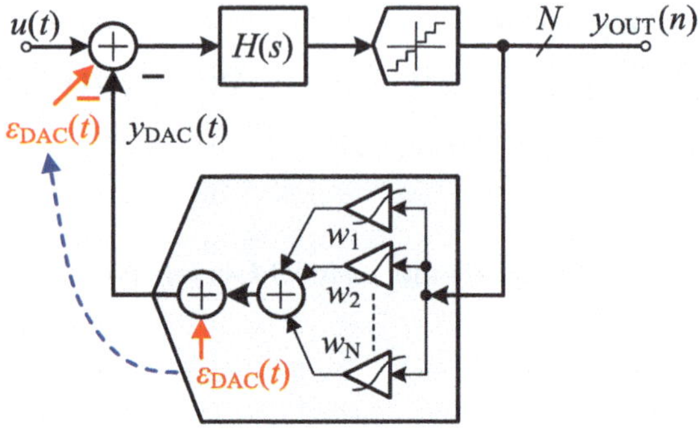

Fig. 6.1 DAC error ϵ_{DAC} within Sigma Delta modulators

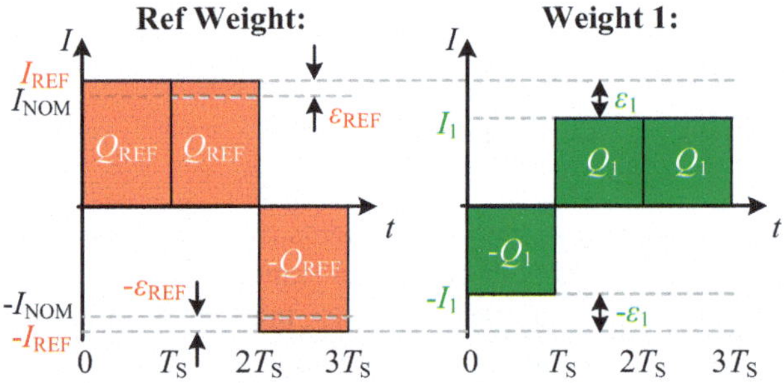

Fig. 6.2 Two DAC unit weights

I_i are adjusted to match the reference weight ($I_i = I_{REF}$), it is still possible that the actual reference weight deviates from the nominal weight defined during the design phase ($I_{REF} \neq I_{NOM}$). It is important to emphasize that only deviations from a common reference weight (I_{REF}) result in DAC nonlinearities. These nonlinearities manifest as distortions in the output signal, which is particularly critical in precision-oriented applications. Therefore, minimizing or compensating for such deviations is essential to maintain the target modulator performance.

As a result, the current produced by the ith unit current source is expressed as

$$I_i = I_{REF}\,(1 + \epsilon_i)\,, \tag{6.1}$$

where ϵ_i represents the relative error with respect to the reference source I_{REF}.

6.2 Active Body-Bias DAC Linearization

The system concept is illustrated in Fig. 6.3. In this design, each unit weight w_i includes an auxiliary DAC (AUXDAC) with B_{AUX} bit resolution. This allows for independent tuning of each weight, with the objective of eliminating static weight errors ϵ_i. Each unit element of the thermometer-coded DAC consists of a flipped-

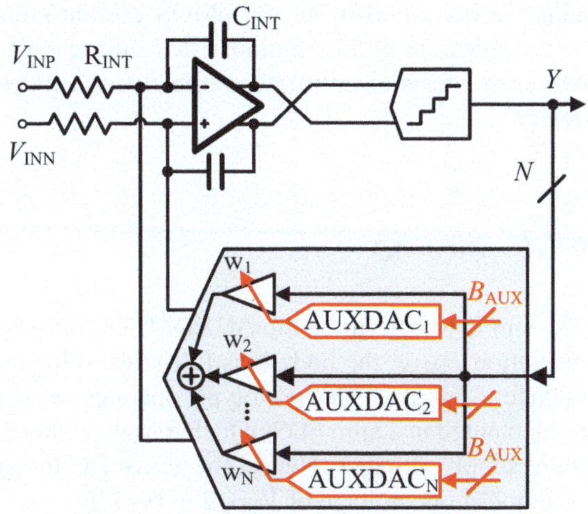

Fig. 6.3 DAC linearization by controlling each DAC weight w_{i} independently

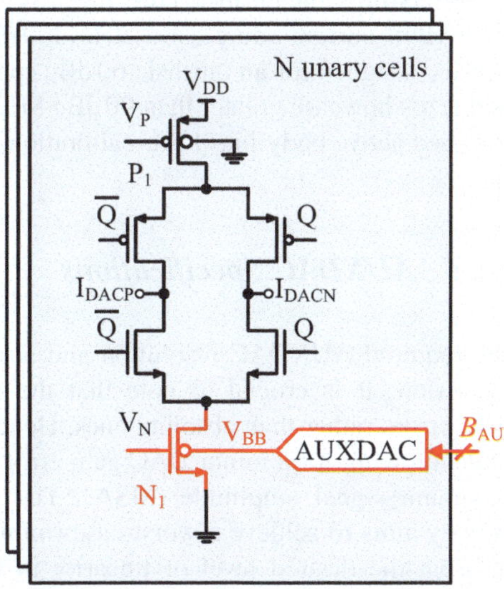

Fig. 6.4 Push–pull DAC unit cell featuring a precisely tuned nMOS body bias

well pMOS and nMOS device, configured as a push–pull current source depicted in Fig. 6.4. The nMOS current source is placed within a separate nWell, facilitating precise tuning of individual DAC weights. The AUXDAC is electrically isolated from the main DAC output nodes, ensuring that its parasitic effects do not compromise the dynamic performance of the main DAC (e.g., bandwidth) or the stability of the closed-loop integrator due to increased virtual ground capacitance. The AUXDAC provides a constant voltage output, which is set initially or periodically updated during error calibration iterations. This design approach optimizes power efficiency, as the AUXDAC is not required to accommodate fast rise and fall times. If the feedback DAC comprises N unit cells, each with B_{AUX} bit, $N \cdot B_{\mathrm{AUX}}$ bits are stored. Alternatively, the current tuning range can be doubled by introducing adaptive body-biasing to the pMOS current sources via a secondary AUXDAC. Nevertheless, as will be demonstrated subsequently, complete compensation for unit cell mismatches of up to 7% is effectively achieved by controlling only the nMOS device.

6.2.1 Strategy

The aim is to determine the AUXDAC circuit parameters, including the AUXDAC resolution B_{AUX}, the body-bias full-scale voltage FS_{AUX} , and the AUXDAC LSB voltage LSB_{AUX}. The prototype operates with a $V_{\text{DD}} = 0.9\,\text{V}$ power supply, implemented in 22 nm FDSOI technology, without requiring negative or greater than 0.9 V supply voltages. Hence, the active RC integrator utilizes an input and output common-mode voltage of $V_{\text{DD}}/2 = 0.45\,\text{V}$.

To maximize the intrinsic DAC linearity, the overdrive voltage V_{OV} for the nMOS and pMOS current sources is set to 400 mV. Following [2–6], the DAC in this work is designed for an intrinsic 60 dBc spurious-free dynamic range (SFDR). The goal is to showcase greater than 90 dBc SFDR post-corrected performance with the proposed active body-bias DAC calibration.

6.2.2 AUXDAC Specifications

The required AUXDAC resolution and accuracy are addressed in the subsequent discussion. It is crucial to note that the focus is on compensating for relative DAC errors rather than absolute ones. Hence, the modulator's linearity is restored, while allowing for a minor DAC gain error that marginally affects the modulator's maximum signal amplitude (MSA). The proposed feedback DAC linearization strategy aims to achieve a worst-case (minimum) post-corrected SFDR of 90 dBc, ensuring the desired level of linearity in the modulator. In [7], the relationship between the DAC integral nonlinearity (INL) and the spectral SFDR is derived with

$$\text{SFDR}_{\text{MIN}} \approx 20\log_{10}\left(\frac{2^B}{\text{INL}_{\text{MAX}}}\right), \tag{6.2}$$

where B denotes the DAC resolution, and INL_{MAX} the worst-case integral nonlinearity. It should be noted that in multi-bit Sigma Delta modulators, the linearity target (90 dBc) exceeds the internal bit width greatly. This design features an internal bit width of 5 bit. Given the 90 dBc SFDR target, a post-corrected 0.1% maximum INL is tolerable for the feedback DAC. Here, only random unit cell mismatch is considered and deterministic errors, e.g., gradients, are neglected.

The authors in [8, 9] establish the relationship between INL and the normalized standard deviation of DAC unit cell mismatch σ_{U}

$$\mathrm{INL}_{\mathrm{MAX}} \approx CI \cdot \sigma_{\mathrm{U}} \frac{\sqrt{2^B}}{2}. \tag{6.3}$$

Here, CI denotes the confidence interval ($CI = 3$ is assumed throughout this chapter). As an example, a 10 µA LSB unit cell current requires a strict matching tolerance of $\Delta I = \pm 3\,\mathrm{nA}(\pm 3\sigma_{\mathrm{U}})$. The FD-SOI process enables active control of the flipped-well nMOS back gate, allowing for precise regulation of the drain current by

$$I_{\mathrm{D}} \approx \beta\left[V_{\mathrm{GS}} - V_{\mathrm{TH}}\left(V_{\mathrm{BB}}\right)\right]^2 \tag{6.4}$$

$$I_{\mathrm{D}} \approx \beta\left[V_{\mathrm{GS}} - V_{\mathrm{TH0}} + \frac{72\,\mathrm{mV}}{\mathrm{V}} V_{\mathrm{BB}}\right]^2. \tag{6.5}$$

Here, β represents the transistor slope parameter, and V_{BB} denotes the actively controlled nMOS bulk-source voltage. With a back-gate sensitivity of 72 mV/V, applying a 1 V bulk-source voltage to transistor N1 reduces its threshold voltage by 72 mV, thereby increasing its drain current. As shown in Fig. 6.4, both the nMOS N1 and the pMOS P1 contribute to the differential current mismatch described by

$$\sigma_{\mathrm{U}} = \sqrt{\sigma_{\mathrm{N1}}^2 + \sigma_{\mathrm{P1}}^2}. \tag{6.6}$$

Therefore, the full-scale voltage and resolution of the AUXDAC must be configured to compensate for the mismatch contributions from both P1 and N1, as discussed in the following sections.

6.2.2.1 AUXDAC Least Significant Bit

Disregarding distance-related mismatch effects between devices, variations in transistor mobility and $C_{\mathrm{OX}}^{'}$, as well as gradient errors, [8] states the unit cell matching of current sources with

$$\sigma_{\mathrm{U}} = \frac{g_{\mathrm{m}}}{I_{\mathrm{D}}} \sigma_{\mathrm{VTH}}, \tag{6.7}$$

where g_{m} denotes the transconductance, I_{D} the drain current, and σ_{VTH} the standard deviation of the mismatch in transistor threshold voltage. As mentioned at the outset, the overdrive voltage V_{OV} is optimized to 0.4 V ($g_{\mathrm{m}}/I_{\mathrm{D}} = 2/V_{\mathrm{OV}}$). The threshold voltage mismatch is derived in [10] with

$$\sigma_{\mathrm{VTH}} = \frac{A_{\mathrm{VTH}}}{\sqrt{W \cdot L}} + S_{\mathrm{VTH0}} D, \tag{6.8}$$

where $W \cdot L$ denotes the transistor area and D the distance between the matching structures. Variables A_{VTH} and S_{VTH0} are process-dependent parameters. In the subsequent analysis, distance-related effects are disregarded ($S_{\mathrm{VTH0}} = 0$), as well as current factor mismatch ($A_{\beta} = 0$). It becomes clear that improving linearity by 1 bit necessitates quadrupling the transistor area.

The proposed body-bias calibration avoids costly area penalties, as the threshold voltage mismatch is now defined by the voltage uncertainty at the back-gate σ_{AUXDAC}, resulting in

$$\sigma_{\mathrm{U}} = \frac{g_{\mathrm{m}}}{I_{\mathrm{D}}} 72 \cdot 10^{-3} \sigma_{\mathrm{AUXDAC}}. \tag{6.9}$$

The voltage uncertainty σ_{AUXDAC} is given by the AUXDAC defining body-bias voltage V_{BB}. Thus, it is interpreted as quantization noise with

$$\sigma_{\mathrm{AUXDAC}} = \frac{LSB_{\mathrm{AUX}}}{\sqrt{12}}. \tag{6.10}$$

Thus, the required AUXDAC LSB voltage LSB_{AUX} is derived by

$$LSB_{\mathrm{AUX}} \leq \frac{\sqrt{12}}{2 \cdot 72 \cdot 10^{-3}} V_{\mathrm{OV}} \cdot \sigma_{\mathrm{U}}. \tag{6.11}$$

To achieve 15 bit linearity, the chosen 400 mV overdrive voltage V_{OV} combined with the target $\sigma_{\mathrm{U}} \leq 0.012\%$, necessitates an

$$LSB_{\mathrm{AUX}} \leq 1.2\,\mathrm{mV}. \tag{6.12}$$

Thus, precise millivolt-level control of the back gate allows for tuning the current source with nanoampere-level accuracy.

6.2.2.2 AUXDAC Full Scale and Resolution

In this design, the AUXDAC full-scale voltage is selected to compensate for $\pm 3\sigma_{\mathrm{U}}$. The intrinsic 60 dBc SFDR performance of the DAC yields a minimum tuning range of $\pm$ 3.3%. To determine the required body-bias full-scale voltage FS_{AUX}, two extreme scenarios are evaluated in the following. Setting the AUXDAC voltage to 0 V results in

$$I\,(0\,\mathrm{V}) = \frac{1}{2} g_{\mathrm{m}} \left[V_{\mathrm{GS}} - V_{\mathrm{TH0}}\right]. \tag{6.13}$$

Table 6.1 Summary of DAC and AUXDAC parameters

DAC specifications	
Internal bit width B	5 bit (therm. coded)
$SFDR_{INIT}$	60 dBc
$SFDR_{POSTCAL}$	>90 dBc
CS overdrive voltage V_{OV}	0.4 V
Supply V_{DD}	0.9 V
Derived DAC parameters and AUXDAC specifications	
INL_{MAX} (post-corrected)	0.1%
max. residual DAC errors σ_U	0.012%
AUXDAC LSB voltage LSB_{AUX}	< 1.2mV
AUXDAC full scale voltage FS_{AUX}	300 mV
AUXDAC resolution B_{AUX}	min. 8 bit (binary coded)

Conversely, applying the full-scale voltage FS_{AUX} to the back gate shifts the nMOS threshold voltage downward, thereby increasing the nMOS current to

$$I\left(FS_{AUX}\right) = \frac{1}{2} g_m \left[V_{GS} - V_{TH0} + 72 \frac{\text{mV}}{\text{V}} FS_{AUX} \right]. \tag{6.14}$$

Given the 60 dBc intrinsic SFDR, the AUXDAC must compensate $\geq 6.6\% (\pm 3\sigma_U)$ current variations yielding

$$\frac{\Delta I}{I} = \frac{I\left(FS_{AUX} - I\left(0\,\text{V}\right)\right) -}{I\left(0\,\text{V}\right)} \approx \frac{1}{V_{OV}} 72 \frac{\text{mV}}{\text{V}} FS_{AUX} \geq 6.6\%. \tag{6.15}$$

Solving the equation for the body-bias full-scale voltage FS_{AUX} yields a minimum required range of 300 mV. This range is critical to ensure the proper tuning of the DAC to 15 bit linearity. Given the target LSB_{AUX} value of ≤ 1.2 mV and the derived FS_{AUX}, it is necessary to achieve an AUXDAC resolution B_{AUX} of at least 8 bit. This resolution ensures that the AUXDAC accurately adjusts the nMOS body-bias voltage to compensate for mismatches in the current sources. Table 6.1 provides a summary of the derived parameters for the DAC and AUXDAC.

6.3 Error Detection Algorithm

The following section discusses the implemented algorithm used to estimate the error ϵ_i for each unit cell in the feedback DAC. The DAC error estimation method is based on the approach detailed in [12], which is reviewed in the following section. To facilitate this, an additional DAC unit cell w_T is incorporated into the feedback DAC, as illustrated in Fig. 6.5.

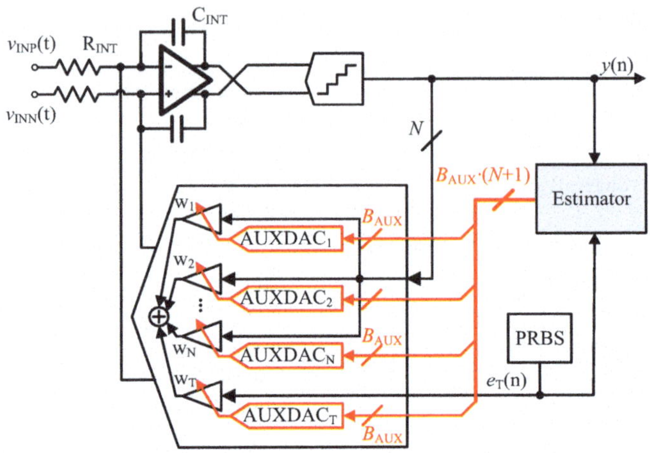

Fig. 6.5 System with extra test-source w_T, PRBS generator, and estimator

This test cell is connected to a digital pseudo random binary sequence (PRBS) source that operates at a reduced bit transition rate($f_S/256$). Due to the slower PRBS update rate, the majority of its signal power is concentrated within the modulator in-band frequencies. The cross-covariance between PRBS test signal $e_T(n)$ and the modulator output $y(n)$ is computed for a single tap m, as outlined in [11].

$$R_i = R_{ey}(m) = \sum_{n}^{N-m-1} \left[\left(e_T(n+m) - \overline{e_T} \right) \left(y(n) - \overline{y} \right) \right], \tag{6.16}$$

where $\overline{e_T}$ and $\overline{y}$ represent the respective mean values. R_i refers to the cross-covariance coefficient corresponding to the ith DAC weight. Subsequently, each unit cell w_i in the feedback DAC is replaced by the test cell w_T , after which the PRBS is reapplied to this substituted source, and the corresponding cross-covariance is computed. The DAC weights contain random errors ϵ_i relative to the designated reference weight w_{REF}, leading to the following dependency

$$w_i = w_{REF}(1 + \epsilon_i). \tag{6.17}$$

As the test source w_T temporarily takes over the signal delivery from each unit source w_i, the modulator continues to function normally, preserving the signal transfer function (STF). By applying this procedure across all N DAC unit sources, including the test source w_T, a total of $N + 1$ cross-covariance coefficients are generated.

$$\mathbf{R} = \left(R_1 \; R_2 \; \ldots \; R_N \; R_{N+1} \right)^T. \tag{6.18}$$

By selecting one DAC cell as the reference source, represented by the cross-covariance coefficient R_{REF}, [12] calculates the estimated relative error using the following method

$$\epsilon_{\text{i}}^{'} = \frac{R_{\text{i}}}{R_{\text{REF}}} - 1. \tag{6.19}$$

This concept becomes particularly evident when the cross-covariance R_{eq} is analyzed in the $\mathcal{Z}$-domain. Under the assumption of a linear time-invariant system, the PRBS $e_{\text{T}}[n]$ is initially processed through the gain stage w_{T}, followed by the negative modulator signal transfer function (STF), resulting in the following expression:

$$S_{\text{ey}}(z) = -w_{\text{T}} \cdot STF(z) \cdot S_{\text{ee}}(z). \tag{6.20}$$

Here, S_{ee} represents the test pattern auto-covariance in $\mathcal{Z}$-domain.

The subtraction of the DAC signal at the modulator input introduces a negative sign. Since both S_{ee} and the STF are independent of the analog DAC weights, the only effect is on the gain difference. The relative mismatch ϵ_{i} results in a gain error between the two cross-covariances. Hence, it is sufficient to calculate the cross-covariance R for a single tap m. Once the test cell w_{T} is swapped with a DAC weight w_{i}, the DAC offset voltage changes, thereby impacting the overall offset behavior of the modulator. Unlike cross-correlation, cross-covariance removes the offset in the samples before determining the correlation, making it a more precise measure for this purpose.

6.3.1 Reference Weight Selection

Unlike in digital correction methods, it is essential to avoid AUXDAC clipping when using the proposed DAC calibration technique. To address this, the cross-covariance is computed for all DAC cells at a specific tap (e.g. $m = 0$). The unit cell with a cross-covariance value closest to the midpoint $(MIN + MAX)/2$ is then designated as the reference weight w_{REF}, as shown in Fig. 6.6. This strategy is designed to maximize the correction range for both positive and negative DAC mismatches, given the fixed AUXDAC full-scale range FS_{AUX}.

If the estimated relative unit cell error $\epsilon_{\text{i}}^{'}$ is positive, it indicates that the corresponding DAC unit cell is delivering a higher output current compared to the reference source. To correct this, as illustrated in Fig. 6.6, the nMOS body-bias voltage V_{BB} should be reduced, which will increase the threshold voltage V_{TH} and consequently decrease the drain current I_{D}. After determining the reference cell wREF, the estimator applies Eq. 6.19 to the remaining N cross-covariance coefficients R_{i} to update the AUXDAC codes, thereby minimizing the relative errors ei across all unit cells.

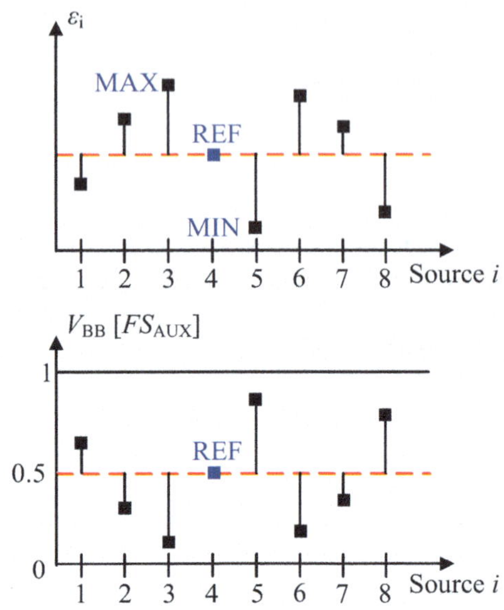

Fig. 6.6 Strategy for selecting reference weight w_{REF}

Initially, all AUXDACs are set to their mid-code to facilitate both positive and negative error corrections during the first iteration. Following the update of each AUXDAC value, the process enters the next iteration. Throughout this process, the AUXDAC associated with the reference weight w_{REF} remains fixed at mid-code. It's important to note that as the iterations progress, the accuracy of the error estimation improves, correlating with the increasing precision of the modulator [12]. Since the PRBS is uncorrelated with the modulator input signal, cross-covariance enables background error estimation and correction [12].

6.3.2 AUXDAC Nonlinearity

The AUXDAC LSB_{AUX} and full-scale FS_{AUX} requirements are determined with a minimum ideal resolution of $B_{AUX} = 8$ bit to achieve a worst-case post-corrected SFDR of 90 dBc. Since the AUXDACs are also prone to gradient and random mismatch, which impacts the accuracy of the DAC calibration, the AUXDAC resolution is increased by 1 bit to 9 bit B_{AUX} with the design goal to maintain 8 bit AUXDAC linearity.

Given $(N+1) = 32$ DAC unit cells that all comprise an AUXDAC with $B_{AUX} =$ 9 bit resolution the finite state machine (FSM) has to store $B_{AUX} \cdot (N+1) = 288$ digital bits.

For example, after estimating the unit cell error ϵ_i using the cross-covariance method and obtaining $\epsilon_i^{'}$, the FSM conducts a binary search with digital AUXDAC weights to reduce $\epsilon_i^{'}$ to zero. This process is repeated for all estimated DAC unit cell errors using their corresponding digital weights.

6.4 Prototype Design

This section validates the proposed DAC linearization method through a detailed analysis of a prototype design, illustrated in Fig. 6.7. The prototype is specifically engineered to meet rigorous requirements: a 30 MHz bandwidth, 12 bit effective number of bits (ENOB), and a 74 dB signal-to-noise and distortion ratio (SNDR), with a post-corrected SFDR exceeding 90 dBc.

Key to achieving these performance targets are the DAC and quantizer, both of which utilize a 5 bit internal resolution. The quantizer is implemented as an asynchronous successive approximation register analog-to-digital converter (SAR-ADC), optimized for high power efficiency. A crucial design feature is the introduction of an excess loop delay (ELD) of 0.5 T_S (half a sampling period). This ELD

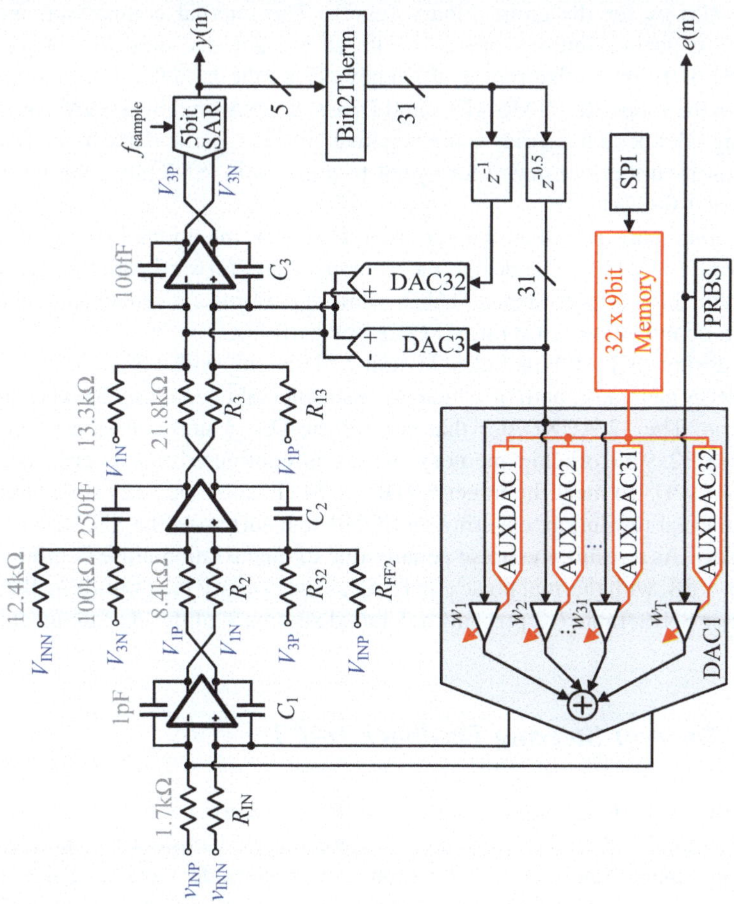

Fig. 6.7 60 MS/s CT Sigma Delta modulator with active body-bias DAC calibration © 2022 IEEE. (Reprinted with permission from [13])

Table 6.2 Modulator core specifications

Parameter	Unit
Bandwidth	30 MHz
SNDR	74 dB
$SFDR_{MIN}$	90 dBc
OSR	10
Int. resolution	5 bit
Full scale	1.2 V_{PPD}

ensures that sampling occurs on one clock edge (e.g., the rising edge), while the feedback DAC signal is updated on the opposite clock edge (e.g., the falling edge). Consequently, the SAR-ADC has less than half of the modulator's clock period to complete the binary search process. Equipped with a 30 fF sampling capacitor, the SAR-ADC is optimized for high-speed operation, achieving an average conversion time of 400 ps for the entire binary search. The internal comparator, including logic and capacitor settling, operates with an average decision time of just 80 ps (12.5 GHz). To maximize power efficiency across the modulator, all components, including the loop filter, SAR-ADC, and DACs, are powered by a 0.9 V core supply. The design deliberately avoids using negative voltages, particularly in the DAC. The modulator is further optimized for signal-to-noise-ratio (SNR) by configuring it for a MSA of 1.08 V_{PPD}.

To maintain the overall modulator order, DAC32 is intentionally delayed by a full clock cycle (z^{-1}) [14]. The design incorporates a feedforward path through R_{FF2} to mitigate the effects of feedback DAC2, while R_{32} forms a resonator that boosts the signal-to-quantization-noise ratio (SQNR) by 8 dB [1].

The design of DAC1 includes 31 unit cells, along with an additional test cell and a PRBS generator, both of which are instrumental in enabling background error estimation. The 32 AUXDACs that control the DAC1 unit cells are programmed through a 32x9 bit on-chip memory, which is configured via a serial peripheral interface (SPI). To meet the target SNDR of 74 dB, the loop filter is designed with a 5 bit internal resolution, ensuring an SQNR that surpasses the SNDR target by at least 10 dB. As a result, the noise penalty due to quantization noise is minimized to less than 1 dB, with thermal noise emerging as the primary factor limiting the overall modulator SNR. Table 6.2 presents a detailed summary of the core parameters.

6.4.1 Current-Steering Feedback DAC1

The feedback DAC employs a push–pull current-steering topology and is specifically designed to function from a 0.9 V core supply. Considering the operational transconductance amplifiers (OTA) input common-mode voltage of 450 mV, the overdrive voltage for the current source transistors P1 and N1 is set at 400 mV. The current source area is scaled to achieve the intrinsic 60 dB worst-case matching performance. Unlike other designs in [15–17], which employ a negative power supply or higher-than-core supply voltages to maximize the current source overdrive voltage, this design eliminates the need for additional power supplies, thereby enhancing the power efficiency of DAC1. To enable positive bulk voltage control of the back gate in transistor N1 (as illustrated in Fig. 6.8a), the nMOS current source is implemented as a flipped-well device, positioned above an nWell. The pMOS transistor P1 is also a flipped-well device, with its bulk connected to ground.

The AUXDAC, illustrated in Fig. 6.8b, operates in a static mode, using a pMOS current source array (I_{AUX}) and a pull-down resistor (R_{AUX}) to establish the body-bias voltage V_{BB}. It is essential that the AUXDAC shares the same local ground reference as the current source N1. By employing local decoupling, any noise present in the local ground is uniformly injected into both the bulk and gate nodes. This approach ensures that the bulk-source and gate–source voltages remain stable to a first-order effect, leading to a data-independent drain current I_{D}. To minimize power consumption, the design aims to maximize R_{AUX} while reducing I_{AUX} as much as possible. The AUXDAC generates continuous-time noise, so incorporating a large capacitance (C_{W}) helps mitigate its noise contribution, provided the noise bandwidth at the AUXDAC output is narrower than the modulator bandwidth. Even if the noise bandwidth is wider than the modulator's, the decimation filter at the modulator's output will only capture noise within the modulator's bandwidth. Therefore, the AUXDAC is designed with an emphasis on minimizing area and power consumption.

This design incorporates a 36 kΩ R_{AUX} with an AUXDAC LSB current step of 17 nA, producing a 600 μV LSB_{AUX} necessary to achieve 90 dBc SFDR correction performance. When the AUXDAC code is set to zero, it draws no power, while the full-scale setting consumes 8.7 μA from the 0.9 V supply. Assuming that deterministic DAC errors are negligible, the chosen reference weight represents the average of all unit cell errors. Consequently, after error correction, the total power consumption of all AUXDACs is estimated by assuming each AUXDAC is set to mid-code (255), resulting in an overall power consumption of 125 μW.

Figure 6.8b shows the layout floorplan of DAC1, where the DAC unit cells are organized in two rows, with the current source transistors N1 located along the edges of DAC1. The strategic placement of nWells at the boundaries facilitates easy access and allows the AUXDACs to be positioned adjacent to DAC1. This design approach leads to the AUXDACs only doubling the overall area of DAC1. Consequently, the 30 dB improvement in SFDR, from 60 dBc to 90 dBc, is achieved with just a 2x increase in area. By contrast, traditional sizing methods following Pelgrom's law [10] would require a significantly larger area to achieve the same performance enhancement.

$$\sigma_{\mathrm{VTH}} = \frac{A_{\mathrm{VTH}}}{\sqrt{W \cdot L}}, \tag{6.21}$$

With A_{VTH} representing the mismatch coefficient and $W \cdot L$ the transistor area, achieving the same performance would require 1000 times the DAC1 area, assuming that β mismatch and spacing effects are neglected.

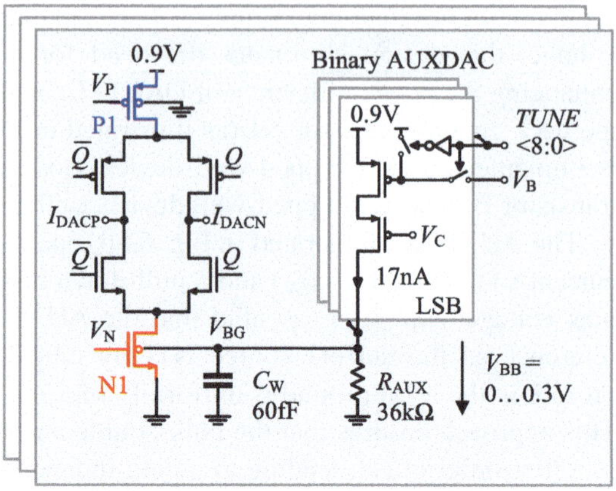

(a)

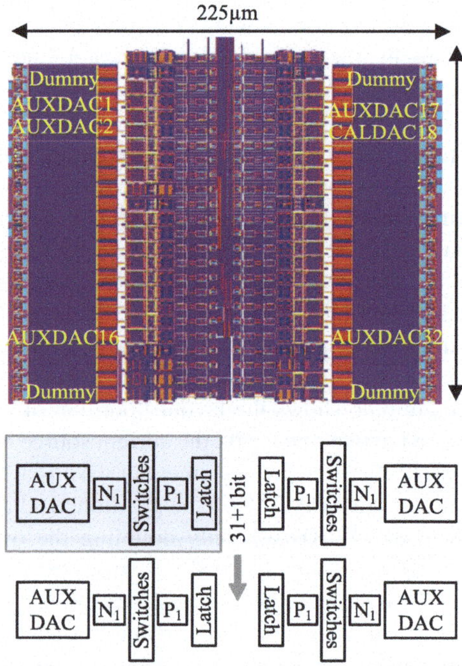

(b)

Fig. 6.8 (**a**) CS unit cell w/ AUXDAC; (**b**) Implemented DAC layout view © 2022 IEEE. (Reprinted with permission from [13])

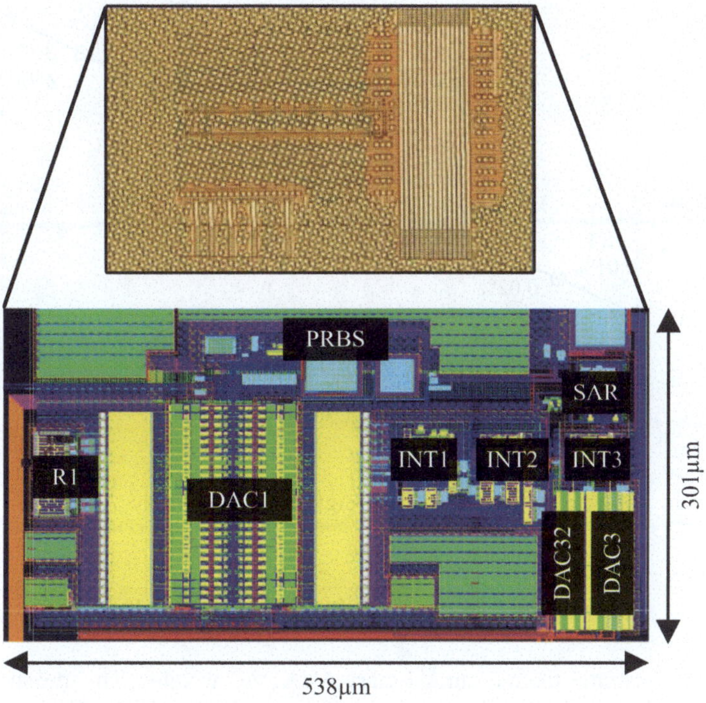

Fig. 6.9 Fabricated prototype © 2022 IEEE. (Reprinted with permission from [13])

6.5 Measurement Results

The prototype, fabricated using 22nm FDSOI CMOS technology, is depicted in Fig. 6.9 and occupies a total area of 538 μm x 301 μm (0.16 mm^2). To minimize parasitic capacitance at the INT1 virtual ground node, DAC1 is strategically placed between the first integrator and the integration resistor R_{INT}.

6.6 AUXDAC Weights and Current-Source Tuning Range

The tunability of a single DAC unit cell is initially evaluated by varying the corresponding AUXDAC code and monitoring the resulting changes in unit cell current. Since direct measurement of the DAC currents for offline high-precision evaluation is not practical, the cross-covariance algorithm is employed to determine the 9 bit AUXDAC weights. The process begins with setting the AUXDAC for the test source to zero, applying the PRBS pattern, and calculating the cross-covariance coefficient $R_{\mathrm{CODE}=0}$. This procedure is then repeated for the binary AUXDAC codes 1, 2, 4, 8, 16, 32, 64, 128, and 256. By referencing each measurement to AUXDAC

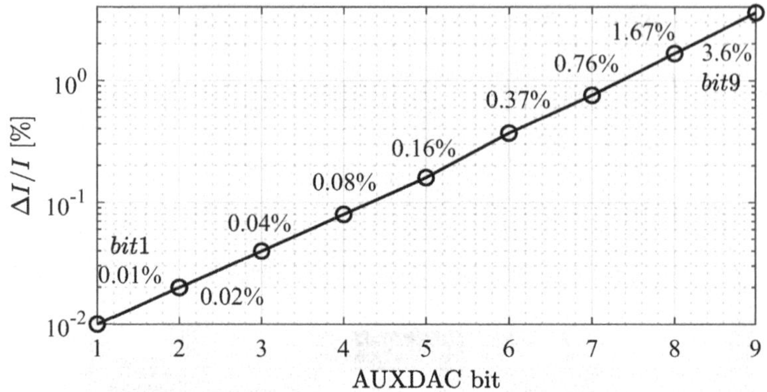

Fig. 6.10 Measured DAC unit-cell current change across AUXDAC bits

code 0, it is possible to estimate the weight of each binary AUXDAC code as follows:

$$bit_1 = \frac{R_{\text{code}=1}}{R_{\text{code}=0}}, \ldots, bit_9 = \frac{R_{\text{code}=256}}{R_{\text{code}=0}}. \tag{6.22}$$

Figure 6.10 illustrates the measured binary AUXDAC weights. The measured 0.1% LSB verifies that the AUXDAC manages to control the 23 μA DAC unit cell with

$$\frac{\Delta I}{I} = bit_1 \cdot I_U = 0.01\% \cdot 23\,\mu\text{A} = 2.3\,\text{nA}. \tag{6.23}$$

Thus, the theory presented in this thesis is validated by the measurement results: A millivolt resolution at the transistor back gate enables nanoamp-level precision in current-source tuning. Adjusting the AUXDAC to its maximum code setting (511) increases the unit cell current I_U by

$$\sum_{k=1}^{9} bit_k = 6.7\%. \tag{6.24}$$

Figure 6.11 shows the source errors of the prototype, estimated through cross-covariance, along with the corresponding AUXDAC values used for correction. Source 32 is chosen as the reference because its cross-covariance coefficient is closest to the average of all computed values. Therefore, the AUXDAC code for source 32 remains fixed at mid-code (255) across all iterations. If a unit cell has a positive mismatch compared to source 32, it indicates an excessive current, requiring a reduction in the nMOS current source's back-gate voltage. This reduction decreases the DAC element current by lowering the forward body bias. Thus, negative relative mismatch errors ϵ_i are compensated by AUXDAC values

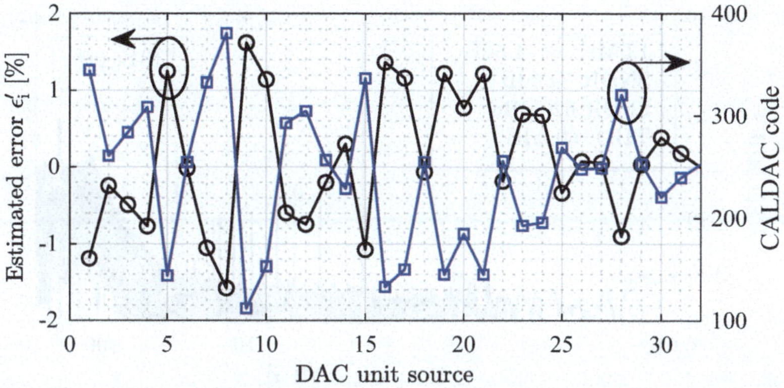

Fig. 6.11 Measured estimated relative DAC errors and applied AUXDAC codes © 2022 IEEE. (Reprinted with permission from [13])

above mid-code (255), while positive errors are compensated by values below mid-code.

6.7 Spectral Performance

Figure 6.12 displays the intrinsic spectral performance of the modulator, which achieves an 61.2 dBc SFDR. With the implementation of the proposed analog correction, as shown in Fig. 6.13, the SFDR is significantly improved to 91 dBc, and the total harmonic distortion THD is reduced to −84.7 dB. The resulting 74.6 dB SNDR aligns closely with the simulation predictions. The peak at 200 MHz corresponds to the division of the 600 MHz data stream into three parallel 200 MHz channels for the logic analyzer. The DR plot, presented in Fig. 6.14, demonstrates a dynamic range of 77 dB. By utilizing analog correction, linearization methods dependent on oversampling, like data weighted averaging, are eliminated. This approach achieves outstanding spectral purity with an SFDR exceeding 100 dB, in particular at low signal amplitudes.

6.8 Comparison with State-of-the-Art

Table 6.3 provides a comparison between the proposed design and other state-of-the-art CT Sigma Delta modulators with comparable bandwidths. The proposed multi-bit loop filter architecture delivers 12 bit ENOB with an oversampling ratio (OSR) of 10, demonstrating competitive linearity for multi-bit designs. A key advantage of this design is its exceptionally low power consumption, achieved

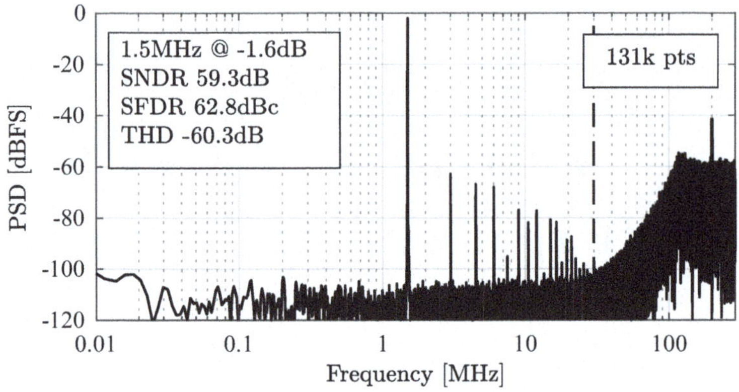

Fig. 6.12 Measured intrinsic spectral performance © 2022 IEEE. (Reprinted with permission from [13])

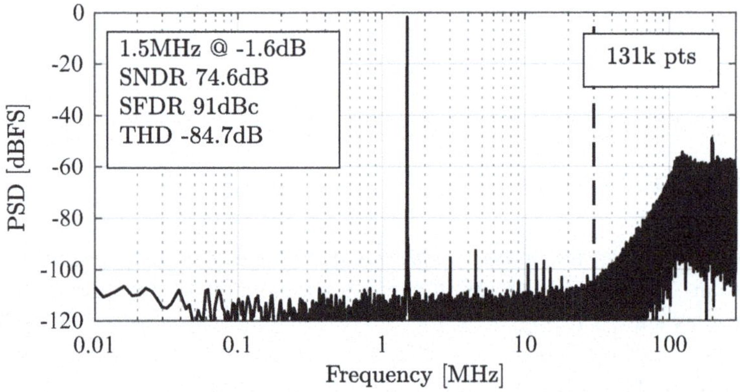

Fig. 6.13 Measured post-corrected spectral performance © 2022 IEEE. (Reprinted with permission from [13])

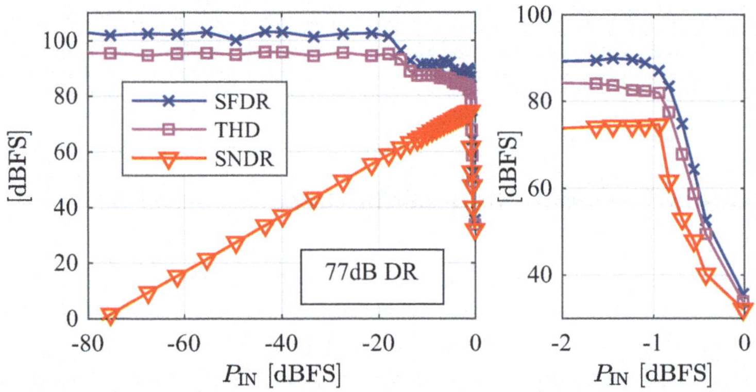

Fig. 6.14 Measured DR © 2022 IEEE. (Reprinted with permission from [13])

Table 6.3 Comparison with state of the art CT Sigma Delta modulators © 2022 IEEE. (Reprinted with permission from [13])

	This work	ISSCC '19 [18]	VLSI '19 [19]	ISSCC '18 [20]	ISSCC '17 [21]	ISSCC '12 [5]
Technology [nm]	22 SOI	7	12	28	130	90
DAC linearization	Back-gate	Mismatch error shaping	DWA	Off-line calibr.	Sizing DAC area	Analog
Supply [V]	0.9	1.3/1	0.8/1/1.5		1.2	1.2
f_S [MHz]	600	400	832	2000	640	500
BW [MHz]	30	25	30	50	15	25
SNDR_{PK} [dB]	74.6	74	71.4	79.8	80.4	67.5
SFDR [dB]	92.7	76.6	87	95.2	87.8	79
DR [dB]	77	79.2	74.4	82.8	82.9	72
Power [mW]	2.87	3.8	3.2	64.3	11.4	8.5
Active Area [mm^2]	0.162	0.056	0.059	0.25	0.17	0.23
FoM_{W} [fJ/conv.]	10.8	18.6	17.6	80.5	24	88
FoM_{SNDR} [dB]	174.8	172.2	171	168.7	171	162.1
FoM_{DR} [dB]	177.1	177.6	174	171.7	176	166.7

through the 125 μW back-gate DAC linearization technique. It's important to note that the power consumption does not include the estimator engine and FSM responsible for updating the AUXDAC values. The design achieves an outstanding 10.8 fJ/conversion-step FoM_{W} and 174.8 dB FoM_{SNDR}, outperforming other state-of-the-art designs in terms of power efficiency.

References

1. Ortmanns, M., & Gerfers, F. (2006). *Continuous-time sigma-delta A/D conversion*, vol. 21. Berlin/Heidelberg: Springer-Verlag. ISBN: 3-540-28406-0.
2. Sarhang-Nejad, M., & Temes, G. C. (1993). A high-resolution multibit sigma delta ADC with digital correction and relaxed amplifier requirements. *IEEE Journal of Solid-State Circuits, 28*(6), 648–660.
3. Kauffman, J. G., Witte, P., Becker, J., & Ortmanns, M. (2011). An 8.5 mW continuous-time $\Delta\Sigma$ modulator with 25 MHz bandwidth using digital background DAC linearization to achieve 63.5 dB SNDR and 81 dB SFDR. *IEEE Journal of Solid-State Circuits, 46*(12), 2869–2881.
4. Lee, S.-C., & Chiu, Y. (2014). A 15-MHz bandwidth 1-0 MASH $\Sigma\Delta$; ADC with nonlinear memory error calibration achieving 85-dBc SFDR. *IEEE Journal of Solid-State Circuits, 49*(3), 695–707.
5. Kauffman, J. G., Witte, P., Lehmann, M., Becker, J., Manoli, Y., & Ortmanns, M. (2014). A 72 dB DR, CT $\Delta\Sigma$ modulator using digitally estimated, auxiliary DAC linearization achieving 88 fJ/conv-step in a 25 MHz BW. *IEEE Journal of Solid-State Circuits, 49*(2), 392–404.

6. Cao, J., Meng, X., Temes, G. C., & Yu, W. (2016). Power-on digital calibration method for delta-sigma ADCs. In *2016 IEEE International Symposium on Circuits and Systems (ISCAS)* (pp. 2002–2005). IEEE. ISBN: 978-1-4799-5341-7.
7. Manganaro, G. (2012). *Advanced data converters*. Cambridge University Press. ISBN: 9781107005570.
8. Razavi, B. (2018). The current-steering DAC [A circuit for all seasons]. *IEEE Solid-State Circuits Magazine, 10*(1), 11–15.
9. Lin, C.-H., & Bult, K. (1999). A 10-b, 500-MSample/s CMOS DAC in 0.6 mm2. *IEEE Journal of Solid-State Circuits, 33*(12), 1948–1958.
10. Pelgrom, M. J. M., Duinmaijer, A. C. J., & Welbers, A. P. G. (1989). Matching properties of MOS transistors. *IEEE Journal of Solid-State Circuits, 24*(5), 1433–1439.
11. Buck, J. R., Daniel, M. M., & Singer, A. C. (2002). *Computer explorations in signals and systems using MATLAB*, 2nd edn. Prentice Hall signal processing series. Upper Saddle River, NJ: Prentice Hall. ISBN: 0130421553.
12. Witte, P., & Ortmanns, M. (2010). Background DAC error estimation using a pseudo random noise based correlation technique for sigma-delta analog-to-digital converters. *IEEE Transactions on Circuits and Systems I: Regular Papers, 57*(7), 1500–1512.
13. Runge, M., Edler, J., Schmock, D., Kaiser, T., & Gerfers, F. (2022). A 30-MHz BW 74.6-dB SNDR 92-dB SFDR CT $\Delta\Sigma$ modulator with active body-bias DAC calibration in 22nm FDSOI CMOS. In *2022 IEEE Custom Integrated Circuits Conference (CICC)* (pp. 01–02). https://doi.org/10.1109/CICC53496.2022.9772799.
14. Mitteregger, G., Ebner, C., Mechnig, S., Blon, T., Holuigue, C., & Romani, E. (2006). A 20-mW 640-MHz CMOS continuous-time $\Sigma\Delta$ ADC with 20-MHz signal bandwidth, 80-dB dynamic range and 12-bit ENOB. *IEEE Journal of Solid-State Circuits, 41*(12), 2641–2649.
15. Shibata, H., Schreier, R., Yang, W., Shaikh, A., Paterson, D., Caldwell, T. C., Alldred, D., & Lai, P. W. (2012). A DC-to-1 GHz tunable RF $\Delta\Sigma$ ADC achieving DR =74 dB and BW =150 MHz at f0 =450 MHz using 550 mW. *IEEE Journal of Solid-State Circuits, 47*(12), 2888–2897.
16. Dong, Y., Yang, W., Schreier, R., Sheikholeslami, A., & Korrapati, S. (2014). A continuous-time 0–3 MASH ADC achieving 88 dB DR with 53 MHz BW in 28 nm CMOS. *IEEE Journal of Solid-State Circuits, 49*(12), 2868–2877.
17. Dong, Y., Zhao, J., Yang, W. W., Caldwell, T., Shibata, H., Li, Z., Schreier, R., Meng, Q., Silva, J. B., Paterson, D., & Gealow, J. C. (2016). A 72 dB-DR 465 MHz-BW continuous-time 1-2 MASH ADC in 28 nm CMOS. *IEEE Journal of Solid-State Circuits, 51*(12), 2917–2927.
18. Lo, T.-Y., Weng, C.-H., Hsieh, H.-Y., Shu, Y.-S., & Chiu, P.-C. (2019). 20.4 an 8x OSR 25MHz-BW 79.4dB/74dB DR/SNDR CT $\Delta\Sigma$ modulator using 7b linearized segmented DACs with digital noise-coupling-compensation filter in 7nm FinFET CMOS. In *2019 IEEE International Solid-State Circuits Conference - (ISSCC)* (pp. 334–336). IEEE. ISBN: 978-1-5386-8531-0.
19. Weng, C.-H., Wei, T.-A., Hsieh, H.-Y., Wu, S.-H., & Wang, T.-Y. (2019). A 71. 4dB SNDR 30MHz BW continuous-time delta-sigma modulator using a time-interleaved noise-shaping quantizer in 12-nm CMOS. In *2019 Symposium on VLSI Circuits* (pp. C228–C229). IEEE. ISBN: 978-4-86348-720-8.
20. He, T., Ashburn, M., Ho, S., Zhang, Y., & Temes, G. (2018). A 50MHZ-BW continuous-time $\Delta\Sigma$ ADC with dynamic error correction achieving 79.8dB SNDR and 95.2dB SFDR. In *2018 IEEE International Solid - State Circuits Conference - (ISSCC)* (pp. 230–232). IEEE. ISBN: 978-1-5090-4940-0.
21. Kim, T., Han, C., & Maghari, N. (2017). 28.2 An 11.4mW 80.4dB-SNDR 15MHz-BW CT delta-sigma modulator using 6b double-noise-shaped quantizer. In L. C. Fujino (Ed.), *2017 IEEE International Solid-State Circuits Conference* (pp. 468–469). IEEE. ISBN: 978-1-5090-3758-2.

Chapter 7
Highly-Linear T&H-Amplifiers Utilizing Bandwidth Boosting for Time-Interleaved ADCs

Enne Wittenhagen and Friedel Gerfers

7.1 Introduction of Bandwidth Boosting T&H-Amplifiers

Future wireless communication standards such as 6G require data rates of several Gbit/s. Therefore, research is focused on ultra-wideband, high-linearity, and low-noise RF ADCs that support these data rates. In addition to throughput, a small footprint is critical for a cost-effective single-chip CMOS solution that can eliminate multiple RF components from the traditional receiver chain, saving power and silicon area. This trend is supported by shrinking CMOS feature sizes that enable high transit frequencies above 400 GHz [1] and low phase noise [2]. Direct sampling of the RF signal, achieved by moving the ADC closer to the antenna, reduces the number of mixers, lowers the noise figure, and improves flexibility and area efficiency. However, achieving high data rates requires RF sampling with ADC sample rates in the GS/s range, a high input bandwidth, and a clock with low phase noise. In addition, a front-end (FE) sampler can be used to reduce the requirements of the sub-ADCs, saving power and reducing time interleave errors. Both approaches require an FE buffer, and a large sampling capacitance, in a Nyquist rate ADC, to achieve high resolution as it accumulates less noise. Consequently, an FE buffer is typically utilized at the input to shield the ADC, providing a wide input bandwidth and isolating the FE sampler to reduce kickback effects. Thus, the FE buffer and FE sampler are among the most critical components that determine the ADC's performance in terms of bandwidth, noise, and linearity. Body-biasing provides benefits such as increasing bandwidth and linearity and serves as a feedback node for a common-mode feedback (CMFB). Two track-and-hold (T&H) amplifiers are presented that utilize the strong body-biasing option of the 22 nm fully depleted silicon-on-insulator (FDSOI) CMOS technology. In the

E. Wittenhagen (✉) · F. Gerfers
Technische Universität Berlin, Berlin, Germany
e-mail: Wittenhagen@tu-berlin.de; friedel.Gerfers@tu-berlin.de

F. Gerfers (ed.), *Bulk-Driven Circuit Techniques for CMOS FDSOI Processes*,
https://doi.org/10.1007/978-3-031-85114-8_7

first part, a 12 GS/s amplifier is analyzed using a small-signal analysis to determine the body-biasing effect on the overall bandwidth, followed by a large-signal analysis to calculate the linearity improvement due to forward body-biasing. In the second part, a highly linear 3-GS/s amplifier is presented using body-biasing for CMFB and dynamic body-biasing to improve the overall bandwidth and linearity of the T&H amplifier.

7.2 Bandwidth Boosting for Top-Plate T&H-Amplifiers

In a standard wireless receiver chain, the incoming signal is first amplified by a low-noise amplifier (LNA) and subsequently by a power amplifier (PA) to enhance the signal power. However, integrated CMOS PAs typically have low output power [3, 4], which often necessitates the use of multiple stages to increase the gain, consequently raising the overall power consumption. The amplified signal from the PA is then digitized using an RF-ADC. To reduce the requirements on the PA, it is essential for the RF-ADC to achieve a high signal-to-noise ratio (SNR) at low input power levels. One way to enhance the SNR at low input amplitudes is by using a larger sample capacitance, which reduces accumulated noise. However, this approach limits the SNR at high frequencies due to a decrease in signal bandwidth. To address this issue, the proposed RF front-end incorporates body-biasing to improve the bandwidth and adapt it inside the sample switch. Therefore, this section proposes a 12 GS/s body-driven top-plate T&H-amplifier for low input amplitudes. The block diagram of the T&H-amplifier is depicted in Fig. 7.1a and comprises a 50 Ω impedance matched front-end buffer, a top-plate sampler, and a back-end buffer driving a high-speed ADC [4]. Figure 7.1b depicts the T&H-amplifier on a schematic level only the single-ended schematic is shown for simplicity, though the actual implementation is fully differential.

The FE buffer is designed with a low output impedance of approximately 8.3 Ω, resulting in a transconductance g_{M1} of 120 mS, providing sufficient input and output bandwidth while minimizing kick-back effects from the T&H. The buffer current is determined by the slew rate required to charge the sample capacitance C_S to the input full-scale voltage of 290 mV$_{ppd}$ (approx. 250 mV$_{ppd}$ FE-buffer output full-scale). A 600 fF capacitance is needed to achieve an overall (FE-buffer/T&H/BE-buffer) low-frequency SNR above 49 dB at the BE-buffer output. This configuration necessitates an FE-buffer current of approximately 13 mA to achieve a 6 ps slewing time, which is 15% of the total track time of around 40 ps.

Transistor M_2 supplies the current for the FE buffer and is designed to have low parasitic capacitance to maximize bandwidth. The FE-buffer output is sampled using transistor M_3, implemented as a PMOS switch with an ON-resistance below 9 Ω, balancing nonlinear resistance, capacitances, and clock driver gate-capacitance ratio.

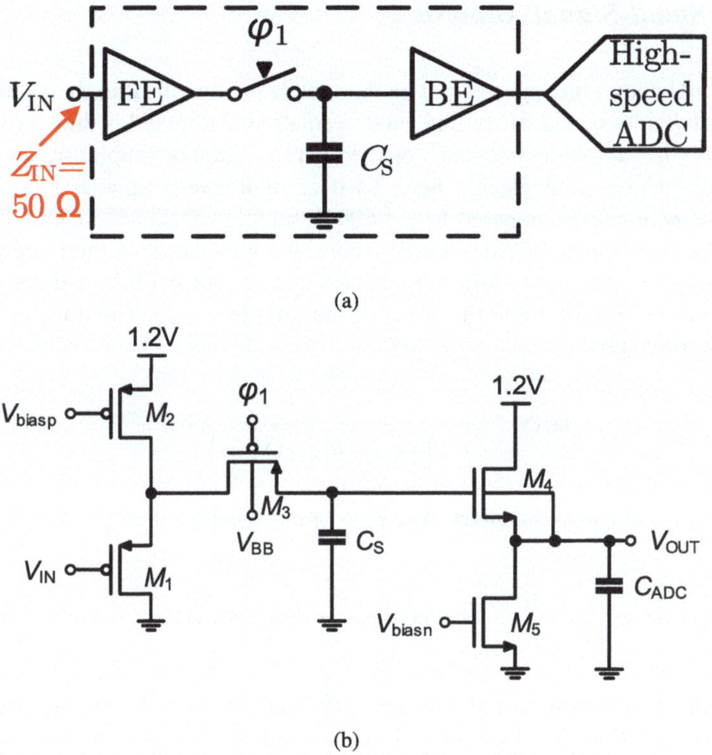

Fig. 7.1 (**a**) Block diagram and (**b**) schematics of the 12 GS/s T&H amplifier

Utilizing the GF22FDX process allows for forward-body-biasing, which can reduce the transistor's threshold voltage by 83 mV/V [5]. Consequently, the ON-resistance of the sample switch M_3 can be decreased by applying a negative voltage to the PMOS body, enhancing overall bandwidth and linearity.

Figure 7.1b illustrates the BE-buffer, implemented as an NMOS source-follower, which drives C_{ADC} and the measurement device. In this configuration, the body of M_4 is connected to its source to enhance the gain of the source-follower. The settling requirements for the BE-buffer are more relaxed compared to the FE-buffer, since it receives an already-settled signal from the T&H. Consequently, the slewing time is doubled, reducing the current demand to 6.5 mA.

Body-biasing helps in terms of bandwidth as well as linearity performance. In order to analyze the tuneability, a small signal is drawn, analyzing the body-bias dependency on the bandwidth. Furthermore, a large-signal analysis follows, discussing the effect of the body-biasing on the linearity.

7.2.1 Small-Signal Analysis

The overall bandwidth depends first on the sample bandwidth after the T&H and the output bandwidth of the BE-buffer. First, a small-signal model is drawn of the FE-buffer with the sampler connected, depicted in Fig. 7.2a. For simplicity, the parasitic capacitance C_{PAR1} is neglected here, as it is significantly smaller (according to simulation estimates) compared to the 600 fF sample capacitance C_S and the BE-buffer input capacitance C_{PAR2}. Furthermore, the impedance of the current source r_{DS2} and r_{DS1} of the source-follower can be neglected since it is higher compared to the g_{M1} of the FE-buffer and the R_{ON3} of the sample switch. The transfer function can be expressed by:

$$H_1(s) \approx \frac{1}{C_1\left(\frac{1}{g_{M1}} + R_{ON3}\left(V_{BB}\right)\right)s + 1} \tag{7.1}$$

with $C_1 = C_S + C_{PAR2} \approx 640\,\text{fF}$. The pole of the system is:

$$p_1 \approx \frac{-1}{2\pi C_1(1/g_{M1} + R_{ON3}(V_{BB}))}.$$

Hereby is g_M the transistor's transconductance, g_{DS} the output conductance, r_{ON} the switch ON-resistance, and V_{BB} the body-bias voltage. As highlighted by the dependency of R_{ON3} on V_{BB} the pole can be moved by applying forward-body-biasing to the switch M_3. This helps in terms of SNR as the signal power at high input frequencies benefits from it. Adding the BE-buffer reduces the overall bandwidth since the strict requirements regarding sampling are only present in the front-end part. Thus, the g_{M4} of the BE-buffer is reduced to reduce the overall power consumption. Figure 7.2b depicts the simplified equivalent small-signal circuit of the BE-buffer. Here, the r_{DS5} of the current source is neglected since the impedance is higher than that of others. Furthermore, the termination impedance R_{TERM} from the measurement device is added to have a reasonable comparison between the measurement and calculation. The transfer function of the output is given by:

$$H_2(s) \approx \frac{1}{C_2\left(\frac{1}{g_{M4}+\frac{1}{r_{DS4}}+\frac{1}{R_{TERM}}}\right)s + 1}, \tag{7.2}$$

with $C_2 = C_{ADC} + C_{PAR3}$. The combined transfer function is depicted in Fig. 7.3a and shows a bandwidth tuning range above 10% by forward body-biasing the switch from 0 to −3 V. The implemented T&H amplifier is measured with a vector network analyzer (VNA), and the output is shown in Fig. 7.3b. Here, the tuning range matched perfectly with the calculations; however, series resistances and parasitic capacitances from the layout reduce the overall bandwidth down to 10.8 GHz (zero body-biasing).

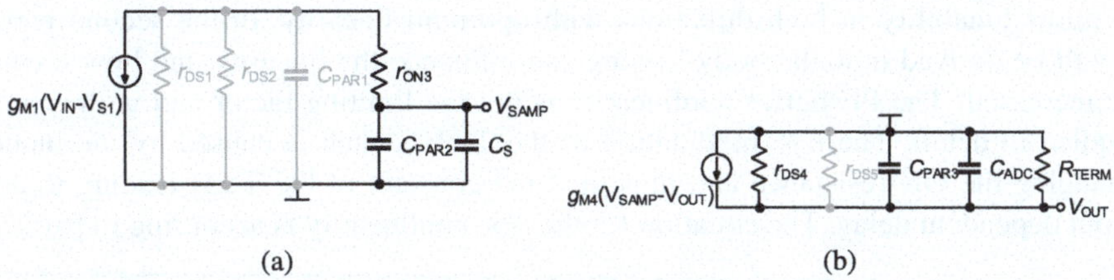

Fig. 7.2 Small-signal equivalent circuit of the T&H amplifier

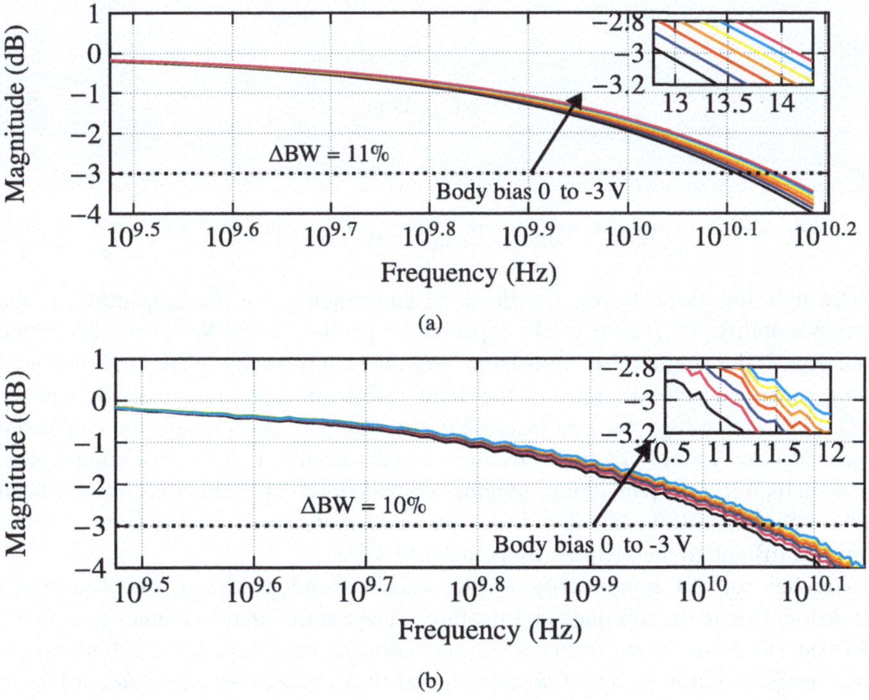

Fig. 7.3 Frequency response of the body-biased T&H amplifier (**a**) calculated and (**b**) measured

7.2.2 *Large-Signal Analysis*

In this section, a large-signal analysis helps to analyze the body-biasing in terms of linearity. Therefore, only the third-order harmonics are analyzed since the circuit is differential and even-order harmonics are eliminated. The source of nonlinearity can be separated into different parts: FE-buffer, T&H, and BE-buffer. Where the T&H nonlinearity is separated into: r_{ON}, charge injection, and timing nonlinearity [6]. In the first part, the nonlinearity of different parts will be calculated as a function of the input frequency and compared to the simulation results. In this part, all calculations and simulations are done with a nominal body-biasing of -0.5 V (pMOS switch) to allow tunability in both directions with optimum linearity. In the second part, it will be derived how the body-biasing can influence the linearity and how it can be increased. The FE-buffer nonlinearity is not the limiting factor and is thus not studied in detail. The r_{ON} modulation of the T&H switch is caused by the input changing the ON-resistance and thus the time constant of the T&H leading to an input dependent delay. The equation for the r_{ON} nonlinearity is according to [6]:

$$HD3_{\text{R}_{\text{ON}}} \approx \frac{n^2 C_{\text{L}} \omega_0}{4\beta V_{\text{OV}}^3} A_0^2 \tag{7.3}$$

with

$$\beta \triangleq \mu C'_{\text{OX}} W/L, \tag{7.4}$$

and

$$V_{\text{OV}} \triangleq V_{\text{DD}} - \mid V_{\text{THp}} \mid - n(V_{\text{DD}} - V_{\text{DC}}), \tag{7.5}$$

where n is the slope factor, C_{L} the load capacitance, A_0 the amplitude, μ the carrier mobility, C'_{OX} gate oxide capacitance per unit area, W the width of the transistor, L the length of the transistor, V_{DD} the supply voltage, V_{THp} the threshold voltage, and V_{DC} the DC-level of the input. All parameters are extracted from the GF22FDX technology and are inserted in the equation. As a result, the calculated r_{ON} nonlinearity across the input frequency is depicted in Fig. 7.4. For comparison, the switch (W/L = 20 µm/20 nm) together with the load capacitance C_1, is simulated in through mode. Both, calculated and simulated, match accurate and show a SFDR above 67 dBc up to the Nyquist frequency of 6 GHz.

Another root of nonlinearity is the signal-dependent charge injection-based distortion. Due to the top-plate architecture, charge stored in the channel goes in the OFF-state to the drain and source node, depending on their impedance at both nodes. The charge injection is signal-dependent and thus causes distortion. According to reference [6], the charge injection modulation can be calculated with:

$$\begin{aligned} HD3_{\text{CI}} \approx & \frac{n^3}{48} \frac{A_0^2}{V_{\text{DD}} V_{\text{OV}}} \frac{C_{\text{G}}}{C_{\text{L}}} \frac{\beta V_{\text{OV}}}{C_{\text{L}}} t_{\text{R}} \\ & \times \left(\frac{V_{\text{OV}}}{V_{\text{DD}}} \frac{\beta V_{\text{OV}}}{C_{\text{L}}} t_{\text{R}} - 1 \right) exp \left(-\frac{1}{2} \frac{V_{\text{OV}}}{V_{\text{DD}}} \frac{\beta V_{\text{OV}}}{C_{\text{L}}} t_{\text{R}} \right), \end{aligned} \tag{7.6}$$

with C_{G} the gate capacitance of the sample switch, and t_{R} the rise time of the clock.

For simplicity, the calculation assumes a zero source resistance as the calculation shows that due to the small amplitude of 72.5 mV (single ended) and the large sample capacitance of 640 fF the effect of charge injection is not the main root of nonlinearity. Therefore, the charge injection-caused distortion is depicted in

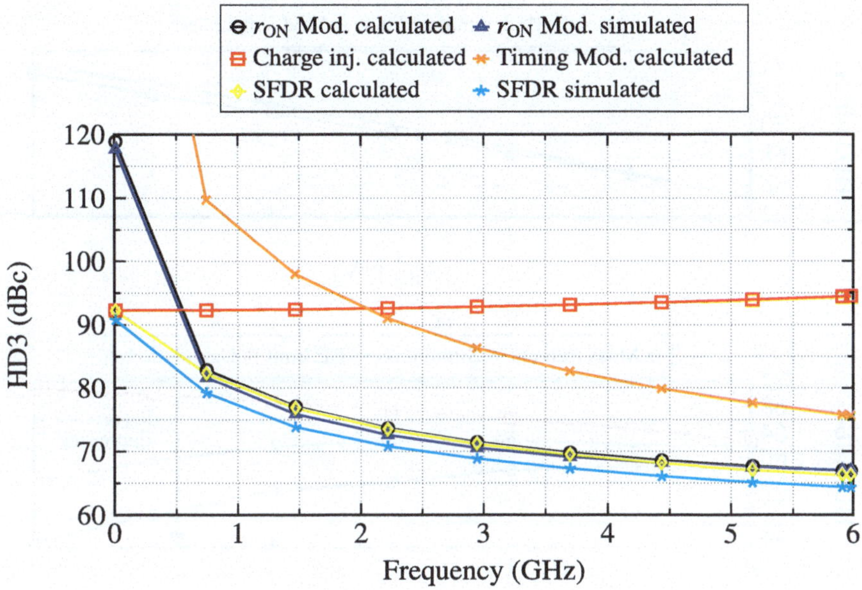

Fig. 7.4 Non-linearity of only the T&H

Fig. 7.4 and is relatively flat up to 3 GHz and slightly decreased due to the lowering amplitude. However, it stays over 90 dBc and is thus not a limiting factor.

The third root of nonlinearity is the timing modulation, which is present in the turn-off point as there is a time shift from the ideal turn-off point. According to [6], the error can be evaluated with:

$$HD3_{\mathrm{TIM}} \approx \frac{n^2}{8}\left(\frac{A_0}{V_{\mathrm{DD}}}\right)^2 (\omega_0 t_{\mathrm{R}})^2 \tag{7.7}$$

Figure 7.4 depicts the timing introduced error across the input frequency. Only the high-frequency SFDR is affected by the error since the input varies faster, and thus, the error increases with the time shift.

Combining all sources of nonlinearity leads to a calculated SFDR, which is close to the actual simulated SFDR with a 2–3 dB variation. As a result, the SFDR of only the T&H is above 64 dBc in the entire Nyquist band and is mainly limited by the r_{ON} nonlinearity.

With the possibility of tuning the ON-resistance, not only the bandwidth can be extended but also the nonlinearity can be reduced according to (7.5) and (7.3). Therefore, the threshold sensitivity of 83 mV/V is inserted in the (7.3) and compared with simulations in a non-switching mode of operation at the Nyquist frequency of 6 GHz. Figure 7.5a shows that the linearity can be improved by 9 dB by applying a forward body-biasing up to −3 V. The calculation also matches with the simulation, revealing a linearity improvement of roughly 8 dB. In the next phase, the FE-buffer

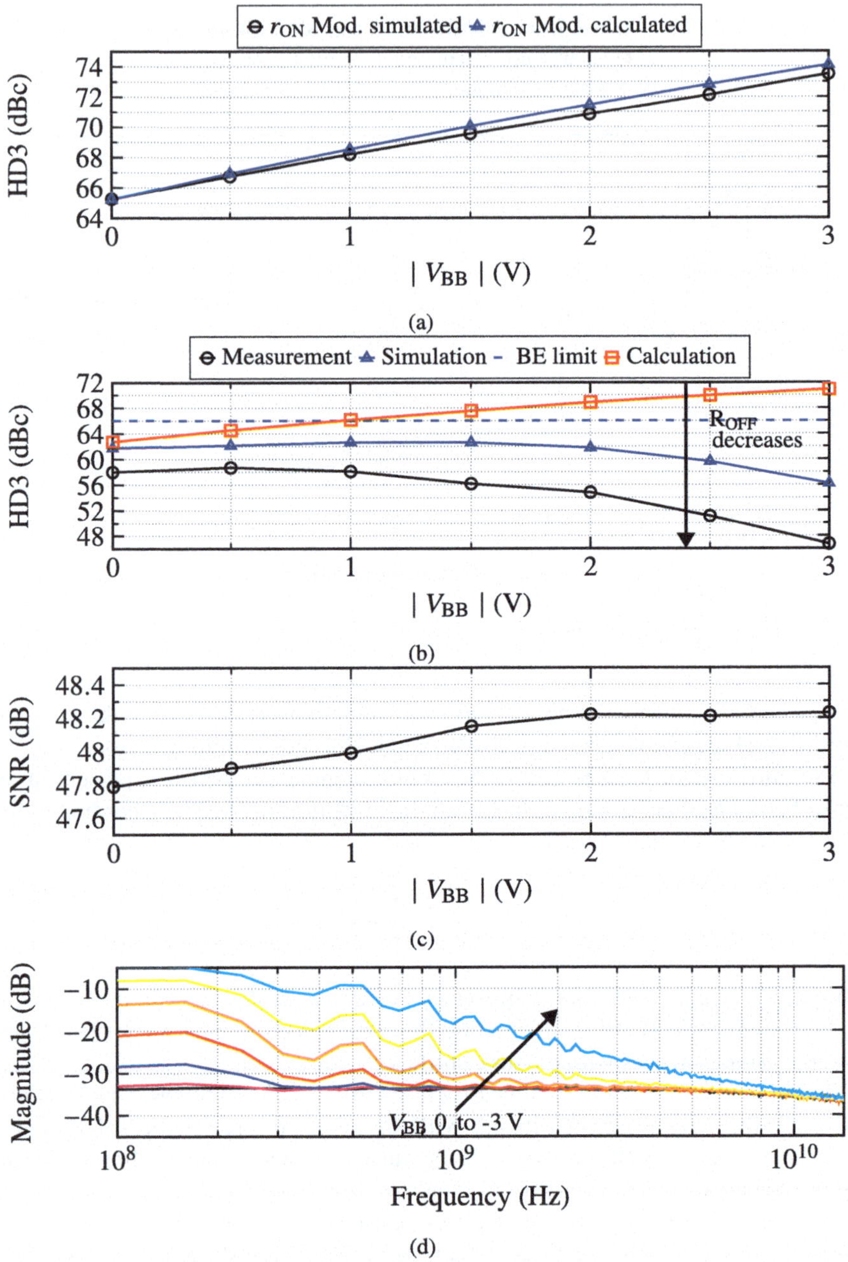

Fig. 7.5 (**a**) r_{ON} nonlinearity, (**b**) overall calculated, simulated, and measured linearity, (**c**) SNR at Nyquist input frequency, and (**d**) the measured hold-mode frequency response of the T&H amplifier

and BE-buffer are added, and the linearity at the output is examined. The calculation assumes an ideal FE- and BE-buffer linearity to visualize the trend by utilizing the body-biasing. Therefore, all Eqs. (7.3)–(7.7) are evaluated in dependency of the body-biasing and are depicted in Fig. 7.5b. Ideally, the overall linearity increases by 7 dB; however, the BE-buffer is designed to provide a linearity of 66 dBc at a Nyquist input frequency. Thus, the influence on the linearity is present up to −1.5 V forward body-biasing, improving the linearity by approximately 1 dB. The linearity decreases below −1.5 V due to a low OFF-resistance of the sample switch since not only the ON-resistance reduces with forward body-biasing but also the OFF-resistance reduces, degrading the overall linearity. Furthermore, a measurement of the overall T&H amplifier shows the same trend, with an early drop-off point at −1 V. Figure 7.5c highlights the simulated SNR improvement since the bandwidth is extended, and thus, the signal power increases. Here, a 0.35 dB SNR improvement from 0 to −1.5 V forward body-biasing is achieved. Figure 7.5d depicts the measured hold-mode frequency response, visualizing the trend of the reduced OFF-resistance at high forward body-biasing voltages V_{BB}.

7.3 Bandwidth Boosting for High Linear T&H-Amplifiers

In this section, a T&H amplifier designed for a higher linearity is presented unlike the T&H amplifier from before this amplifier utilizes dynamic as well as static body-biasing improving the linearity, maximizing the SNR, and regulating the common mode. Furthermore, the T&H amplifier is designed for a higher input swing of 1.2 V full-scale voltage. Figure 7.6 shows the circuit diagram of the T&H amplifier [7]. For the sake of clarity, the single-ended version is shown, in contrast to the fully differential integration. The input buffer is differentially terminated to 100,Ω to achieve adequate return loss (RL). The body-biased wideband FE buffer drives the time-interleaved dynamic body-biased 1.5 GS/s T&Hs. The T&Hs operate on

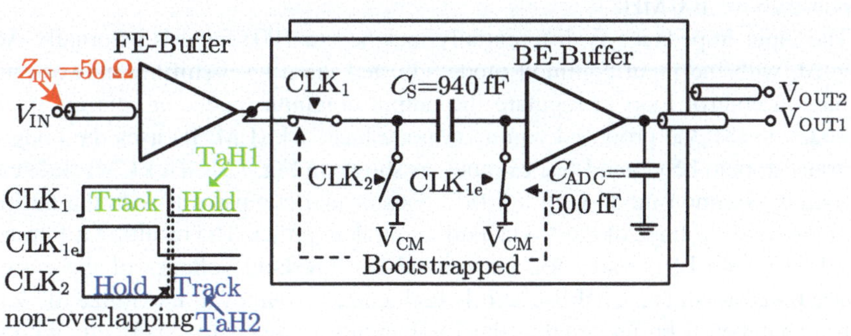

Fig. 7.6 Schematic of 3 GS/s T&H amplifier © 2022. This work is openly licensed via CC BY 4.0 from [7]

the ping-pong principle, with one T&H sampling while the other holds its sample. Bottom-plate sampling is used to achieve high linearity and to prevent signal-dependent charge from flowing into the sampler due to the high-impedance node on the sampler side. Three different clock phases are required per T&H for sampling: CLK1 and the early clock CLK1e determine the sampling phase by connecting the sampling capacitance C_S to the output of the FE buffer. The input signal is sampled with respect to the common-mode voltage, which is buffered by a two-stage closed-loop amplifier. In the next phase, the early clock CLK_{1e} opens just before the top plate switch to form a high-impedance node to mitigate signal-dependent charge injection. During the hold mode, CLK2 connects the top plate to the common mode V_{CM} (Fig. 7.6) and transfers the charge to the back-end buffer (BE). Normally, a series of SAR ADCs or a pipeline ADC would convert the sampled and buffered output voltage to the digital domain. However, in this design, the ADC resampling capacitance is modeled with a statically connected 500 fF MOM capacitance C_{ADC} as shown in Fig. 7.6. Transmission lines connect the output to the pads.

The FE-buffer is depicted in Fig. 7.7a, employing a pseudo-differential push–pull architecture with bootstrapped drain nodes [7–10]. This design effectively minimizes the output impedance of the FE-buffer to: $R_{OUT} \approx 1/(g_{M3} + g_{M4})$.

The cascodes ensure a high-linearity for the increased signal swing as they reduce the nonlinear r_{DS} of the main source-follower M_3 and M_4. Regarding the body-biasing of all source-followers, one solution would be to connect the body nodes to the output to increase the DC gain; however, that would reduce the overall bandwidth since the FE-buffer has to drive the additional well including all dummy devices from the layout. Thus, all source-followers are forward-biased to allow a higher-voltage headroom for the level shifters. Furthermore, all source-followers M_1 - M_6 require a DC biasing to ensure the right operating point. Therefore, level shifters are required to provide the DC biasing and the AC signal for M_1–M_6. Figure 7.7b depicts the two lower level-shifter LS_{1n} and LS_{2n}. Two current sources supply a constant current of 100 μA to control the IR drop across the resistor R_{BIAS}. A 1 pF forward capacitance C_{FEED} is added in parallel to create an AC path. Together, R_{BIAS} and C_{FEED} form a pole/zero pair that produces a flat frequency response above 100 MHz.

The input impedance is differentially matched to 100 Ω and is externally AC-coupled, with the input common mode adjusted via a level-shifting network that requires a control loop to regulate the output common mode. In contrast to the approach in [8], the proposed common-mode feedback (CMFB) uses the body of the transistors in LS_{1n} and LS_{2n} as input, as shown in Fig. 7.7b. This CMFB detects the output common mode with a 50 kΩ resistor and compares it with the desired common-mode voltage of 0.6 V via a differential amplifier. The resulting difference is fed back into LS_{1n} and LS_{2n} to regulate the threshold voltages of the current source transistors. The CMFB controls both current sources to maximize the loop gain so that the FE-buffer and thus the T&H amplifier can be driven directly without the need for a bias T (no DC voltage level required) on the source side. Figure 7.8 depicts the CMFB loop stability; here only the DC common mode is of importance.

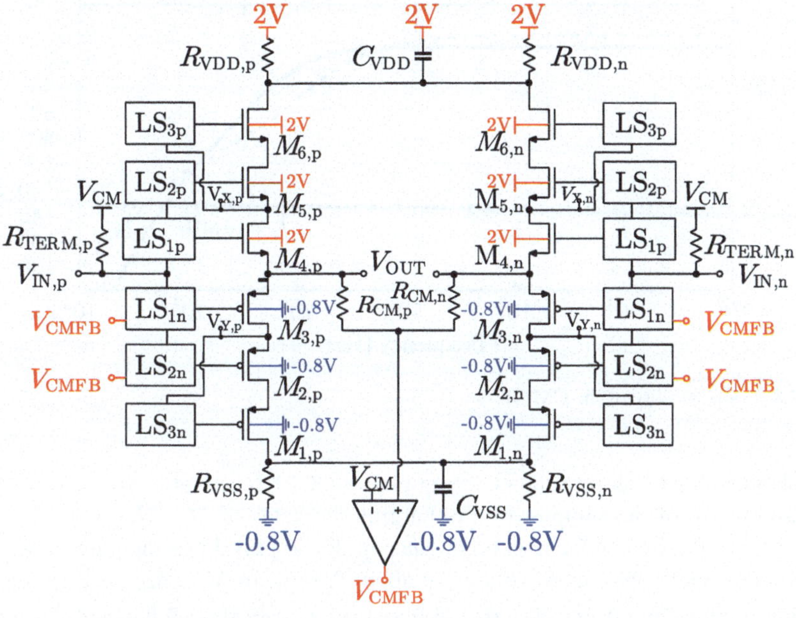

(a)

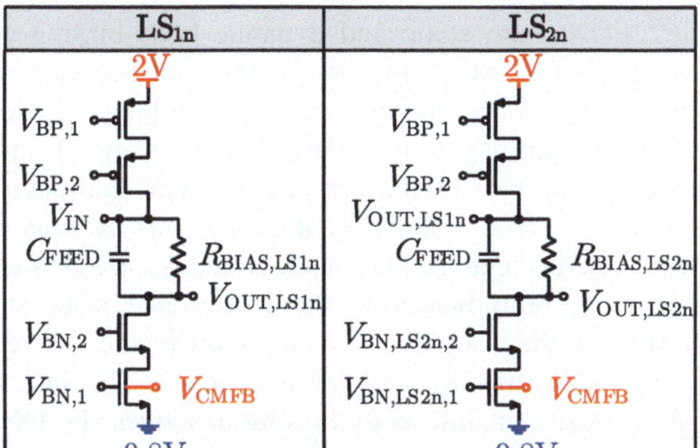

(b)

Fig. 7.7 Schematic of (**a**) the FE-buffer and (**b**) the body-biased level-shifters © 2022. This work is openly licensed via CC BY 4.0 from [7]

Hence, the CMFB has a unity gain frequency of 16 MHz and a gain of 28 dB together with a phase margin of 62°.

In order to achieve high linearity and a high bandwidth, the switches on the top and bottom plates are implemented as bootstrapped switches, as shown in Figs. 7.6 and 7.9a. In this configuration, M_3 and M_{12} charge the bootstrap capacitance C_3, while M_4 raises the gate of M_8 and disconnects the bootstrap capacitance from the

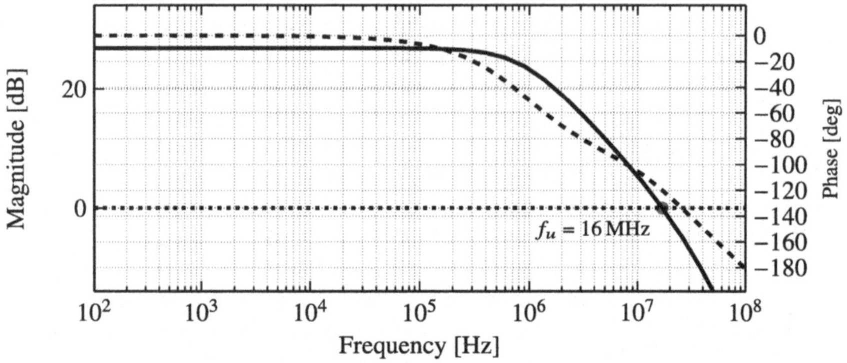

Fig. 7.8 Loop stability of the CMFB

sample switch M_{11} during the charging process. CLKb activates M_{10} and initiates the hold phase of the sampler by discharging the gates of M_9, M_{11} and M_{13}. The track phase starts when CLK is high, causing M_5 to pull down the gate of M_8 and apply a constant gate-source voltage of about $0.9V_{DD}$ to M_{11}. M_6 ensures that the floating node between M_7 and M_{10} is biased to V_{DD} to reduce leakage during the track phase.

Thanks to GF22FDX, both static and dynamic body-biasing can be applied within the bootstrap circuit, which means a significant increase in performance compared to conventional bootstrapping methods. The body of M_8 is zero-body-biased (ZBB) to GND resulting in an isolation of the well capacitance from the signal path improving the V_{GS} of the sampling switch. Although this typically introduces a latch-up risk, SOI technology mitigates this by separating the wells from the drain and source regions [11]. Consequently, M_8 is ZBB in the hold phase and FBB (forward-body-biased) in the track phase, since all G/D/S nodes are shifted by $0.9V_{DD}$ in the track phase, which reduces the ON-resistance of the transistor while the OFF-resistance remains unchanged. The sample switch M_{11} and the switch M_9 utilize dynamic body-biasing to reduce the ON-resistance, to reduce the distortion of the resistance modulation and to increase the bandwidth during the track phase. In the hold phase, these switches are ZBB, which increases the OFF-resistance and thus improves linearity, unlike the T&H amplifier discussed in Sect. 7.2. To achieve such a behavior, the bootstrapped clock controls both the gate and the well of the transistors. Simulations of the T&H connected to the FE buffer show an improvement in tracking bandwidth of more than 10% compared to ZBB, as shown in Fig. 7.9b. The drop of −1.2 dB at low frequencies is due to the bulk transconductance g_{MB} of the FE-buffer. In addition, simulations show a sixfold reduction in leakage in hold mode compared to static FBB at the FF 125 °C corner. Simulations of the T&H amplifier with various body-bias configurations show that using dynamic body-biasing for the sample switch enhances overall linearity while maintaining a similar SNR compared to the static body-biasing approach, as shown in Fig. 7.10. Here, an SFDR benefit above 2 dB is gained up to 2.5 GHz. Above

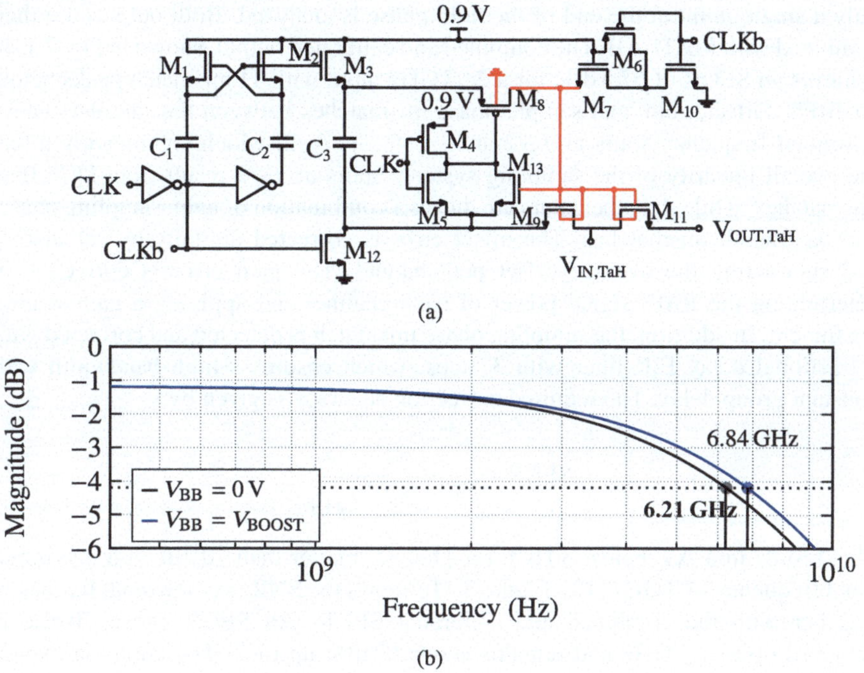

Fig. 7.9 (**a**) Schematic of the modified bootstrap circuit and (**b**) the resulting bandwidth benefit © 2022. This work is openly licensed via CC BY 4.0 from [7]

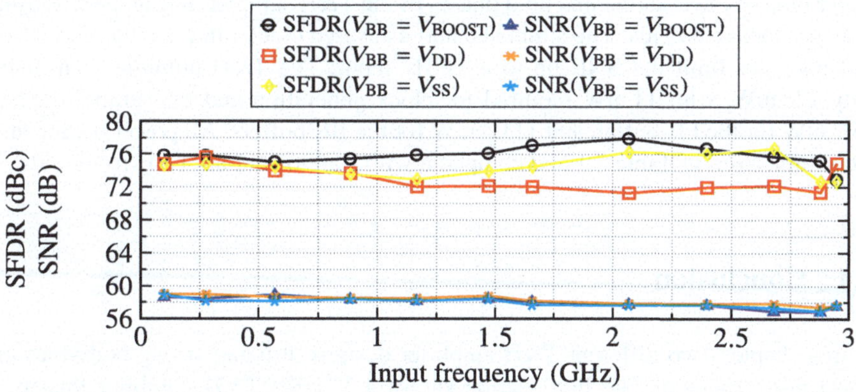

Fig. 7.10 Simulated linearity enhancement of the different body-biasing options of the T&H © 2022. This work is openly licensed via CC BY 4.0 from [7]

2.5 GHz, the timing distortion increases due to the slower falling edges compared to the ZBB case, and thus, both are in the same range.

The output of the T&H amplifier is measured using a sampling scope and a spectrum analyzer, as depicted in Fig. 7.11a. For the sampling scope measurements, only a single point at the end of the hold phase is analyzed. Both outputs are then combined with MATLAB. The combined and calibrated output, shown in Fig. 7.11b, achieves an SFDR of 67.5 dBc for a 2.97 GHz input with TI mismatch peaks below 75 dBFS. Offset, gain, and sample phase mismatches between the samplers cause unwanted frequency spurs at $f_S/2$ and $\pm f_{IN} + f_S/2$, which significantly affect the overall linearity of the sampling system. Spurs at $f_S/2$ result from DC offset mismatches, while the other spurs are due to a combination of gain, sampling phase, and bandwidth mismatches. The offset error is corrected digitally by calculating and subtracting the average offset per channel. The gain error is corrected by determining the RMS signal power of each channel and applying a gain scaling coefficient. In addition, the sampling phase mismatch is detected and corrected with a fractional delay FIR filter with 37 taps, which ensures a high bandwidth with constant group delay. The required correction accuracy is given by:

$$\mathrm{SFDR} = \frac{1}{\pi^2 \cdot \Delta T^2 \cdot f_{\mathrm{IN}}^2},$$

This results in a ΔT below 30 fs for a TI spur smaller than 70 dBc at a maximum input frequency of 3 GHz [12]. Figure 7.11c shows the SNR over the input frequency together with the combined and calibrated SFDR. An SFDR above 70 dBc is achieved up to 2.5 GHz and remains above 67 dBc up to 3 GHz, while an overall SNR/SNDR above 55 dBFS is maintained. Simulation results show an SFDR difference of about 3 dB to 5 dB due to mismatch and the limited linearity of the sampling scope, with an SNR difference of 0.5 dB to 2.5 dB caused by the integration of the tracking and hold phases of the T&H sampler during the spectrum analyzer measurements. The simulated and measured clock jitter is only 45 fs RMS and does not limit the SNR up to a 3 GHz input. The T&H amplifier consumes only 178 mW, with 11 mW required for clock generation and bootstrap circuitry, 55.7 mW for the FE-buffer, and 111.4 mW for the BE-buffers. All power parameters are summarized in Table 7.1 and compared with state-of-the-art T&H front-ends.

7.4 Conclusion

In this chapter, two different T&H amplifier designs utilizing strong body-biasing effect are presented. The first section shows a 12 GS/s T&H amplifier that uses forward body-biasing inside the sample switch to improve the bandwidth above 10% and to increase the linearity by 1 dB. Therefore, a small-signal analysis is done and compared with the measurement results. Furthermore, a large-signal linearity analysis is performed to deeply understand the benefit of forward body-biasing on the linearity. As a result, the bandwidth can be extended by more than 10% resulting in an SNR improvement of 0.35 dB with a 1 dB linearity increase. Forward-body-biasing above −1.5 V results in a reduced linearity, since the OFF-resistance decreases. The second part addresses the issue of the finite OFF-resistance and

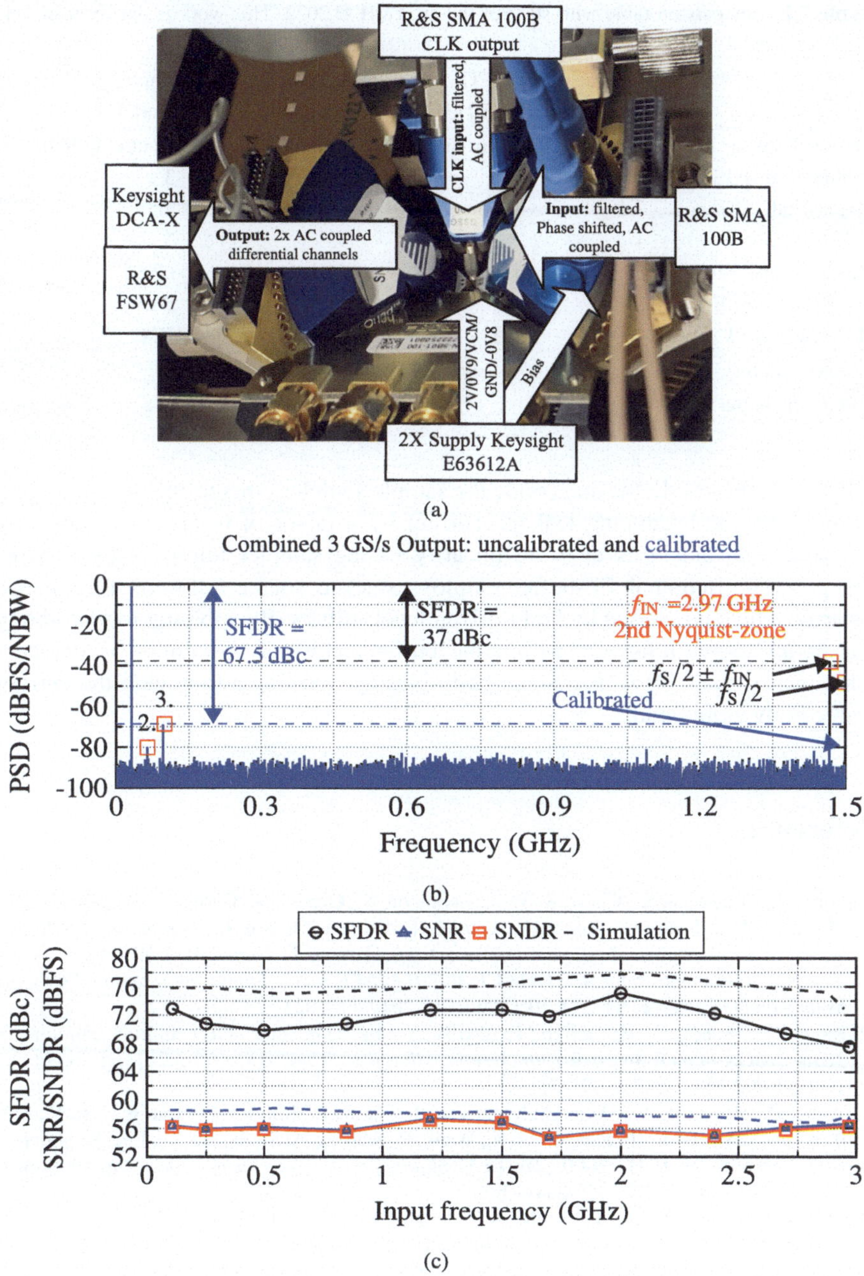

Fig. 7.11 (**a**) The measurement setup and (**b**) spectrum of the calibrated and uncalibrated T&H amplifier output with a resolution bandwidth of 732 kHz and (**c**) SFDR and SNR across the input frequency © 2022. This work is openly licensed via CC BY 4.0 from [7]

Table 7.1 Comparison table with state-of-the-art T&H © 2022. This work is openly licensed via CC BY 4.0 from [7]

	ISSCC 2017 [8]	ISSCC 2017 [13]	ISSCC 2020 [14]	This work [7]
Technology	28 nm	16 nm FinFET	16 nm FinFET	22 nm FD-SOI
Sample-rate (GS/s)	10	4	12	3
Bandwidth (GHz)	7.4	–	18	4.5
f_{IN} (GHz)	4	2	4	3
SFDR at f_{IN} (dBc)	65.5	67	61	67.5
SNDR at f_{IN} (dB)	55	57.3	53	56.3
FE-Power (mW)	400[a]	282	220	178
Power supplies	3	–	2	3

[a] Only FE-buffer power [15]

utilizes dynamic body-biasing inside the sample switches, increasing the bandwidth by over 10% and reducing leakage current by a factor of 6. However, not only the bandwidth enhances also the linearity of the sampler improves by 2–3 dB. The pseudo-differential FE-buffer employs stacked source-followers to enhance linearity, which are biased by body-driven level shifters. By applying forward body bias to all source-followers within the buffers, power consumption is reduced. Additionally, the CMFB uses static body-biasing, forming a loop with the current source body as an input.

References

1. Watts, J., Sundaram, K., Chew, K. W. J., Lehmann, S., Ong, S. N., Chow, W. H., Chan, L. H., Mazurier, J., Schwan, C., Andee, Y., Feudel, T., Pirro, L., Erben, E., Nowak, E., Smith, E., Bazizi, E. M., Kammler, T., Taylor, R., Rice, B., & Harame, D. (June 2017). RF-pFET in fully depleted SOI demonstrates 420 GHz FT. In *2017 IEEE Radio Frequency Integrated Circuits Symposium (RFIC)* (pp. 84–87). 10.1109/RFIC.2017.7969023.
2. Runge, M., Scholz, P., & Gerfers, F. (2019). A 44fs RMS jitter 6GHz limiting amplifier in 22nm CMOS FDSOI. In *2019 12th German Microwave Conference (GeMiC)* (pp. 111–114). 10.23919/GEMIC.2019.8698183.
3. Wang, H., Huang, T.-Y., Mannem, N. S., Lee, J., Garay, E., Munzer, D., Liu, E., Liu, Y., Lin, B., Eleraky, M., Jalili, H., Park, J., Li, S., Wang, F., Ahmed, A. S., Snyder, C., Lee, S., Nguyen, H. T., & Smith, M. E. D. (2000). *Power Amplifiers Performance Survey 2000-Present.* https://gems.ece.gatech.edu/PA%5C_survey.html.
4. Wittenhagen, E., Kurth, P., Lotfi, N., Hecht, U., Edler, J., Scholz, P., & Gerfers, F. (2023). A 12 GS/s RF-sampler employing inductive peaking in 22 nm FD-SOI CMOS. In *2023 18th European Microwave Integrated Circuits Conference (EuMIC)* (pp. 317–320). 10.23919/EuMIC58042.2023.10288655.
5. Wittenhagen, E., Runge, M., Lotfi, N., Ghafarian, H., Tian, Y., & Gerfers, F. (2021). Advanced mixed signal concepts exploiting the strong body-bias effect in CMOS 22FDX®. *IEEE Transactions on Circuits and Systems I: Regular Papers, 68*(1), 57–66. 10.1109/TCSI.2020.3023077.

6. Iizuka, T., Ito, T., & Abidi, A. A. (2018). Comprehensive analysis of distortion in the passive FET sample-and-hold circuit. *IEEE Transactions on Circuits and Systems I: Regular Papers, 65*(4), 1157–1173 10.1109/TCSI.2018.2797987.
7. Wittenhagen, E., Artz, P. J., Scholz, P., & Gerfers, F. (2022). A 3-GS/s RF track-and-hold amplifier utilizing body-biasing with >55-dBFS SNR and >67-dBc SFDR up to 3 GHz in 22-nm CMOS SOI. *IEEE Open Journal of the Solid-State Circuits Society, 2*, 135–143. 10.1109/OJSSCS.2022.3217019.
8. Devarajan, S., Singer, L., Kelly, D., Kosic, S., Pan, T., Silva, J., Brunsilius, J., Rey-Losada, D., Murden, F., Speir, C., Bray, J., Otte, E., Rakuljic, N., Brown, P., Weigandt, T., Yu, Q., Paterson, D., Petersen, C., & Gealow, J. (Feb. 2017). A 12b 10GS/s interleaved pipeline ADC in 28nm CMOS technology. In *2017 IEEE International Solid-State Circuits Conference (ISSCC)* (pp. 288–289). 10.1109/ISSCC.2017.7870374.
9. Wittenhagen, E., Artz, P., Scholz, P., & Gerfers, F. (2021). A 3 GS/s >55 dBFS SNDR time-interleaved RF track and hold amplifier with >67 dBc SFDR up to 3 GHz in 22FDX. In *2021 IEEE Radio Frequency Integrated Circuits Symposium (RFIC)* (pp. 139–142). 10.1109/RFIC51843.2021.9490479.
10. Linnhoff, S., Buballa, F., Reinhold, M., & Gerfers, F. (2020). A 12 bit 8 GS/s time-interleaved SAR ADC in 28nm CMOS. In *2020 27th IEEE International Conference on Electronics, Circuits and Systems (ICECS)* (pp. 1–4). 10.1109/ICECS49266.2020.9294903.
11. Carter, R., Mazurier, J., Pirro, L., Sachse, J., Baars, P., Faul, J., Grass, C., Grasshoff, G., Javorka, P., Kammler, T., Preusse, A., Nielsen, S., Heller, T., Schmidt, J., Niebojewski, H., Chou, P., Smith, E., Erben, E., Metze, C., Bao, C., Andee, Y., Aydin, I., Morvan, S., Bernard, J., Bourjot, E., Feudel, T., Harame, D., Nelluri, R., Thees, H., M-Meskamp, L., Kluth, J., Mulfinger, R., Rashed, M., Taylor, R., Weintraub, C., Hoentschel, J., Vinet, M., Schaeffer, J., & Rice, B. (Dec. 2016). 22nm FDSOI technology for emerging mobile, Internet-of-Things, and RF applications. In *2016 IEEE International Electron Devices Meeting (IEDM)* (pp. 2.2.1–2.2.4). 10.1109/IEDM.2016.7838029.
12. Razavi, B. (2013). Design considerations for interleaved ADCs. *JSSC, 48*(8), 1806–1817. 10.1109/JSSC.2013.2258814.
13. Vaz, B., Lynam, A., Verbruggen, B., Laraba, A., Mesadri, C., Boumaalif, A., Mcgrath, J., Kamath, U., De Le Torre, R., Manlapat, A., Breathnach, D., Erdmann, C., & Farley, B. (2017). A 13b 4GS/s digitally assisted dynamic 3-stage asynchronous pipelined-SAR ADC. In *2017 IEEE International Solid-State Circuits Conference (ISSCC)* (pp. 276–277). 10.1109/ISSCC.2017.7870368.
14. Ali, A. M. A., Dinc, H., Bhoraskar, P., Bardsley, S., Dillon, C., Kumar, M., McShea, M., Bunch, R., Prabhakar, J., & Puckett, S. (2020). A 12b 18GS/s RF sampling ADC with an integrated wideband track-and-hold amplifier and background calibration. In *2020 IEEE International Solid-State Circuits Conference - (ISSCC)* (pp. 250–252). 10.1109/ISSCC19947.2020.9063011.
15. Devarajan, S., Singer, L., Kelly, D., Pan, T., Silva, J., Brunsilius, J., Rey-Losada, D., Murden, F., Speir, C., Bray, J., Otte, E., Rakuljic, N., Brown, P., Weigandt, T., Yu, Q., Paterson, D., Petersen, C., Gealow, J., & Manganaro, G. (2017). A 12-b 10-GS/s interleaved pipeline ADC in 28-nm CMOS technology. *IEEE Journal of Solid-State Circuits, 52*(12), 3204–3218. 10.1109/JSSC.2017.2747758.

Chapter 8
High-Speed Flash ADC Using Bulk-Driven Flash Reference Generation Technique

Nima Lotfi and Friedel Gerfers

8.1 UWB Track-and-Hold

This section presents the design and implementation of an ultra-wideband track-and-hold (UWB T&H) operating at a sampling speed of 18.5 GS/s. The T&H is intended to be used in a 4X-time-interleaved (4X-TI) flash ADC to get the target sampling rate of 74 GS/s for a baseband ADC as part of a W-band receiver. The down-converted signals in the IF (Intermediate Frequency) band have a bandwidth of 1–36 GHz. Hence, a sampling rate of 74 GS/s is chosen for the IF-to-digital converter. The total ENOB target for the baseband ADC including the UWB T&H and the 5-bit flash ADC is more than 4.5 bits. Since a resolution of 5 bits is considered for the ADC, an accuracy of better than 5.5 bits is required for the T&H to meet the total ENOB target. To prove the UWB T&H performance on silicon, a test chip as a stand-alone sampler has been fabricated and characterized prior to the flash ADC design. The highlighted area in Fig. 8.1 shows the implemented cores. The measurement results reveal an ENOB of better than 5.5 bits up to the 5th Nyquist zone of the ADC (fin =39 GHz) with a 3-dB bandwidth of 44 GHz. Hence, the measured ENOB and bandwidth meet the targeted specs.

8.1.1 System-Level Architecture of the 4X-TI Flash ADC

The block diagram of the 4X-TI flash ADC with the proposed front-end T&H and clock-phase generator is depicted in Fig. 8.1. The total required sampling rate is 74 GS/s to digitize the down-converted IF band (1–36 GHz) Furthermore, the

N. Lotfi (✉) · F. Gerfers
Technische Universität Berlin, Berlin, Germany
e-mail: nima.lotfi@tu-berlin.de; friedel.Gerfers@tu-berlin.de

F. Gerfers (ed.), *Bulk-Driven Circuit Techniques for CMOS FDSOI Processes*,
https://doi.org/10.1007/978-3-031-85114-8_8

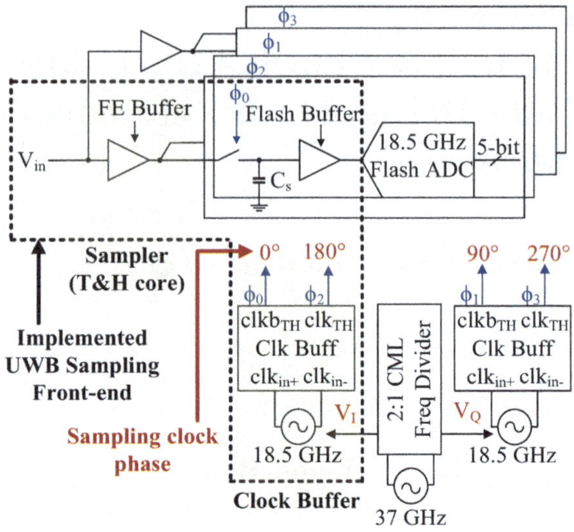

Fig. 8.1 Block diagram of the 4X-TI flash with proposed sampling channels and quadrature-phase clock generator © 2022 EuMA. (Reprinted with permission from [1])

input buffer along with the T&H requires an accuracy of better than 5.5 bits. If a 2X-TI T&H architecture with 37 GS/s per lane is adopted, a settling time of only t_s =9 ps is available assuming that 2/3 of the track phase is available, as acquisition time due to finite rise and fall times as well as non-overlapping sampling clocks and slewing time. As a result, the T&H according to Eq. 8.2 requires a 3 dB bandwidth of more than 80 GHz to meet the target settling accuracy, which is not achievable given the sampling capacitance value of 30 fF to meet the THD (Total Harmonic Distortion) target. Therefore, a 4X-TI architecture is adopted for the T&H relaxing the minimum bandwidth requirement to 40 GHz. Using a double input buffer front-end, as depicted in Fig. 8.1, the overall linearity, isolation, and kickback performance are enhanced. Two T&H lanes share an input buffer as the clock phases of the respective ADCs are 180° apart. Hence, the two FE input buffers for the 4X-TI implementation consume only 84 mW from the 1.2 V supply. This way, also the input capacitance of the FE is reduced by half enhancing the input bandwidth for the TI sampler. The required quadrature-phase clocks for the sampling switches are generated by a clock-phase generator. It takes a half-rate sinusoidal signal (f_{in} = 37 GHz) as input and provides the clock buffers with an in-phase and a quadrature-phase (V_I, V_Q) differential signal at 18.5 GHz. The V_I and V_Q from the CML divider are AC-coupled to the input of the clock buffers to enable common-mode voltage shift. Finally, the sampling clock phases are generated by clock buffers, and a 5-bit sample phase tuning using a CDAC is considered. In order to avoid second rank time interleaving for the ADC, the ADC conversion rate is chosen to be the same as the channel sampling speed (f_s = 18.5 GS/s). Hence, a flash ADC as the fastest candidate is used.

8.1.2 Single-Channel UWB T&H Design

Sampling, as the first step in the A/D conversion, is a nontrivial task, and the performance of the ADC is often limited by the SNR or SFDR of the front-end track-and-hold (T&H) specifically at high input frequencies. A low-jitter clock path is a must for sampling at high input frequencies. Otherwise, jitter, as the dominant noise contributor, will limit the total SNR. There are three noise sources contributing to the total SNR in an ADC, which are quantization noise, thermal noise (kT/C), and jitter. The SNR due to clock jitter ($\mathrm{SNR_j}$) is given as

$$\mathrm{SNR_j} = -20log_{10}(2\pi f_{in}\sigma_j). \tag{8.1}$$

where f_{in} is the frequency of the sinusoidal input applied to the T&H and σ_j is the standard deviation of the sampling clock jitter. As the input frequency increases, the uncertainty of the sampling instant as a result of external or local jitter in the clock path reduces the SNR. Another parameter affecting the linearity and therefore the overall ENOB is the T&H settling behavior. In order to avoid performance loss due to incomplete settling, a 5-bit settling accuracy (n =5) according to Eq. 8.2 has to be obtained.

$$\begin{aligned} e^{-t_s/\tau_{TH}} &< \frac{1}{2^{n+1}} \\ \tau_{TH} &< \frac{t_s}{(n+1)ln(2)} \\ f_{3dB,TH} &= \frac{1}{2\pi\tau_{TH}}. \end{aligned} \tag{8.2}$$

where n is the target resolution τ_{TH} is the T&H time constant, t_s is the available settling time for the T&H (acquisition time) and is inversely proportional to the T&H sampling speed. Hence, an ultralow-jitter clock path and a small T&H time constant prove essential for high-frequency T&Hs, which are addressed in this design.

An ultralow-jitter clock path is necessary in order to maintain 5.5-bit ENOB at 39 GHz input signal due to jitter. The schematic of the proposed ultralow-jitter clock buffer is depicted in Fig. 8.2. The transistors M_0 and M_1 form a self-biased CMOS amplifier, which is designed for an output DC common-mode voltage of 450 mV ($V_{DD}/2$) and an input impedance of 50 Ω. It has an LF gain of 18 dB and a bandwidth of 30 GHz. The amplifier uses a self-biased scheme employing the resistor R_f as a feedback resistor. The cross-coupled inverter sets the crossing point to approximately 450 mV, so that a phase shift of 180° between the complementary clock phases is achieved. The last three inverters are scaled with a fan-out of 3 to drive the large sampling switch (W = 30 μm) and as emphasized in Fig. 8.2 are optimized for a sharp rising edge at the output. Therefore, the first and the last inverters are sized with sharp rising clock edges, while the middle one has a fast falling edge. This way, supply and jitter noise effects as well as sample switch

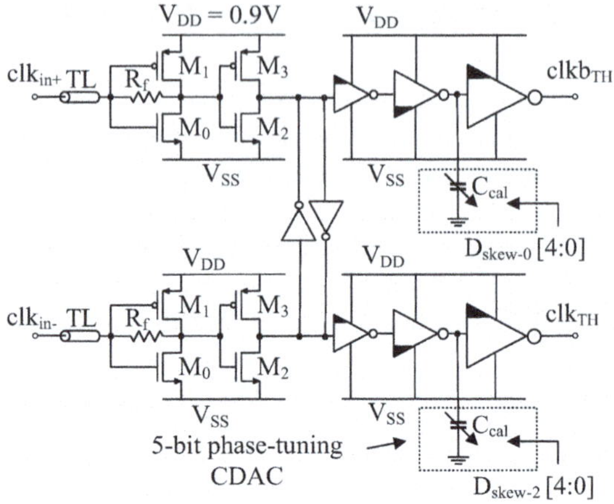

Fig. 8.2 Schematic of the clock path (clock buffer) © 2022 EuMA. (Reprinted with permission from [1])

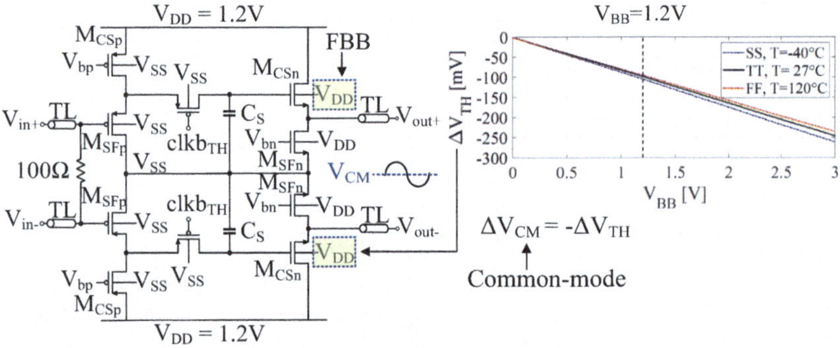

Fig. 8.3 Schematic of the T&H with pMOS and nMOS source followers © 2022 EuMA. (Reprinted with permission from [1])

offset sensitivity are reduced. The sampling clocks ($clkb_{TH}$ and clk_{TH}) can be used for a 2X-TI (TI channels at complementary clock phases). However, in this single-channel design, only one clock output is used by the T&H. To correct the sampling clock timing skew within $\pm 3\sigma$ range, a 2 ps tuning range is required. Hence, a 5-bit phase tuning CDAC with timing LSB of 30 fs will be considered for the 4X-TI T&H, which is good enough for a sample phase accuracy to achieve 49 dBc SFDR at 37 GHz input frequency, as explained in [2].

A detailed schematic of the track-and-hold building blocks is shown in Fig. 8.3. The input port is differentially 100 Ω-terminated for a good return loss. The output port is designed to drive the flash comparator array, hence targeting a low output impedance. The FE or input buffer, which is a pMOS source follower (SF), provides a source resistance of $R_S = 1/gm_{SFp} \approx 6\,\Omega$, and the flash buffer decouples the load of the flash comparator array from the T&H. The use of these two buffers provides a time constant of less than 4 ps to meet the dynamic settling requirement of the

T&H. Input parasitic capacitance of an SF includes a nonlinear and a linear term originating from C_{gd} and C_{gs}, respectively. Sampling just on the input capacitance of the SF without a dedicated sampling capacitor C_S is possible resulting in higher bandwidth while the targeted kT/C noise budget is also met without C_S. However, the achieved linearity is not sufficient. Therefore, an additional sampling cap of $C_S = 30\,\text{fF}$ is added to increase the overall linearity and achieve better than 36 dB THD, while maintaining a bandwidth of higher than 40 GHz.

A common problem of having several SF buffers in series is the common-mode (CM) voltage shifts limiting the usable signal swing. The first SF shifts the CM from 350 mV at the input to an output CM of approximately 700 mV in TT corner. Since no further positive CM shift is possible, the output SF is an nMOS-based SF. However, the minimum input common-mode voltage for the flash comparator (400 mV) is not met. To solve this issue, forward-body bias (FBB) is applied to the back gate of M_{SFn} as shown in Fig. 8.3. Using the supply voltage $V_{DD} = 1.2\,\text{V}$ as FBB, and thanks to the flipped-well option of the 22FDX process, nMOS threshold voltage is reduced by approximately 100 mV solving the minimum output CM issue. The transistor M_{SFn} is a supper low-threshold voltage (SLVT) device. The SLVTs in 22 nm FDSOI benefit from a flipped-well architecture [3], which allows FBB for both SLVTs without any latch-up limitation, which needs a positive and negative body bias for the SLVT nMOS and pMOS, respectively.

8.1.3 UWB T&H Measurement Results

The complete 18.5 GS/s sampling channel as depicted in Fig. 8.1 has been implemented in 22 nm FD-SOI. The fabricated test chip microphotograph and the time-domain measurement setup are shown in Figs. 8.4 and 8.5. The core T&H along with the clock buffer consumes a total core area of 75 µm × 62 µm while the test chip with pads occupies 650 µm × 740 µm. In order to calibrate differential phase imbalance of the setup due to cable mismatch, two tunable phase shifters are added to the signal path in Fig. 8.5. The spectral results are taken from a

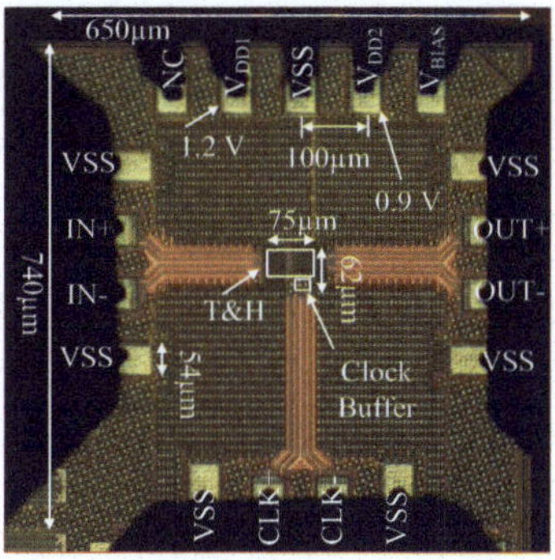

Fig. 8.4 Die microphotograph of the stand-alone implemented T&H © 2022 EuMA. (Reprinted with permission from [1])

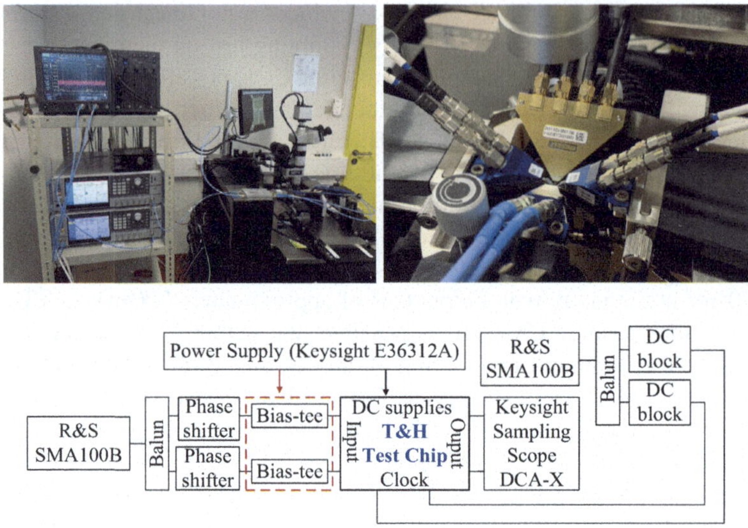

Fig. 8.5 Time-domain T&H measurement setup © 2022 EuMA. (Reprinted with permission from [1])

1024-point FFT applied to the sub-sampled data from the 80 GHz sampling scope. Dynamic performance of the T&H, as highlighted in Fig. 8.7, is verified across input frequency. The proposed FE+T&H achieves a peak SFDR and ENOB of 62 dBc and 6.9 bits at $f_{in} = 198$ MHz while the SFDR and ENOB at $f_{in} = 37$ GHz (4th Nyquist zone) are 38.6 dBc and 5.6 bits, respectively. Furthermore, the performance across sampling frequency with a low input frequency ($f_{in} = 198$ MHz ... 300 MHz) and an input amplitude of 1 dBFS, is shown in Fig. 8.8. The measurement results reveal an ENOB of better than 6.7 bits up to $f_s = 27$ GHz. Figure 8.9 shows the spectral results at $f_{in} = 198$ MHz (LF) and $f_{in} = 37$ GHz (HF) versus input amplitude, which is normalized respective to the full-scale range ($V_{in} = 0.6\,V_{ppd}$). The FFT plots of the single-tone input at $f_{in} = 39$ GHz and two-tone input at $f_{in1} = 1.5$ GHz and $f_{in2} = 36.1$ GHz are given in Fig. 8.6. The measured S_{11} and S_{21} during the track phase are given in Fig. 8.10 highlighting the 44 GHz 3dB-BW of the T&H. Table 8.1 shows the state-of-the-art sampler comparison. The T&H core consumes 65 mW from the 1.2 V supply while the clock buffer consumes 10 mW from the 0.9 V supply (75 mW in total).

8.2 Bulk-Driven Flash ADC

The growing demand for high-speed data converters in deep sub-micron technologies continues to increase their design challenges as well as tighten the constraints. Time interleaving is the most common technique to increase the

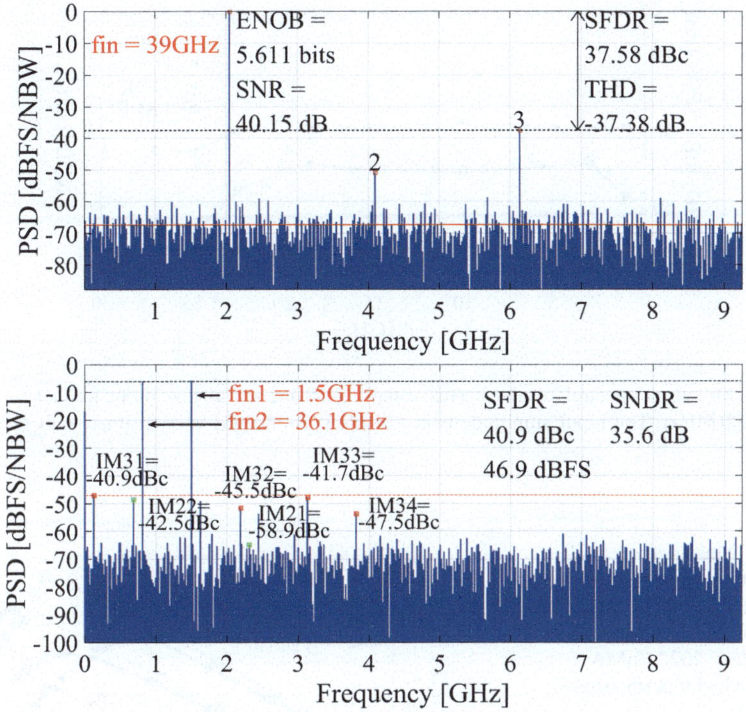

Fig. 8.6 Measured spectral results at $f_{in} = 39$ GHz (fifth Nyquist zone) with 0 dBFS ($V_{in} = 0.6\,V_{ppd}$) input amplitude and two-tone measurement with $f_{in1} = 1.5$ GHz and $f_{in2} = 36.1$ GHz at −6 dBFS (Samples=1024) © 2022 EuMA. (Reprinted with permission from [1])

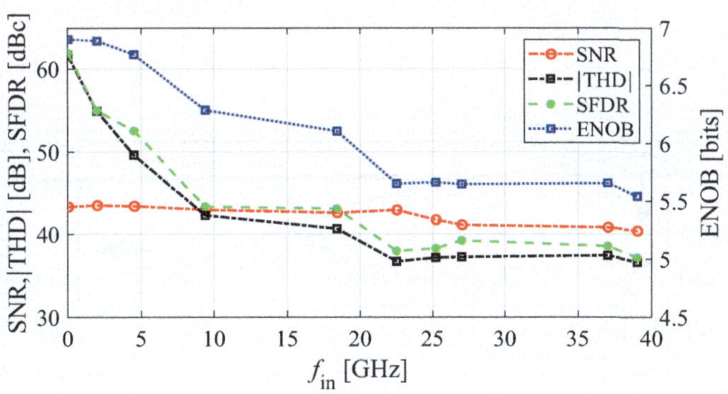

Fig. 8.7 Measured spectral results across input frequency, whereas the input amplitude is always adjusted to 0 dBFS ($V_{in} = 0.6\,V_{ppd}$) © 2022 EuMA. (Reprinted with permission from [1])

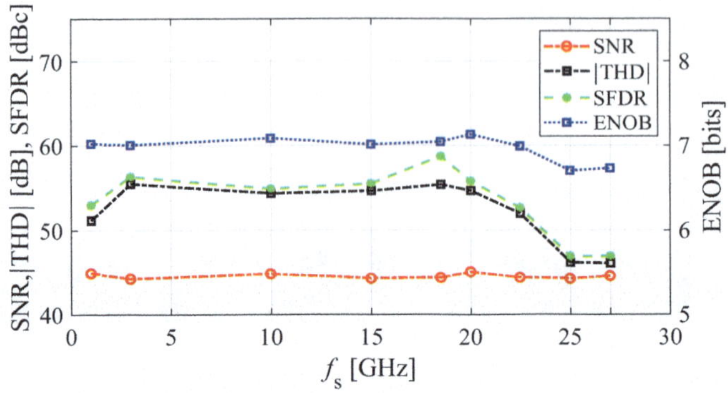

Fig. 8.8 Measured spectral results versus sampling frequency at low input frequency (f_{in} = 198 ... 300 MHz). The input amplitude is at 1 dBFS © 2022 EuMA. (Reprinted with permission from [1])

Fig. 8.9 Measured spectral results at f_{in} = 198 MHz (LF) and f_{in} = 37 GHz (HF) versus normalized input amplitude © 2022 EuMA. (Reprinted with permission from [1])

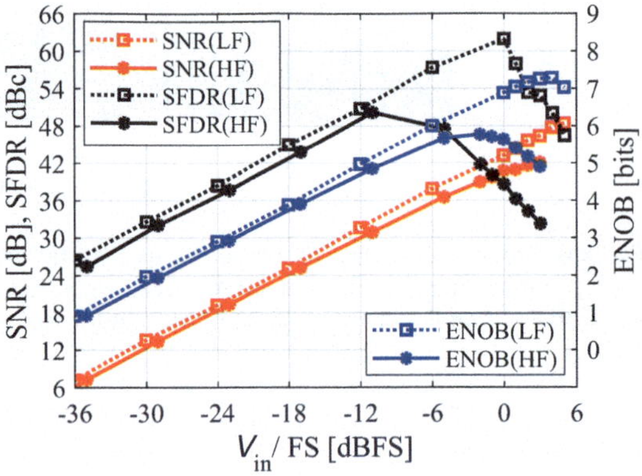

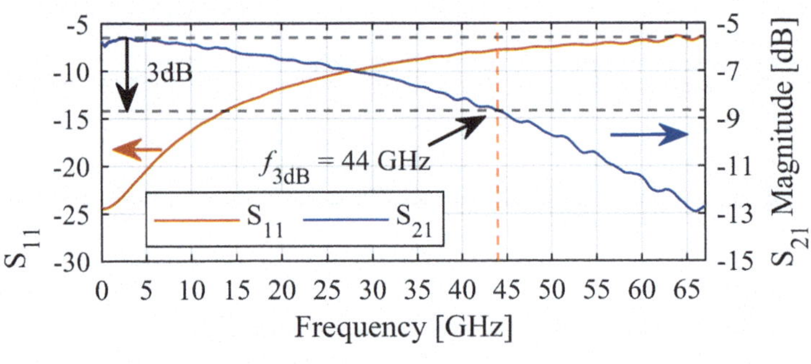

Fig. 8.10 Measured S_{11} and S_{21} during the track phase of the T&H © 2022 EuMA. (Reprinted with permission from [1])

Table 8.1 State-of-the-art sampler comparison © 2022 EuMA. (Reprinted with permission from [1])

Reference	TMTT '14 [4]	IMS '16 [5]	TCASII '16 [6]	BCICTS '18 [7]	This Work
Process	65 nm	180 nm	28 nm	130 nm	22 nm
	CMOS	SiGe	CMOS	BiCMOS	FDSOI
Supply [V]	3.3	2.5	1.75, 1.4	5.5, 3.5	1.2, 0.9
Input Swing [Vppd]	0.16	0.5	0.4	0.5	0.6
f_{sample} [GHz]	32.5	16	25	25.6	18.5
BW [GHz]	19.3	8	55	40 (1dB)	44
THD [dB]	−37@6	−45@7	−32@50	−39.4@40	−36@39
SNDR [dB] f_{in} [GHz]	–	–	29@50	36.7@40	35@39
P_{DC} [mW]	211	132	73	913	75
Die area [mm^2]	0.36	1.16	0.53	1.5	0.48
(Active)	0.07	–	0.03	–	0.004

sampling rate. However, highly time-interleaved ADCs with 32 or more lanes suffer from significant drawbacks [2] such as reduced input bandwidth, increased input capacitance, low-jitter multi-phase clock generation, cross talk and kickback penalties, reference voltage distribution, and finally significant increased effort to realize a robust TI calibration scheme. On the other side, the maximum achievable speed for a certain ADC architecture defines if for the overall target speed, a single-channel implementation or reducing the interleaving factor is feasible. Flash ADCs have been known as the fastest solution for medium-resolution high-speed ADCs. Hence, a 5-bit flash architecture has been adopted to realize a single-channel ADC at the target lane speed of 18.5 GS/s. This ADC is to the author's best knowledge the fastest single-core 5-bit flash ADC among single-core ADCs reported so far in the CMOS process. It makes realization of the targeted 74 GS/s ADC possible with only an interleaving factor of 4, which substantially reduces the calibration and design complexity as well as increasing the effective ADC bandwidth.

Flash ADCs are generally categorized as power-hungry converter types. Therefore, a novel bulk-driven flash reference generation technique utilizing the dedicated comparator offsets as flash reference levels are introduced and adopted in this design, which is discussed further in this section. This technique substantially reduces the power consumption of the flash comparator array and low-impedance R-ladder as two of the most power-hungry blocks in a flash ADC. Finally, this design has been implemented and proven on silicon, which is discussed in this section as well.

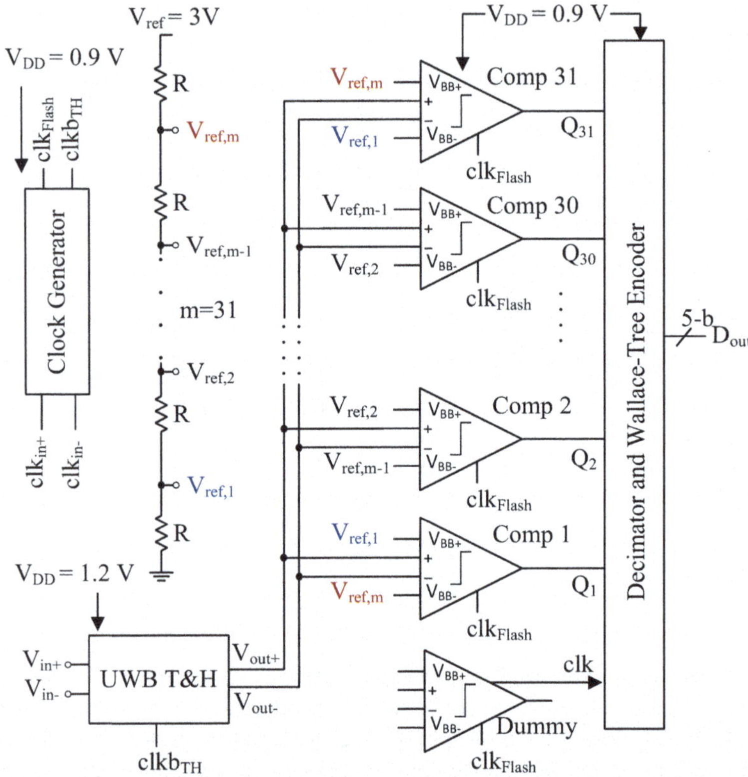

Fig. 8.11 Overall architecture of the implemented flash ADC © 2023 EuMA. (Reprinted with permission from [8])

8.2.1 Overall Architecture of the Flash ADC

The overall architecture of the implemented flash ADC is shown in Fig. 8.11. The main difference of this architecture compared to the conventional flash ADCs is providing the R-ladder reference levels to the back gate of the preamplifier used in the comparator instead of applying the references directly to the classical DDA. This bulk-driven reference generation concept and how it defines the ADC full-scale range and LSB voltage are further discussed in Sect. 8.2.2. The R-ladder uses an off-chip 3 V reference and consumes only 450 uW. The T&H core is discussed in the previous section and is the only core block, which operates from a 1.2 V supply. The rest of the cores use a standard 0.9 V supply. Since an external 3V reference is available in this design, it has been used as forward-body-bias (FBB) for the nMOS-based source-follower to further increase the output common mode of the T&H to 550 mV. The clock-phase generator is the same clock buffer already discussed in Sect. 8.1.2. It takes the 18.5 GHz sine-wave and provides the clock for the T&H

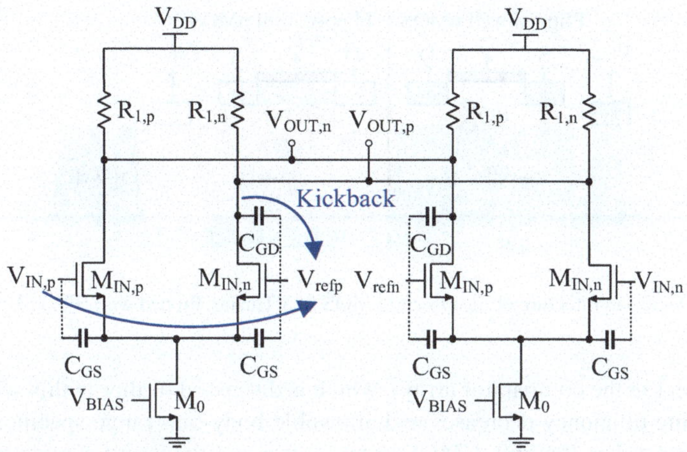

Fig. 8.12 Schematic of a classical differential-difference amplifier (DDA)

core ($clkb_{TH}$) and flash comparator array. However, the complementary clock phase taken from the clock buffer (clk_{TH}) requires duty cycle correction and certain buffering to drive the clock load of the comparator array. Hence, a comparator-array clock driver is used to generate the required clock phase (clk_{Flash}). The thermometer code from the comparator array is down-sampled by the decimator generating an output data rate of 144 MHz (decimation by 128x) or 96 MHz (decimation by 192x), which is selectable using a control bit. The thermometer-to-binary conversion is finally performed by the decimator at the target output speed using a Wallace-tree encoder.

8.2.2 Bulk-Driven Reference Generation Concept and Comparator Design

The state-of-the art flash ADCs typically use a standard differential-difference amplifier (DDA) as preamplifier. The DDA requires two differential pairs, as shown in Fig. 8.12 to compare the differential input signal with the flash references. The kickback to the reference voltages, as highlighted in Fig. 8.12, is a result of the existing parasitic capacitances C_{GS} and C_{GD}. Hence, a low-impedance and power-hungry R-ladder is required as a result of the introduced kickback to achieve a small RC time constant and maintain enough settling accuracy and linearity. The proposed body-biased comparator technique remedies the kickback issue and results in 100x higher R-ladder unit resistors. Apart from power saving in the R-ladder, implementation of calibration methods such as reference trimming is possible due to negligible kickback and relaxed design. Furthermore, it employs only one differential pair in each comparator resulting in significant power saving (50% less

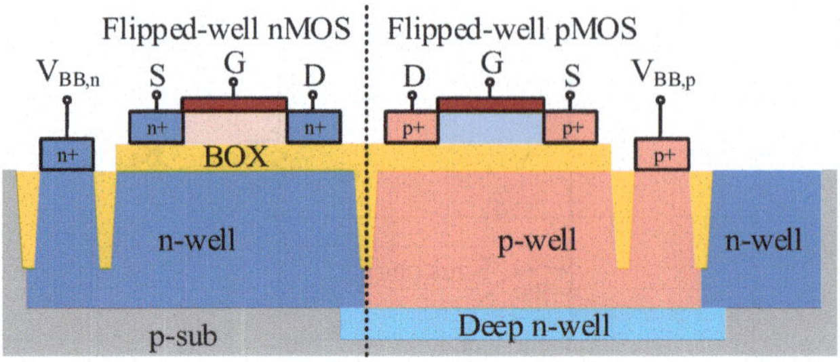

Fig. 8.13 Device architecture of nMOS and pMOS SLVT using flipped well (FW) [3]

static current in the comparator array), which is discussed further in this section. As body-biasing efficiency decreases and allowable body-bias range specifically FBB is highly limited in the bulk CMOS process, it is not the case for modern FDSOI technologies. A cross section of a super-low-threshold voltage (SLVT) nMOS and pMOS of the used FDSOI technology using flipped well (FW) is shown in Fig. 8.13. The device architecture of the depicted SLVTs in FW allows FBB for both SLVTs without any latch-up limitation, which needs a positive and negative body bias V_{BB} for the nMOS and pMOS, respectively ($V_{BB,n} > 0$ and $V_{BB,p} < 0$). Apart from the constantly decreasing efficiency of the body biasing and its small allowed range in the bulk CMOS process, another issue is nonlinear junction capacitances C_{DB} and C_{SB}. As the device cross section of the SLVT devices depicted in Fig. 8.13 shows, the mentioned nonlinear junction caps no longer exist. Hence, linearizing the MOS devices leaves a linear dependency between the threshold voltage (V_{TH}) and the back-bias voltage (V_{BB}). This linear behavior of the threshold voltage of an SLVT nMOS as a function of V_{BB} is also discussed in [9] across process corners and confirmed using simulation extracted plots. As a result, another degree of freedom is gained even for analog designs based on body biasing, which has been the motivation for a different flash comparator design in this work.

The novel bulk-driven flash reference generation concept adopted in this design is shown in Fig. 8.14. The DDA, as discussed earlier, requires two differential pairs, one for the actual input signal and the other one for the reference voltages. The differential pair for the reference voltages is impaired by parasitic capacitances (C_{GD}, C_{GS}) causing high-frequency coupling and kickback to the R-ladder. Our proposed technique eliminates the differential pair for the reference voltages therewith the low impedance requirement of the reference ladder to guarantee settling of the R-ladder is alleviated. Moreover, the power consumption of the power-hungry preamplifier is reduced by using a single differential pair. The references from the R-ladder, as shown in Fig. 8.14, are applied to the back gate of the preamplifier input devices, generating the actual flash ADC reference levels by altering the input transistor threshold voltages. As discussed earlier, V_{TH} reduces linearly by increasing the V_{BB} as FBB voltage applied to the back gate of the SLVT nMOS devices. Therefore, considering this linear dependency, and applying uniformly spaced references from the R-ladder to the input transistor back gate (V_{BB+}, V_{BB-}), uniformly spaced

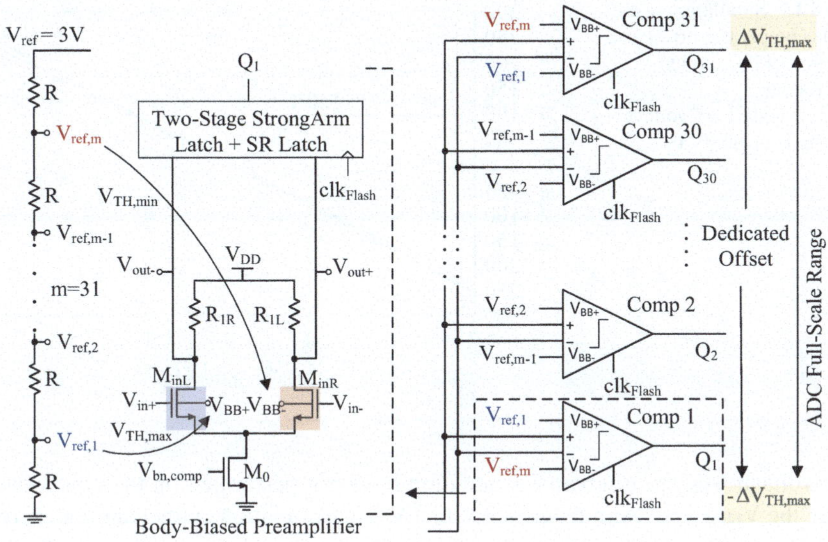

Fig. 8.14 Bulk-driven flash concept

comparator offsets are obtained correspondingly. Therefore, each comparator has a dedicated offset respective to the applied differential FBB, which is the actual flash reference level in this architecture. Each comparator consequently compares the differential input from the T&H to its built in offset. The body-biased preamplifier is followed by a two-stage StrongArm latch as the comparator core and a final S-R latch.

The flash reference levels are defined by dedicated comparator offsets. As a result, the full-scale range of the ADC (FSR) is the difference between the minimum and maximum comparator offsets in the comparator array. Since the R-ladder of the flash is placed between the ground and $V_{ref} = 3\,V$, this difference (FSR) is twice the corresponding V_{TH} change of the input devices for the variation of the back-bias V_{BB} from 0 to V_{ref}, which are represented as $V_{TH_{max}}$ and $V_{TH_{min}}$ respectively in Fig. 8.14. The FSR and LSB voltage are therefore calculated as

$$FSR = 2 \times (V_{TH_{max}} - V_{TH_{min}}) = 2 \times S_{BB} \times V_{ref}$$
$$\Delta = \frac{FSR}{2^n - 1} \tag{8.3}$$

Where $S_{BB} = \frac{dV_{TH}}{dV_{BB}}$ is the tuning gain of the V_{TH} respective to V_{BB}. Since the adopted technology provides a threshold voltage tuning gain of $S_{BB} \approx 80\,\text{mV/V}$. By using a 3 V V_{ref} as external reference, a nominal full-scale range of approximately 480 mV_{ppd} is achieved. Variation of the S_{BB} across PVT according to simulation results given in [9] affects the ADC full-scale range only by +/−7%. To verify the dedicated comparator offsets as a function of the applied differential body-

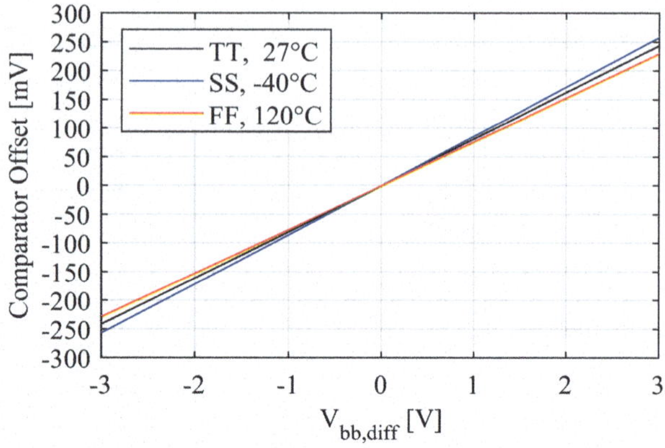

Fig. 8.15 Simulation result of the comparator offset as a function of the applied differential forward-body-bias Voltage ($V_{bb,diff}$) across PVT

bias voltage $V_{bb,diff}$, a separate simulation has been performed on the comparator, while the $V_{bb,diff}$ is swept from − 3 V to 3 V. The simulation extracted offset across PVT, as shown in Fig. 8.15, confirms both the linear behavior of the comparator offset respective to the applied $V_{bb,diff}$ and also the +/−7% variation of the FSR across PVT. The minimum and maximum offsets in the plot represent the offset of the first and last comparators in the flash comparator array. Hence, the FSR of the flash or the dedicated offset range across PVT, as discussed before, can be simply extracted by subtraction of these maximum and minimum achieved comparator offset voltages, while $V_{bb,diff}$ changes from − 3 V to 3 V. Although the ADC FSR is altered across PVT, only a linear gain error is introduced to the transfer function of the ADC, and this gain error can be calibrated out by trimming the 3 V reference or setting the external 3 V reference.

Accuracy of the flash reference levels depends on the matching of the R-ladder and comparator offsets. The resistor mismatch in the reference ladder is typically small [10], The comparator offset on the other hand is the main design challenge in flash ADCs. The offsets in the comparators are categorized into three types, which are systematic dynamic and random offset. Systematic offset is predictable and can be remedied by adopting proper circuit design and layout techniques. Dynamic offset can occur as a result of kickback or memory effect of the comparator core (StrongArm latch). Putting a preamplifier before the StrongArm latch mitigates the dynamic offset. The random offset is a result of process variation during the fabrication and can lead to large static offsets. Since a preamplifier is used in this design, the major contribution to the random offset is made by the input devices of the preamplifier (M_{inL}, M_{inR}). The variance of the random offset caused by the threshold voltage mismatch at the input of the differential pair $\sigma^2_{V_{TH}}$ can be approximated as [11].

$$\sigma^2_{V_{TH}} \approx \frac{A^2_{V_{TH}}}{W \cdot L} \tag{8.4}$$

Where $A_{V_{TH}}$ is a process-dependent coefficient and W · L is the gate area of the input devices. Therefore, in order to reduce the random offset, the size of the input transistors needs to be made larger. However, the penalty is more capacitive load on the T&H resulting in reduced bandwidth or increasing the power consumption and input capacitance of the input buffer to maintain the same bandwidth. Statistical Monte Carlo simulation with n =1000 samples shows a total input referred comparator offset of $\sigma \approx 6.2$ mV, while as the major contributor $\sigma_{V_{TH}} = 5.1$ mV for this design. To get a fabrication yield of 99.7%, 3σ needs to be less than 0.5 LSB. Considering 1 LSB $\approx$ 15 mV, this condition is met for 1.2σ in the current design. Therefore, the yield is reduced to 77%. In order to further increase the yield without increasing the size of the input transistors and maintain the ADC linearity, offset calibration is required. Since the random offset is static, foreground calibration methods with lower complexity compared to background calibration can be used. The other advantage of the bulk-driven technique is the possibility of reference trimming for the static offset calibration since the introduced kickback to the references is negligible. Therefore, a 3-bit reference trimming scheme, as suggested in [12], is considered for the next design as our future work to tune each comparator offset within a tuning range of ± 1.5 LSB using a tuning step of 0.5 LSB. Hence, based on the measured INL/DNL, a calibration code will be obtained off-chip, and the single comparators offset will be tuned using a simple reference trimming (7 trimming states utilizing a 3-bit code per comparator). This way, the entire mismatch range of $\pm 3\,\sigma$ for the static offset will be calibrated improving the yield and linearity of the ADC. The complete block diagram of the comparator used in the comparator array is shown in Fig. 8.16. The aforementioned body-biased preamplifier, as shown in this figure, is followed by a two-stage StrongArm latch as the comparator core. In order to remedy high metastabilty rate at the core speed of 18.5 GHz and improve bit-error rate (BER), a two-stage StrongArm latch is considered in the comparator architecture. The second-stage latch samples the output of the first-stage during regeneration. Hence, a higher gain is achieved aleviating the metastability. Furthermore, each latch stage utilizes forward-body-bias to 3 V for the implemented nMOS SLVT devices as highlighted in the Fig. 8.16 reducing the device threshold voltage enhancing the comparator speed. A detailed analysis of StrongArm latch is given in [13]. Finally, the output data are stored in the SR-latch, while the output of the second StrongArm latch resets. The SR-latch is followed by a D-FF, which is time-aligned with the first D-FF used in the decimator operating at the same speed with half a clock cycle latency to relax the timing. The comparator array includes 31 comparators resulting in a high capacitive load on the clock. Therefore, a local clock buffer or clock tree is used to generate the complementary clock phase required for the two-stage latch represented as $\boldsymbol{\phi}$ and $\bar{\boldsymbol{\phi}}$, as well as reducing the total clock load on the flash clock clk_{Flash}. To verify the comparator speed regarding metastability, a transient simulation at SS corner (worst case) has been performed, while the inputs of the comparator are shorted. The obtained eye diagram at the output of the first and second stage StrongArm latch is shown in Fig. 8.17. The eye diagram of the second latch reveals no metastability for 1000 clock cycles even at SS-corner.

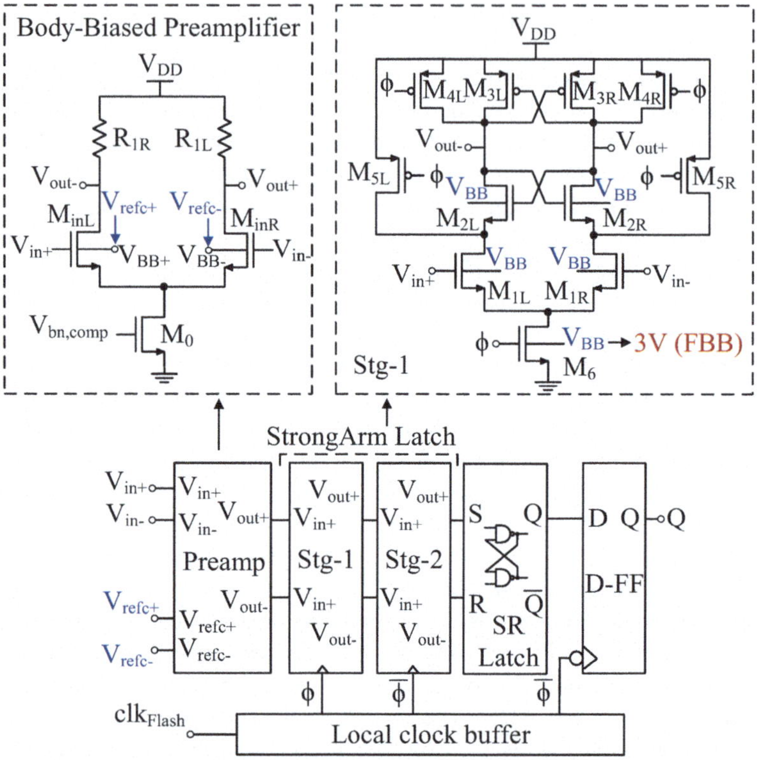

Fig. 8.16 Schematic of the body-biased comparator with two-stage latch

8.2.3 Decimator Design

Due to high speed of the flash, including a thermometer-to-binary encoder at the lane speed (18.5 GS/s) for 5-bit resolution is not easily implementable using a standard Wallace-tree encoder. Furthermore, considering 5-bit data at the core speed of 18.5 GS/s results in 92.5 Gb/s. Hence, a decimator as shown in Fig. 8.18 has been implemented after the comparator array to simplify both data read out and thermometer-to-binary conversion. The clock phases are generated from sequential 2:1 CMOS clock dividers (D-FF with Qb shorted to D) and all D-FFs operating at f_s and $f_s/2$ are designed as dynamic flip-flops. To relax the timing and avoid setup and hold-time violations, one every other flip-flop array is triggered with the opposite clock edge. Due to the high clock rate, the decimation utilizes two consecutive levels of 2x decimation to relax the timing. This is followed by a selectable third level 32x/48x decimation using a control bit. The final binary output is generated using a Wallace-tree encoder at data rates of $f_s/128$ $f_s/192$. The input clock reference of the decimator is taken from a dummy comparator, which has the same clock path as

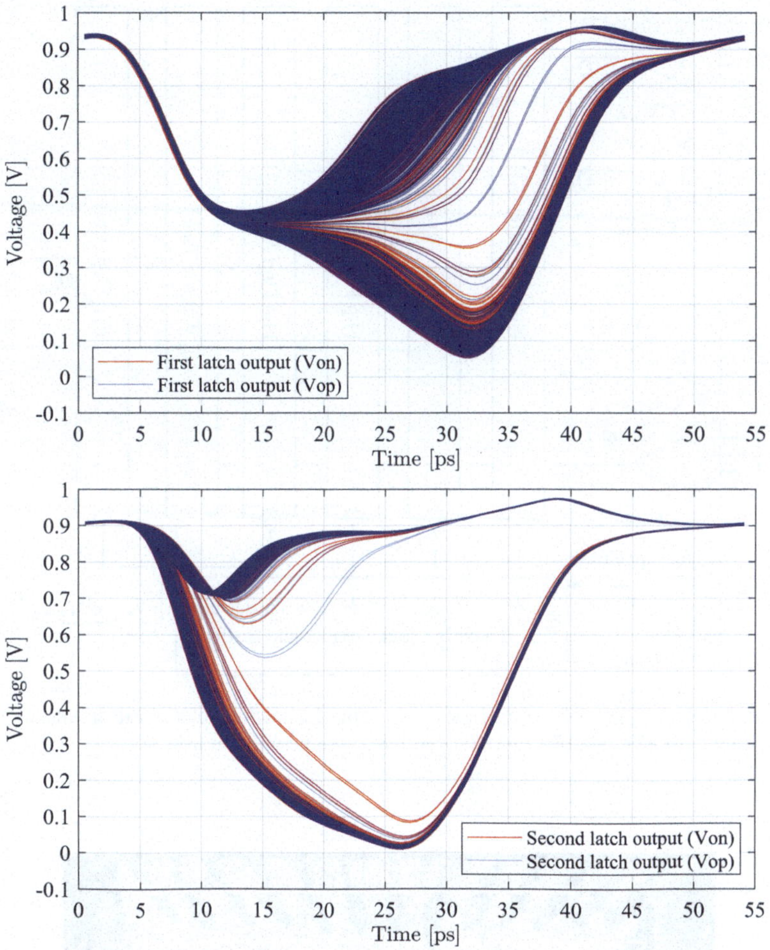

Fig. 8.17 Eye diagram at the output of the first and the second latch, while input of the comparator is shorted (SS corner—T = 27°C)

the other comparators resulting in a matched delay. The further delay for the time alignment is tuned digitally.

8.2.4 Flash ADC Measurement Results

The implemented ADC, as the die micro photograph in Fig. 8.19 shows, consumes an area of 1.01 mm^2 with pads. The measurement setup is shown in Fig. 8.20. The high-frequency clock and ADC input have been wafer probed, while the glued chip is bonded to the PCB for the DC supplies and low-frequency digital I/Os. The input

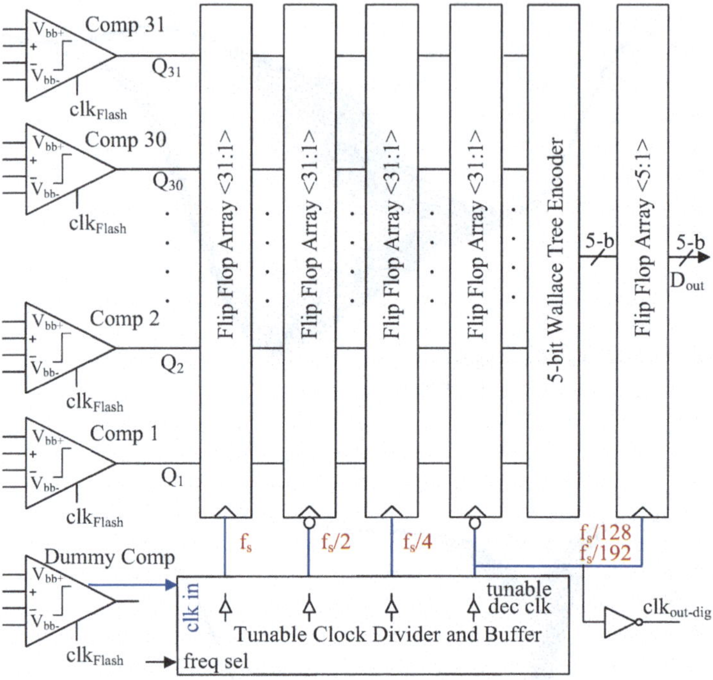

Fig. 8.18 Block diagram of the decimator with tunable decimation ratioOverall architecture of the implemented flash ADC © 2023 EuMA. (Reprinted with permission from [8])

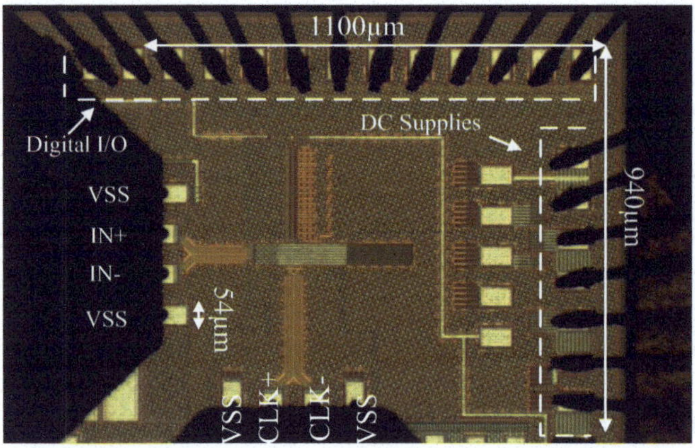

Fig. 8.19 Die microphotograph of the implemented flash ADC © 2023 EuMA. (Reprinted with permission from [8])

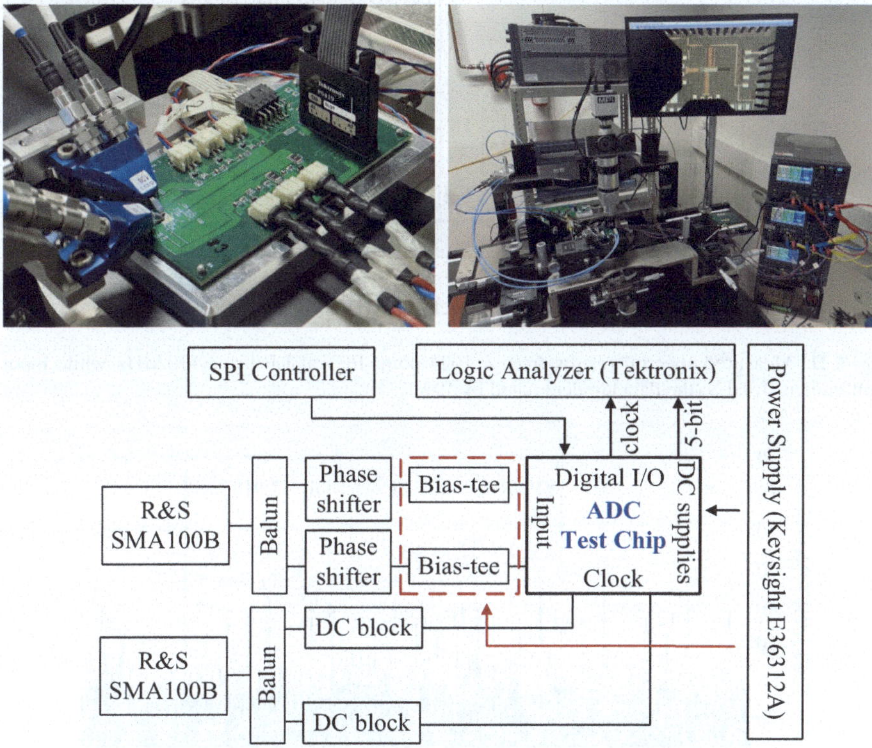

Fig. 8.20 Flash ADC measurement setup © 2023 EuMA. (Reprinted with permission from [8])

common mode of the ADC is applied using two bias-Ts at the input side. The clock input is AC-coupled. Hence, two DC blocks have been used. The SPI controller controls an on-chip daisy chain used to control the drive strength of some buffers, clock skew tuning, and the decimation ratio. The decimated ADC output is given to a logic analyzer, and the recorded data are processed in Matlab. The 1024-point FFT for LF sine-wave input and Nyquist are shown in Figs. 8.21 and 8.22.

The spectral results reveal a Nyquist SFDR and SNDR of 33.6 dBc and 25.3 dB (ENOB =3.9 bits) respectively using a full-scale single-tone sine wave input (f_{in} = 9.1 GHz). The LF (f_{in} = 170 MHz) SFDR and SNDR are 35.9 dBc and 26.1 dB (ENOB=4 bits), respectively. Since an UWB T&H is used in this design, the ADC performance is characterized up to the 4th Nyquist zone of the ADC. Spectral results across input frequency are given in Fig. 8.23 for an input amplitude overall adjusted to − 0.5 dBFS. The measurement reveals an SNDR better 24.3 dB (ENOB=3.75 bits) up to 37.2 GHz. The analysis of the ADC output code range across input frequency shows an amplitude loss or roll-off of about − 1.8 dB at f_{in} = 37.2 GHz with a − 3dB bandwidth of above 40 GHz. The INL/DNL plots are shown in Fig. 8.24. The measured INL and DNL using a 10 MHz sine-wave input

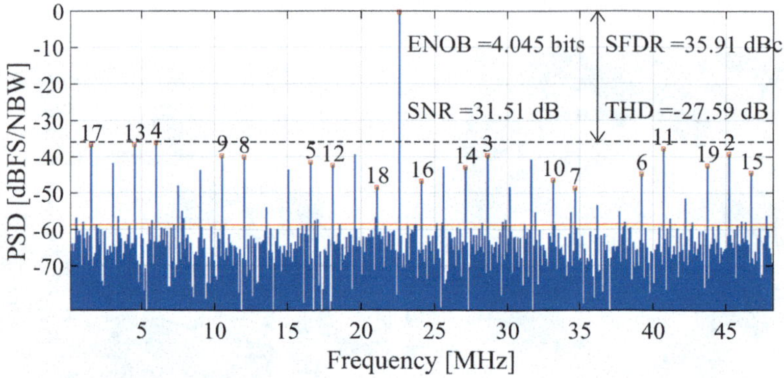

Fig. 8.21 Measured spectral results from a 1024-point FFT at LF fin =170 MHz while input amplitude is 0 dBFS (the data are decimated by 192)

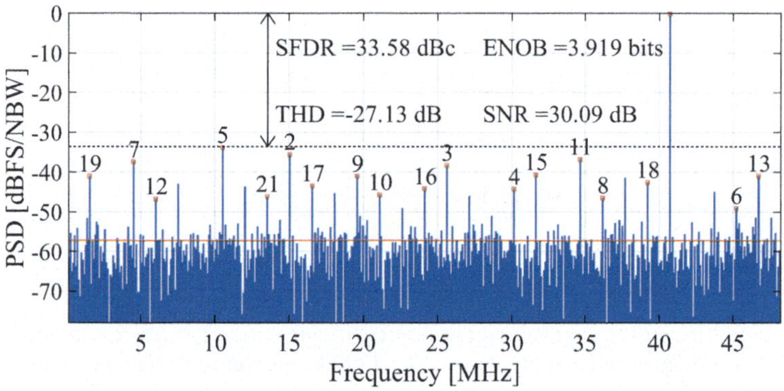

Fig. 8.22 Measured spectral results from 1024-point FFT at Nyquist fin =9.2 GHz while input amplitude is 0 dBFS (the data are decimated by 192)

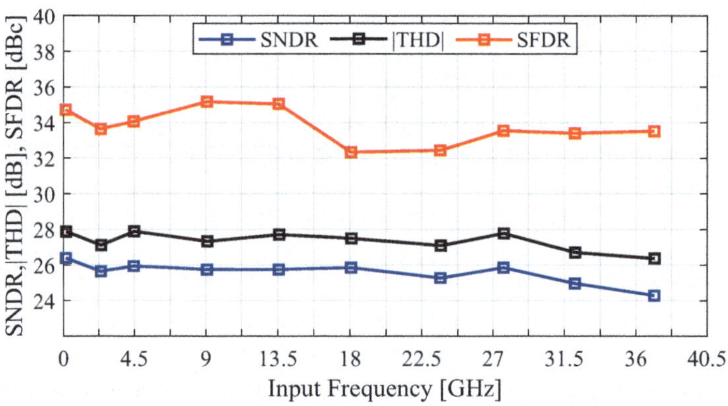

Fig. 8.23 Measured spectral results across input frequency up to the 4th Nyquist zone of the ADC. The input amplitude is overall adjusted to -0.5 dBFS, and output data are decimated by 192x using 1024 spectral points

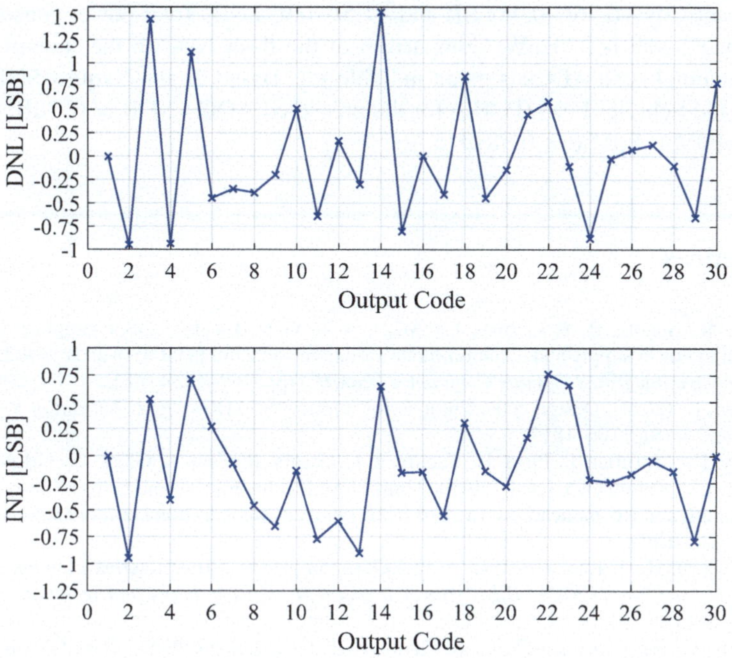

Fig. 8.24 Measured INL/DNL of the ADC using a 10.5 MHz sine-wave input

Table 8.2 State-of-the-art single-channel flash ADC comparison

Reference	EUMIC '10 [14]	JSSCC '14 [15]	RFIC '15 [16]	BCTM '17 [17]	This work
Process	65 nm CMOS	130 nm SiGe	28 nm CMOS	130 nm SiGe	22 nm FD-SOI
	CMOS	SiGe	CMOS	SiGe	FD-SOI
Architecture	Flash	Flash	Flash	Flash	
Resolution [bit]	3	6	3	4	5
f_{sample} [GHz]	20	20	24	40	18.5
ENOB at Nyquist [bit]	2	3.7	2.2	2.8	3.9
Power [W]	3.1	1	0.41	3.5	0.27
Die area [mm^2]	5.2	3.61	2.4	1.3	0.06
Active area [mm^2]	0.16	0.56	0.12	–	0.06
FoM_W [pJ/c-s.]	38.75	3.85	3.72	12.56	0.98
FoM_S [dB]	109	124	120	116	130

are respectively +0.76/−0.95 LSB and +1.5/−0.95 LSB. Total power consumption of the ADC core is 270 mW. Comparison of the flash ADC to the state-of-the-art single-channel flash ADCs is given in Table 8.2. Based on the Nyquist SNDR and ENOB, the Schreier FoM (FoM_S) for the presented ADC SNDR is 130 dB, and the Walden FoM is $FoM_W = 0.98\,pJ/c - s$.

References

1. Lotfi, N., Scholz, P., & Gerfers, F. (2022). A 44 GHz-BW 18.5 GS/s sampling front-end robust to power supply and common-mode variations in 22 nm FDSOI. In *2022 17th European Microwave Integrated Circuits Conference (EuMIC)* (pp. 169–172). IEEE.
2. Razavi, B. (2013). Design considerations for interleaved ADCs. *IEEE Journal of Solid-State Circuits, 48*(8), 1806–1817.
3. Carter, R., Mazurier, J., Pirro, L., Sachse, J. U., Baars, P., Faul, J., Grass, C., Grasshoff, G., Javorka, P., Kammler, T., et al. (2016). 22nm FDSOI technology for emerging mobile, Internet-of-Things, and RF applications. In *2016 IEEE International Electron Devices Meeting (IEDM)* (pp. 2–2). IEEE.
4. Ma, S., Yu, H., & Ren, J. (2014). A 32.5-GS/s sampler with time-interleaved track-and-hold amplifier in 65-nm CMOS. *IEEE Transactions on Microwave theory and Techniques, 62*(12), 3500–3511.
5. Lin, Y.-A., Yeh, Y.-C., Liu, Y.-C., & Chang, H.-Y. (2016). A 55-dB SFDR 16-GS/s track-and-hold amplifier in 0.18 μm SiGe using differential feedthrough cancellation technique. In *2016 IEEE MTT-S International Microwave Symposium (IMS)* (pp. 1–4). IEEE.
6. Tretter, G., Fritsche, D., Khafaji, M. M., Carta, C., & Ellinger, F. (2015). A 55-GHz-bandwidth track-and-hold amplifier in 28-nm low-power CMOS. *IEEE Transactions on Circuits and Systems II: Express Briefs, 63*(3), 229–233.
7. Du, X.-Q., Grözing, M., & Berroth, M. (2018). A 25.6-GS/s 40-GHz 1-dB BW current-mode track and hold circuit with more than 5-ENOB. In *2018 IEEE BiCMOS and Compound Semiconductor Integrated Circuits and Technology Symposium (BCICTS)* (pp. 56–59). IEEE.
8. Lotfi, N., Scholz, P., & Gerfers, F. (2023). The fastest CMOS single-channel 5-bit flash ADC operating at 18.5 GS/s in 22 nm FD-SOI. In *2023 18th European Microwave Integrated Circuits Conference (EuMIC)* (pp. 121–124). IEEE.
9. Wittenhagen, E., Runge, M., Lotfi, N., Ghafarian, H., Tian, Y., & Gerfers, F. (2020). Advanced mixed signal concepts exploiting the strong body-bias effect in CMOS 22FDXR®. *IEEE Transactions on Circuits and Systems I: Regular Papers, 68*(1), 57–66.
10. Noulis, T. (2018). *Mixed-signal circuits*. CRC Press.
11. Pelgrom, M. J. M., Duinmaijer, A. C. J., & Welbers, A. P. G. (1989). Matching properties of MOS transistors. *IEEE Journal of Solid-State Circuits, 24*(5), 1433–1439.
12. Chang, H.-Y., & Yang, C.-Y. (2015). A reference voltage interpolation-based calibration method for flash ADCs. *IEEE Transactions on Very Large Scale Integration (VLSI) Systems, 24*(5), 1728–1738.
13. Razavi, B. (2015). The StrongARM latch [A circuit for all seasons]. *IEEE Solid-State Circuits Magazine, 7*(2), 12–17. ISSN: 1943–0582. https://doi.org/10.1109/MSSC.2015.2418155.
14. Ferenci, D., Grözing, M., Lang, F., & Berroth, M. (2010). A 3 bit 20 GS/s flash ADC in 65 nm low power CMOS technology. In *The 5th European Microwave Integrated Circuits Conference* (pp. 214–217). IEEE.
15. Ritter, P., Le Tual, S., Allard, B., & Möller, M. (2014). Design considerations for a 6 bit 20 GS/s SiGe BiCMOS flash ADC without trackand-hold. *IEEE Journal of Solid-State Circuits, 49*(9), 1886–1894.

16. Tretter, G., Khafaji, M., Fritsche, D., Carta, C., & Ellinger, F. (2015). A 24 GS/s single-core flash ADC with 3 bit resolution in 28 nm low-power digital CMOS. In *2015 IEEE Radio Frequency Integrated Circuits Symposium (RFIC)* (pp. 347–350). IEEE.
17. Du, X.-Q., Grözing, M., Buck, M., & Berroth, M. (2017). A 40 GS/s 4 bit SiGe BiCMOS flash ADC. In *2017 IEEE Bipolar/BiCMOS Circuits and Technology Meeting (BCTM)* (pp. 138–141). IEEE.

Part IV
RF & mmWave Circuit Techniques

Chapter 9
RF Switches in CMOS FDSOI Process—from Circuit Concepts to Implementation

Oğuzhan Özdamar, Valentyn Solomko, Semen Syroiezhin, and Robert Weigel

9.1 Introduction

Solid-state FET-based linear RF switches in dedicated FD-SOI CMOS technologies are extensively used in wide range of applications, including RF front-ends of cellular user equipment [1], WiFi terminals [2], wireless infrastructure equipment [3], to mention only a few. Among the strongest drivers for rapid development of the aforementioned technologies and RF switches based thereupon were 4G and 5G cellular technologies, requiring significant amount of antenna switches [2], tuning circuits [4], switchable directional couplers [5] (all based on CMOS switches) for reconfigurable front-ends. Antenna tuning applications in cellular phones created a strong momentum for rapid development of high-voltage ($V_{RF} > 45\,V$), linear ($IIP_3 > 80\,dBm$) RF switches in specialized SOI-CMOS switch technologies, making them de-facto standard in smartphones with integrated antennas.

Among the benefits of SOI-CMOS stacked switches over other competing technologies (including GaAs, RF-MEMS, PCM, diode-based) are low manufacturing cost, monolithic integration capabilities, overall competitive small- and large-signal RF performance at sub-GHz frequencies, outstanding intrinsic electrostatic discharge (ESD) robustness and low $R_{on}C_{off}$ figure-of-merit (FOM), sufficient for most demanding use cases.

O. Özdamar (✉)
formerly with Infineon Technologies AG, Neubiberg, Germany

V. Solomko · S. Syroiezhin
Infineon Technologies AG, Neubiberg, Germany
e-mail: valentyn.solomko@infineon.com; semen.syroiezhin@infineon.com

R. Weigel
Institute for Electronics Engineering, University of Erlangen-Nuremberg, Erlangen, Germany
e-mail: robert.weigel@fau.de

F. Gerfers (ed.), *Bulk-Driven Circuit Techniques for CMOS FDSOI Processes*,
https://doi.org/10.1007/978-3-031-85114-8_9

Since the introduction of dedicated SOI-CMOS technologies for RF switch applications back in 2008 [6], the performance of its core device—switch MOSFET—has undergone significant improvement, evidenced by a threefold decrease in $R_{on}C_{off}$ FOM over a span of 15 years. Many innovative concepts on a circuit design level have been proposed, particularly for high-voltage, highly stacked switches. Noteworthy examples include capacitive equalization [7, 8], active biasing [9], switching time acceleration [10, 11], and back-gate biasing [12] techniques.

9.2 Design of a Single-Transistor RF Switch in SOI-CMOS

9.2.1 Features of Dedicated CMOS FDSOI Switch Technologies

SOI-CMOS technology has been a predominant force in the antenna switch market for over a decade, commanding over 85% market share [13]. Notably, in 2015, the technology witnessed a noteworthy improvement of 20% in the $R_{on}C_{off}$ product (FOM), and since then, continuous advancements have led to a more than double reduction of the FOM [14]. This technology utilizes both n-type and p-type MOS transistors, isolated from the substrate by a buried oxide (BOX). This isolation serves to minimize the parasitic drain–source capacitance, a benefit derived from the lower permittivity of BOX compared to silicon. The galvanic isolation between transistors in SOI-CMOS contributes to a reduction in terms of leakage current, enhancing overall device performance and reducing current consumption.

In general, RF switches based on SOI-CMOS technologies are manufactured on high-resistivity (HR) silicon substrate materials, with resistance values well exceeding 1 kΩ·cm, thus reducing the substrate losses at high frequencies [15, 16]. However, challenges arise when HR silicon wafers are oxidized, leading to parasitic surface conduction (PSC), where fixed charges within the oxide trigger free carriers near the substrate/BOX interface, degrading the nonlinear performance of the transistors and its sensitivity to dc voltage variations. Introducing a trap-rich layer between the BOX and the HR silicon handle wafer helps to reduce PSC. This layer efficiently captures free carriers at the substrate/BOX interface, allowing the substrate to recover to its nominal resistivity. This not only improves the transistors nonlinear performance and reduces sensitivity to dc voltage variations but also significantly diminishes RF losses and crosstalk [17].

A switch MOSFET is constructed similarly to a general-purpose MOS device, however, optimized for the operation in cut-off and triode, because switches operate in these two regions in off and on states, respectively. A hybrid approach featuring combination of fine-size lithography process generations for FEOL with the older generations of BEOL comprising thick metal layers allows achieving low $R_{on}C_{off}$ FOM and minimizing the layout size of RF switches [18].

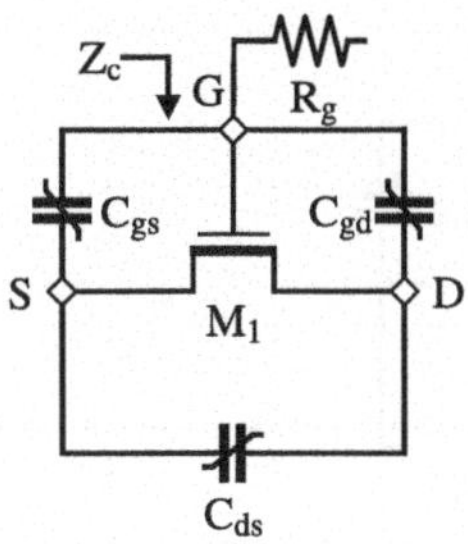

Fig. 9.1 Basic linear RF switch

9.2.2 Single-Transistor RF Switch

A linear RF switch is constructed with a high-ohmic resistor connected in series to the gate terminal of a MOSFET transistor as shown in Fig. 9.1. The gate resistor is crucial in defining the nonlinear behavior and switching time of the RF switch. The switching time is quantified by a time constant described in (9.1) and is directly proportional to the gate resistor. Theoretically, if the gate resistor is removed and the energy of a control source is high enough, the switching time could approach some nanoseconds range. However, such a scenario would introduce significant linearity issues due to uneven voltage division at the gate–source and gate–drain terminals.

$$\tau = R_G(C_{gd} + C_{gs}) \tag{9.1}$$

In order to maximize the linearity, the gate resistor is maintained at a significantly higher value compared to the impedance $Z_C = 1/j\omega(C_{\mathrm{gs}} + C_{\mathrm{gd}})$ of the gate–source overlap capacitance, ensuring that the RF current predominantly flows through the capacitances C_{gs} and C_{gd}. It is important to note that the switching time, linearity, and power losses of an RF switch are interdependent and can be adjusted by varying the gate resistor. In particular, RF switches for communication applications have stringent requirements for linearity, while the acceptable range for the switching time typically ranges between 2 μs and 20 μs. Balancing these factors is crucial to meet specific performance demands for RF switches [19].

9.2.3 Key Characteristics of a Single-Transistor RF Switch

An equivalent circuit diagram and corresponding first-order parasitic elements of an off-state RF switch with a high-ohmic resistor $R_G \gg Z_C$ connected at the gate terminal are illustrated in Fig. 9.2. For simplicity, the resistor is removed from the circuit under the specified bias condition. The transistor operates in the off-state when the dc voltage at the gate terminal is well below the threshold voltage V_{th}

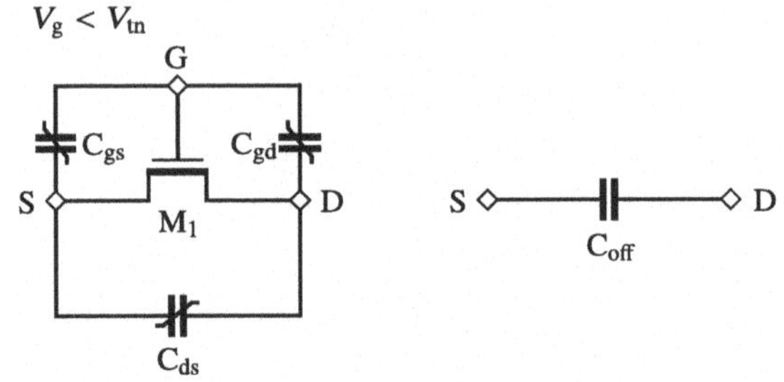

Fig. 9.2 Equivalent circuit of a single-MOSFET RF switch in off-state

(cut-off region). If the gate voltage exceeds the threshold, the channel is inverted, allowing a current flow between the drain and source terminals [20]. The off-state operation resembles a capacitor in a signal path, as depicted in Fig. 9.2. The primary contributors to the off-state capacitance C_{off} are the nonlinear parasitic overlap capacitances C_{gd}, C_{gs}, and the drain–source capacitance C_{ds}. The contribution of the junction capacitances C_{sb} and C_{db} to the off-state capacitance in a single-transistor switch is negligible, primarily due to the use of high-ohmic substrate material. Even small bias voltage levels at the body of the RF switch transistor generates a large depletion region due to low body doping, thus reducing the impact of the junction capacitances dramatically. The off-capacitance is calculated as the parallel connection of C_{gs}, C_{gd}, and C_{ds} as shown in Fig. 9.2, assuming $C_{gs} = C_{gd}$.

Given that the gate oxide is very thin, the overlap capacitance is notably larger than the substrate capacitance. Consequently, for the calculation of C_{off}, the substrate capacitance is not considered. The drain–source capacitances C_{ds} are formed between the source–drain junctions, separated by the depleted p-region. An analytical derivation of the source–drain capacitance can be found in literature [21]. This capacitance is described using an electric field generated from the potential between the source–drain junctions and, according to [21], can be described as:

$$C_{ds} = \frac{\varepsilon_{\mathrm{Si}} \varepsilon_0 \xi W}{L}. \tag{9.2}$$

The depth ξ of the positive charge attracted to the channel between the drain–source diffusions and the variation of the drain–source capacitance with the terminal voltage levels are crucial aspects. These factors contribute to the behavior and performance of the transistor.

In on mode of operation, the gate voltage of an RF switch transistor is maintained above the threshold voltage, causing the channel between the drain and source to invert into an n^+-region. The transistor exhibits different inversion modes—weak, moderate, or strong—depending on the applied gate voltage. Weak inversion, also known as subthreshold operation, occurs when the gate voltage is within 100 mV or higher below the threshold voltage. Conversely, strong inversion is observed when the gate voltage is 100 mV or more above the threshold voltage. Its noteworthy that

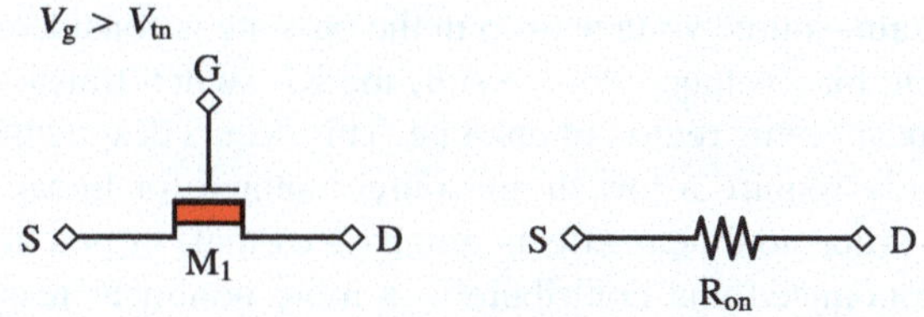

Fig. 9.3 Equivalent circuit of an RF switch transistor in on-state

in RF applications, particularly in RF switches, a majority of transistors operate in strong inversion, as highlighted by the considerations in this chapter [20]. As the gate voltage surpasses the threshold voltage, a channel is formed between the drain–source junctions, enabling the flow of current from drain to source in the on-state.

A first-order equivalent circuit and cross section of a single RF transistor in the on-state, as depicted in Fig. 9.3, illustrate the representation of the RF switch transistor by R_{on} between the drain–source terminals. Examining the operation modes within the on-state, the transistor operates in triode region when the drain–source voltage V_{ds} is less than or equal to the difference between the gate–source voltage V_{gs} and the threshold voltage V_{th}. In this region, the drain current follows a square law. However, during the switch operation, the ac component of V_{gs} is half of V_{ds}, meaning $V_{gs} = V_{gs,H0} + \frac{V_{ds}}{2}$, where $V_{gs,H0}$ (dc gate–source component) represents the bias point of the switch and V_{ds}—the voltage across the switch transistor with zero dc component. Given the response of a MOSFET in a triode region, the drain current can be calculated as:

$$\begin{aligned} I_d &= \mu_n \frac{\varepsilon_{ox}}{t_{ox}} \frac{W}{L} \left[\left(V_{gs,H0} + \frac{V_{ds}}{2} - V_{th} \right) V_{ds} - \frac{V_{ds}^2}{2} \right] \\ &\approx \mu_n \frac{\varepsilon_{ox}}{t_{ox}} \frac{W}{L} \left[\left(V_{gs,H0} - V_{th} \right) V_{ds} \right] \end{aligned} \tag{9.3}$$

The elimination of the qudratic term in (9.3) indicates linear response of the switch to applied RF voltage. The on-channel resistances R_{on} of a single RF switch can be quantitatively described by differentiating the (9.3). This implies that by taking the derivative of the relevant equation, one can obtain an expression that directly relates the on-channel resistance to the operation parameters of the RF switch, providing a detailed understanding of how the resistance varies with changes in key factors, such as applied voltage, transistor size:

$$R_{on} = \left(\frac{\partial I_d}{\partial V_{ds}} \right)^{-1} = \frac{1}{\mu_n \frac{\varepsilon_{ox}}{t_{ox}} \frac{W}{L} (V_{gs,H0} - V_{th})} \tag{9.4}$$

Typically, RF switch transistors have total gate width in the range of several millimeters, resulting in very low on-channel resistance, often below 1 Ω.

Usually, the drain–source voltage drop in the on-state is considerably lowed than the gate–source dc bias voltage. As a result, the RF switch transistor consistently operates in the linear ohmic region in on-state. This occurs despite the fact that (9.4) might not explicitly require a low drain–source voltage for linear operation. The mention of high-order nonlinear effects emphasizes that, in practice, such effects become more pronounced and contribute to a more nonlinear response when the drain–source voltage is not kept low. Ensuring a constant on-channel resistance across the entire range of applied RF excitations is critical. This is achieved by incorporating a large gate resistor, causing the entire transistor to swing with the RF voltage while maintaining a constant gate–source voltage over the complete RF period.

A notable concern is addressed regarding the negative RF swing, during which the source–body diode may be biased in the forward direction, by applying a negative voltage to the body terminal to prevent the undesired current flow through the diode [22].

The drain–source breakdown voltage, which defines voltage handling capabilities and nonlinear distortions of a single transistor, is another key characteristic of a switch MOSFET. RF switches experience high RF voltage levels operating in off-state, resulting in a significant voltage drop across the channel of the transistor. The breakdown voltage of these transistors defines the voltage at which the transistor briefly transitions from the off to the on state. This transition generates substantial harmonics and may lead to breakdown due to the excessive thermal stress. An illustrative example is provided in Fig. 9.4. The magnitude of ac drain–source voltage at which instantaneous transition into the triode region (characterized by the formation of a conductive inversion layer under the gate oxide) occurs is called breakdown voltage and in a first approximation is defined as:

$$V_{RF,BR} = 2(V_{th} - V_{gs,H0}) \tag{9.5}$$

Important to mention that V_{tn} depends on its terminal voltages and may vary due to the drain-induced barrier lowering effect [23]. Figure 9.4 shows the time-domain waveforms of the transistor when it operates in off-state and reaches the breakdown point. Before breakdown, the current waveforms are 90 degrees out of phase with the applied drain–source voltage, indicating a capacitive load as the excitation source. The breakdown occurs when V_{gs} and V_{gd} cross the threshold level V_{th} at positive and negative excitation voltage peaks, respectively. This results in nonlinear drain current peaks in phase with the V_{ds}. It is essential to note that the average power dissipation on the transistor is zero below the breakdown point and increases rapidly once the breakdown point is crossed, leading to excessive heating of the device.

The primarily nonlinear contributors of single off-state RF transistor are influenced by the nonlinear capacitors C_{gs}, C_{gd}, C_{ds}, and the substrate capacitance C_{sub}. Odd-order nonlinearities arise from the nonlinear overlap capacitances C_{gs}, C_{gd}, and C_{ds}. Given that these capacitances are proportional to the width of the transistor, effective sizing during the design-in phase allows for a trade-off in

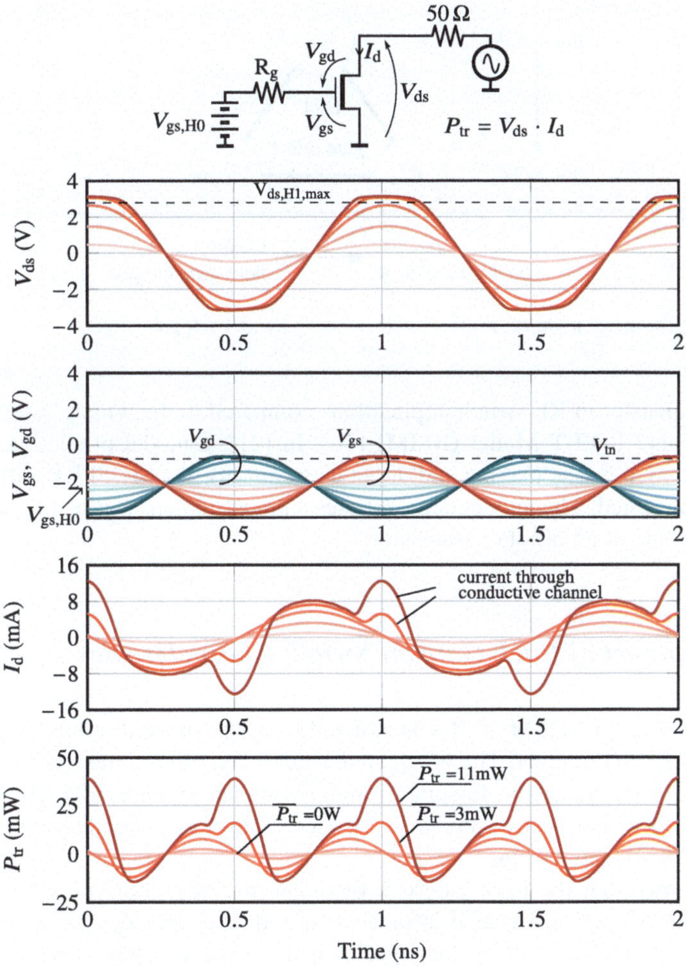

Fig. 9.4 Time-domain response of the single-transistor off-state switch with the transition into breakdown

odd-order nonlinearities. On the other hand, even-order nonlinear products are influenced by the substrate capacitance C_{sub}. Enhancements in this aspect can be achieved through the use of high-resistivity substate materials and trap-rich silicon substrates, as explored in studies [24] and [25]. The large-signal properties of high-resistivity silicon substrates have been investigated in [26], revealing that static harmonic distortion models exhibit simulation errors greater than 15 dB for coplanar transmission lines on HR substrates [27]. Despite efforts to model the nonlinear behavior of HR substrates [27], it is acknowledged that nonlinear substrate modeling is still in a premature state according to [27]. The Penn State Philips (PSP) model of the transistor proves to be more suitable for simulating the nonlinear behavior of

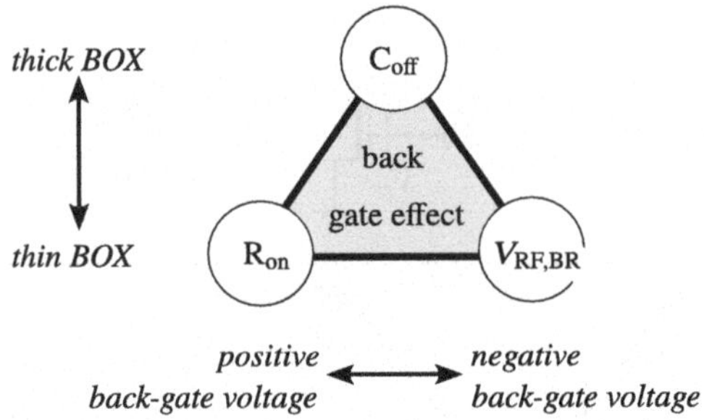

Fig. 9.5 Performance trade-off in RF switch associated with back-gate effect

a single transistor in RF switch applications compared to the widely used Berkeley Short-Channel IGFET Model (BSIM) [28]. In [21], a model predicting the third harmonic based on the overlap and direct capacitances of a MOSFET with Volterra series is presented. This model demonstrates a prediction error of up to 5 dB for the third harmonic of an off-state transistor.

9.2.4 Impact of Back Gate on Switch Characteristics

The body of an FDSOI MOSFET is isolated from the substrate (bulk silicon) by a buried oxide (BOX) layer. The conductive silicon layer under the BOX serves as a back gate for the transistor. In spite of high resistance of the handle silicon, the dc potential under the transistor body can be set to a desired value, thus exploiting the benefits of back-gate biasing.

It is known that the back gate can modulate the threshold voltage of an FET device [29]. Since V_{th} variable is present in both R_{on} and $V_{RF,BR}$ characteristics according to (9.4) and (9.5), this mechanism can be utilized to reduce the on-state resistance and improve the ac voltage handling of the switch. As indicated in [29] and [12], the strength of the back-gate effect if inversely proportional to the thickness of the BOX, becoming significant only for ultrathin BOX layer (thickness below 10 nm, as, e.g., in [30]) due to small body factor [29]. Therefore, utilization of a BOX layer with the thickness comparable with the gate oxide of a switch MOSFET (meaning, in the range of 5 nm to 20 nm) is essential for practically relevant RF performance improvement enabled by the back-gate effect. However, this requirement contradicts the approach aimed at utilizing the thick BOX with the dielectric constant three times below the silicon (4.1 versus 11.9) for minimizing the C_{ds} contributor to the off-state capacitance [18]. The drawing in Fig. 9.5 portrays a tradoff associated with the use of back-gate effect.

9.3 Design of Stacked RF Switches

9.3.1 *Bias Topologies*

High-voltage RF switches are composed of stacked series RF transistors as shown in Fig. 9.6a. The number N of stacked devices is directly proportional to the maximum RF voltage handling $V_{RF,\mathrm{max}}$. The R_{on} and C_{off} are scaled with the number of stacked devices:

$$R_{on,\mathrm{total}} = R_{on} \cdot N \tag{9.6}$$

$$C_{off,\mathrm{total}} = \frac{C_{off}}{N} \tag{9.7}$$

Although R_{on} and C_{off} are linearly scaled with the transistor width, the $R_{on}C_{off}$ ratio, in first approximation, remains constant and independent of the stack size and transistor width. However, practical considerations such as routing, packaging, and PCB parasitic contribute to higher $R_{on}C_{off}$ ratio in high-voltage ICs. The equivalent circuit in off- and on-state of the stack is shown in Fig. 9.6b and c respectively.

In some applications, for example, antenna tuning in cellular handheld devices, the RF switches are designed to reach RF peak voltage levels of above 80 V. Typical values for R_{on} range from 1 Ω to 2 Ω, while C_{off} varies between 200 fF and 140 fF with a typical R_{off} around 30 kΩ at 1 GHz.

The circuit parameters such as cutoff frequency, switching speed, and the uniformity of the RF voltage distribution can be optimized by proper selecting the bias network. Two common topologies include star and series bias configurations as shown in Fig. 9.7. In the star topology, each RF transistor is connected to a common bias supply via a gate resistor, while in the series bias network, the resistors of equal values are used for gate-to-gate, drain-to-source, and body-to-body biasing.

For a series switch design, the GND terminal is replaced by a second RF port. However, special attention should be paid to ESD protection in series RF switch devices. The industry ESD compliance level of above 1 kV (HBM) [31] can be achieved by implementing an auxiliary shunt RF switch connected to the series RF branch. Shunt RF switches are self-protected from the ESD stress.

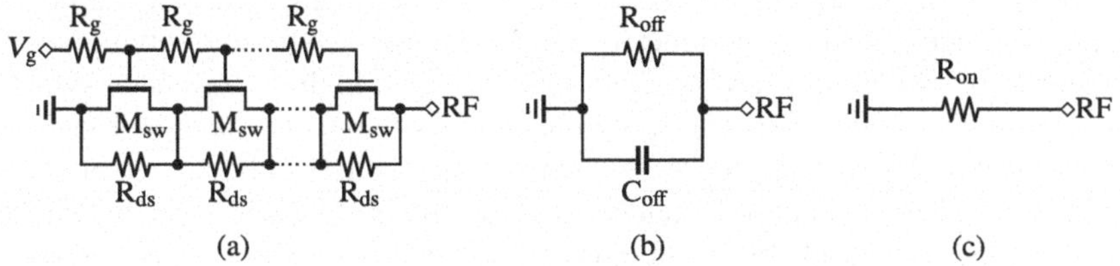

Fig. 9.6 (**a**) Shunt RF switch with equivalent circuit in (**b**)—off-state and (**c**)—on-state

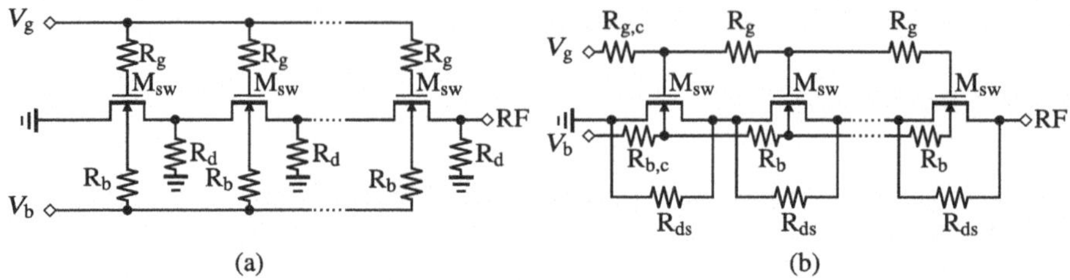

Fig. 9.7 Biasing topologies for RF switch design—(**a**) shunt biasing—(**b**) series biasing

9.3.2 Switching Time Aspects

The switching time is proportional to the time constants of the stacked switches defined by the gate oxide capacitance and resistance of the bias devices [32]. Referring to Fig. 9.7:

$$\tau_{shunt} = (R_g + R_d)(C_{gd} + C_{gs}), \tag{9.8}$$

$$\tau_{series} = \left(\frac{N}{2}R_{g.c} + \frac{(N-1)(N+1)}{2}(R_g + R_{ds})\right)\left(C_{gd} + C_{gs}\right). \tag{9.9}$$

While C_{gs} and C_{gd} are defined by the transistor width, meaning R_{on} specification for the switch, and are constraint in the design, the only way to improve switching time is to reduce the bias resistors. However, bias resistors have a direct impact on the ohmic power losses in off-state switch, defined as:

$$P_{loss,shunt} \approx \frac{V_{RF}^2}{2(\frac{3}{N})R_g \parallel R_d \parallel R_b} \tag{9.10}$$

$$P_{loss,series} \approx \frac{V_{RF}^2}{2N \cdot R_g \parallel R_{ds} \parallel R_b} \tag{9.11}$$

assuming that $R_{g,c} \approx R_g$.

Equations (9.8)–(9.11) reveal a well-known trade-off between the switching time and power losses in stacked MOSFET switch. In many practical cases, a healthy compromise between these parameters can be achieved without applying special techniques for switching speed acceleration. However, high-performance switches may require utilization of special design techniques known in the art for breaking said trade-off and improving power efficiency of the switch without switching time degradation. These techniques include, among others, frequency-dependent bias networks [33], fast discharge techniques [11], basic bypassing the bias resistors during switching optimized for low-stacked configurations and may compromise linearity in high-voltage RF switches [2, 34]. An implementation appropriate for the use in highly stacked shunt switches and efficient in terms of silicon is utilized, and achieved performance benefits are demonstrated in Sect. 9.4.

9.3.3 Substrate Capacitance and Its Compensation

Since the RF switch is composed of stacked transistors, the impact of the substrate capacitance C_{sub} becomes crucial for high-voltage RF switch design, influencing the even-order nonlinear products, the uniformity of the voltage division in the stack, power, as well as voltage handling capability of the switch [7]. Figure 9.8 shows a layout view of a shunt RF switch, where the rectangular shapes indicate individual transistors in stack [8].

The substrate capacitance originates between the switch cells and ground due to the electric field penetrating the high-ohmic substrate, which acts as an insulator with a relative permittivity $\varepsilon_{r,\mathrm{si}}$ of 11.65 at room temperatures and frequencies below 10 GHz. Since the direct measurement of the substrate capacitance is very challenging and prone to errors, C_{sub} can be determined with 3D-EM simulations or analytical evaluations based on microstrip transmission line models for wire-bonded and flip-chip RF switches. Three common analytical approaches for calculating C_{sub} are the models proposed by Wheeler, Schneider, and Hammerstad-Jensen [35–37].

First, we consider the wire-bonded configuration as shown in Fig. 9.9a. The switch itself is represented by a conductive layer with the width W deposited on the silicon substrate with the relative permittivity $\varepsilon_{r,\mathrm{si}}$ and thickness $d_{\uparrow}$ on the infinite ground plane. The surrounding dielectric layer is described by the permittivity $\varepsilon_{r,\uparrow}$ and might be a packaging material or air.

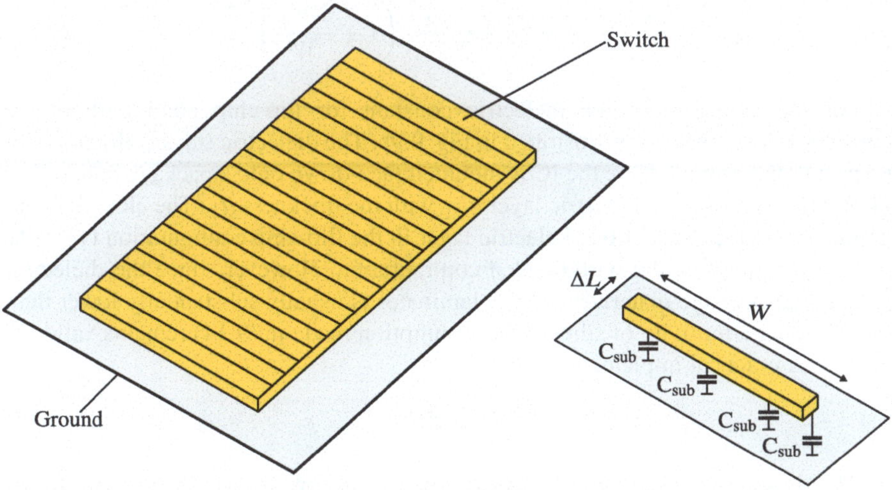

Fig. 9.8 Layout representation of a RF switch deposited on a high-ohmic substrate material

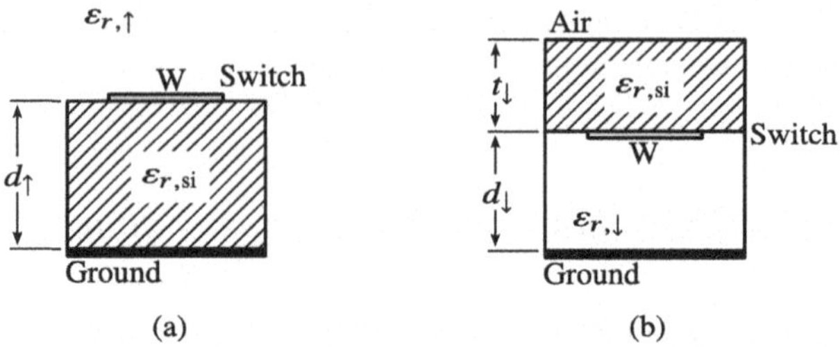

Fig. 9.9 Cross section of (**a**) wire-bonded and (**b**) flip-chip RF switch

The thickness of the conductive layer is assumed to be zero. The substrate capacitance per unit length (measured in F/m) of the RF switch can be, e.g., obtained with Wheelers formula [35]:

$$C_{\text{sub,W}} = \frac{4\pi\varepsilon_0\varepsilon_{\text{eff}}}{\ln\left[1 + \frac{1}{2}\frac{8d}{W}\left(\frac{8d}{W} + \sqrt{\left(\frac{8d}{W}\right)^2 + \pi^2}\right)\right]}, \tag{9.12}$$

Alternative formulas considering Schneider and Hammerstad–Jensen models can be found in literature [22]. All formulas are valid for both wire-bonded and flip-chip mounded RF switch ICs. The effective dielectric constant ε_{eff} depends on the configuration, since the dielectric constant of the insulating layers above and below the conductive layer differences between the wire-bonded and flip-chip configurations. For the wire-bonded configuration, the effective dielectric constant can be determined using the formula for standard microstrip transmission lines [36].

$$\varepsilon_{\text{eff},\uparrow} = \frac{\varepsilon_{r,\text{Si}} + \varepsilon_{r,\uparrow}}{2} + \frac{\varepsilon_{r,\text{Si}} - \varepsilon_{r,\uparrow}}{2}\left(1 + \frac{10d_\uparrow}{W}\right)^{-\frac{1}{2}}. \tag{9.13}$$

For the deviation of the dielectric constant for flip-chip configuration, we consider a cross section as illustrated in Fig. 9.9b. The dielectric stack is flipped with respect to the ground. Here the IC is mounted face-down onto a package material or PCB, represented by a dielectric layer $\varepsilon_{r,\downarrow}$ with the thickness $d_\downarrow$. The electrical and geometrical properties of the dielectric layer in the flip-chip configuration typically differ from those of the wire-bonded configuration. However, since the dielectric constant of package materials or PCB laminates is usually substantially lower than the relative permittivity of silicon, the assumption made in (9.14) remains valid for almost all practical implementations [8]:

$$\varepsilon_{r,\downarrow} < \varepsilon_{r,\text{Si}} \qquad \text{and} \qquad \varepsilon_{r,\uparrow} < \varepsilon_{r,\text{Si}} \tag{9.14}$$

The flip-chip configuration is comparable to standard inverted microstrip transmission line model. The calculation of the substrate capacitance is performed by using quasi-static model as described in literature [38]:

Table 9.1 Material stack parameters for the numerical evaluation of the models

	d	ε_r	$\varepsilon_{r,\mathrm{si}}$	t
Flip-chip	200 μm	3.4	11.65	200 μm
Wire-chip-bonded	750 μm	1	11.65	∞

$$\varepsilon_{\mathrm{eff},\downarrow} = \left(\sqrt{\varepsilon_{r,\downarrow}} + \frac{\sqrt{\varepsilon_{r,\mathrm{Si}}} - \sqrt{\varepsilon_{r,\downarrow}}}{c_0 + (W/d_\downarrow)c_1}\right)^2, \tag{9.15}$$

where

$$c_0 = 2.359 - 0.097 \ln \varepsilon_{r,\mathrm{Si}} - 0.00577 \ln^2 \varepsilon_{r,\mathrm{Si}} + 0.0114 \ln^3 \varepsilon_{r,\mathrm{Si}} + \left[0.04856 - 0.03408 \ln \varepsilon_{r,\mathrm{Si}} + 0.15296 \ln^2 \varepsilon_{r,\mathrm{Si}} - 0.02418 \ln^3 \varepsilon_{r,\mathrm{Si}}\right](d_\downarrow/t_\downarrow) \tag{9.16}$$

$$c_1 = 0.219 - 0.253 \ln \varepsilon_{r,\mathrm{Si}} + 0.208 \ln^2 \varepsilon_{r,\mathrm{Si}} - 0.027 \ln^3 \varepsilon_{r,\mathrm{Si}} + \left[0.915 + 0.338 \ln \varepsilon_{r,\mathrm{Si}} - 0.253 \ln^2 \varepsilon_{r,\mathrm{Si}} + 0.04 \ln^3 \varepsilon_{r,\mathrm{Si}}\right](d_\downarrow/t_\downarrow). \tag{9.17}$$

The substrate capacitance models were evaluated with the provided material stack parameters summarized in Table 9.1. These parameters are determined by the available options for hardware implementation. In practical applications, dielectric layers with constant permittivity values between 1 and 4, and thicknesses up to 1000 μm, may be encountered [8].

The numerical evaluation of the capacitance per 1 μm layout length over the width W is plotted in Fig. 9.10. The curves for the wire-bonded and flip-chip configurations are almost identical, with the average difference being below 0.6% and 0.7%, respectively. For validation of the numerical evaluation, EM simulation was conducted using Sonnet Lite EM solver. The difference between the capacitance curves obtained from the EM simulations and the calculated curves is below 8.7%, for the wire-bonded configuration and 3.5% for the flip-chip configuration.

The substrate capacitance calculation based on transmission line models effectively captures all first-order substrate capacitive coupling effects. The models, despite its simplified material stack for an RF switch and strong abstraction, meets practical RF switch engineering requirements within acceptable tolerances. However, for more accurate modeling, especially if higher precision is required, a full 3D-EM simulation is recommended [8].

As mentioned earlier, the substrate capacitance significantly impacts the uniformity of voltage division within an RF switch stack. The imbalance gets stronger with increasing C_{sub}. The voltage distribution according to the equivalent circuit as shown in Fig. 9.11a is determined as:

$$V_{\mathrm{ds,i}} = \sqrt{\frac{C_{\mathrm{sub}}}{C_{\mathrm{off}}}} \frac{\cosh\Big((2(i-1)+1)\alpha\Big)}{\sinh(2N\alpha)} V_{\mathrm{RF,peak}}, \tag{9.18}$$

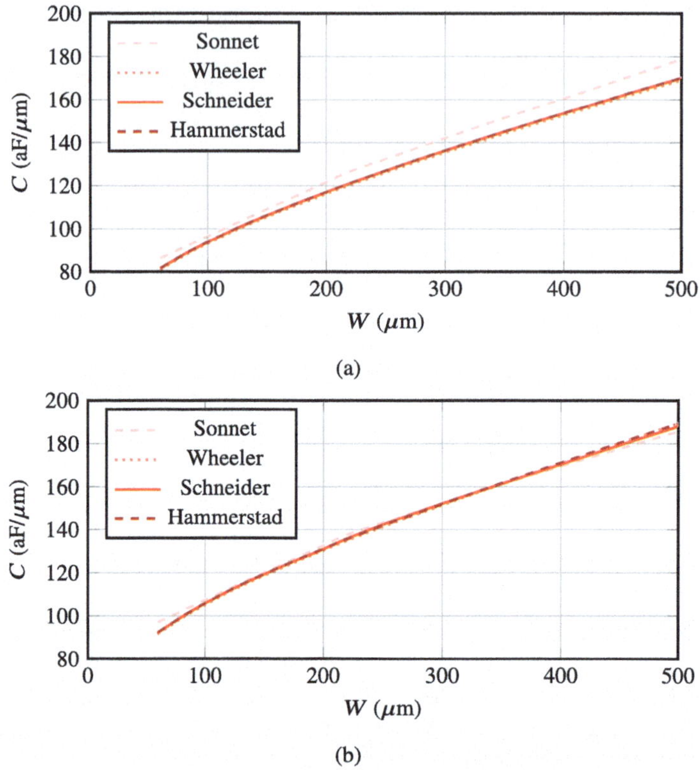

Fig. 9.10 Numerical evaluation of empirical models and respective EM simulation results of the substrate capacitance for (**a**) wire-bonded and (**b**) flip-chip configuration

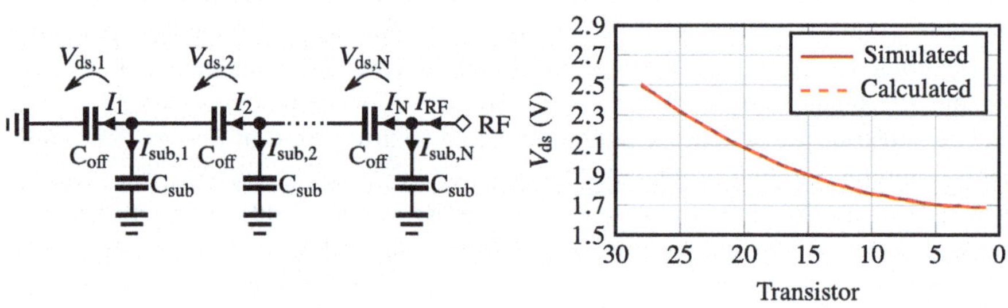

Fig. 9.11 Equivalent circuit of an off-state RF switch and evaluation of (9.18)

The numerical evaluation and circuit simulation result for the wire-bonded RF switch with 28 stacked transistor ($C_{sub} = 2.95$ fF and $V_{RF,peak} = 54.8$ V) is illustrated in Fig. 9.11b. The RF switch transistor located at the RF input is exposed to the highest voltage. Conversely, the transistor located at the grounded side of the switch sees the lowest voltage. This imbalance can be eliminated using equalization capacitors connected parallel to drain and source terminals of each transistor in stack according to Fig. 9.12a. Considering that, after equalization, the voltage drop V_{ds} at each RF switch transistor is the same, the value of the equalization capacitance at each transistor can be calculated as [39]:

$$C_{eq,i} = \frac{i(i-1)}{2} C_{sub}. \tag{9.19}$$

The values of the equalization capacitors and the resulting voltage distribution are shown in Fig. 9.12b. The capacitor values decrease from RF input to the grounded-side of the switch. The RF voltage distribution along the stack is 54.8 V/N = 1.97 V, which is much below the RF breakdown voltage of a single transistor in stack. This leads to an increase of the power handling at the input by about 30% [22].

9.3.4 Back-Gate Biasing in Stacked Switches

The utilization of back-gate effect for performance improvement of MOSFET-based RF switch has been reported in literature [12], where an SPDT switch in 22 nm FDSOI technology with the body and BOX thicknesses of 12 nm and 20 nm respectively, demonstrated the improvement in insertion loss of some 5 %, isolation of around 1 dB and linearity in terms of IIP_3 of 4 dB achieved by applying the back-gate bias voltages of ±3 V. In spite of relatively thin BOX of 20 nm, the improvement brought by the back-gate biasing is marginal.

It is worth noting that biasing voltage sign depends on the switch state. If a monolithic integrated circuit comprises multiple switches operating in different states, local substrate islands under the throws must be implemented for galvanical isolation to be able to have different back-gate biasing for individual throws. In [12], triple wells under the BOX are utilized for this purpose.

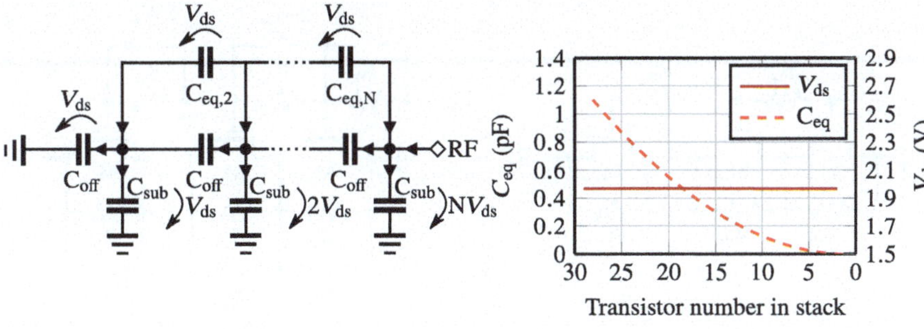

Fig. 9.12 Equivalent circuit of an equalized off-state RF switch and balanced voltage distribution

9.4 Case Study: A 80 V Shunt RF switch with Sub-2 μs Switching Time in FD SOI CMOS Process

The implementation of a high-voltage switch IC in a fully depleted 65 nm SOI-CMOS technology according to the presented design considerations is demonstrated in this section. The utilized technology, especially MOSFET devices and substrate, is optimized for RF switch applications, meaning targeting the lowest $R_{on}C_{off}$ FOM and power losses in the high-ohmic bulk material.

The IC comprises four shunt SPST (single-pole single-throw) switches of two different types. The first type is a stacked switch with a state-of-the-art series resistive bias network and equalization capacitors. The second type is based on the first one, additionally featuring switching time acceleration network speeding up the switching transient from 19.2 μs to 1.6 μs in off-to-on direction and from 0.6 μs to 0.2 μs in on-to-off direction while sustaining low power losses and nonlinear distortion. For convenience, the aforementioned switch types are called "conventional" and "with acceleration."

9.4.1 Circuit Design

A simplified schematic diagram of the implemented conventional switch type is shown in Fig. 9.13a. The circuit comprises 26 MOSFETs M_{sw} arranged in series and a bias network made of R_g and R_{ds}. The high-ohmic resistive connections between the gate terminals as well as the drains and sources of the stacked MOSFETs allow the switch to attain low power losses, high linearity, and the capability to handle high RF voltages. Among the elements not included in the diagram are equalization capacitances in parallel to each device in stack for compensating the influence of substrate on the uniformity of ac voltage division in stack and local body biasing

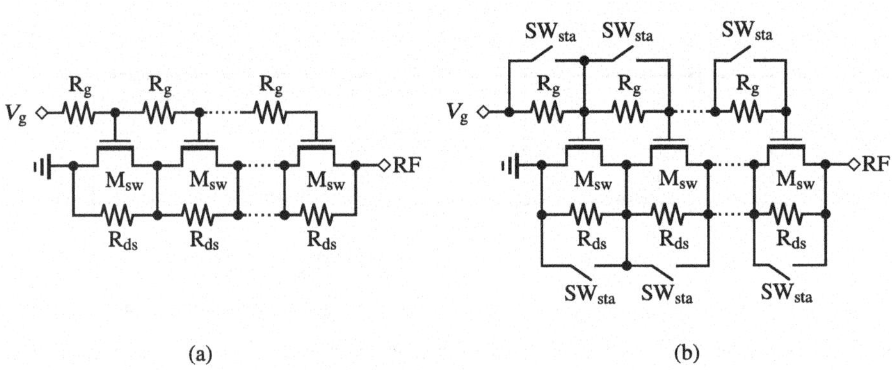

Fig. 9.13 Simplified circuit diagrams of the implemented (**a**) conventional RF switch and (**b**) switch with the acceleration feature

of the switch transistors M_{sw} by utilizing a circuit selecting the lowest voltage between the gate and source–drain nodes and applying it to the body terminal of the respective switch transistor [40, 41].

As mentioned previously, the switch contains 26 stacked elements. The channel size of the switch transistor M_{sw} is $W_{Msw}\,/\,L_{Msw} = 7.3\,\text{mm}\,/\,0.15\,\mu\text{m}$. It is by far the largest device in switch, occupying around 92% of the total stack area. The gate and drain–source bias resistors are configured as follows: $R_g = R_{ds} = 10\,\text{k}\Omega$. The body terminals are shorted to the gate terminals of the respective transistor in off-state; therefore, they are biased from R_g. In on-state, the bodies are dc-coupled to the conductive channel of stacked MOSFET.

The switches in Fig. 9.13 comprise equalization capacitors (not shown in the circuit diagrams), which are calculated in exact accordance to (9.19) with substrate capacitances extracted as described in Sect. 9.3.3.

9.4.2 Switching Time Acceleration

The bias resistors, as well as the gate-channel and metal interconnect capacitances of the transistors, result in a time constant that causes delays in switching events. Simply resizing these components to reduce the delay is not a viable solution because it results in significant penalties in critical RF parameters such as linearity, insertion loss, bandwidth, and others.

There are various strategies documented in literature to resolve the aforementioned difficulty in creating a design that balances competing needs. One possible solution is to use an inductor-based biasing network, as proposed in [33]. This approach successfully tackles the challenge of switching speed, but unfortunately leads to considerable power loss, linearity and bandwidth degradation, and inefficient use of silicon area when applied to high-stacked switch devices. Another solution, presented in [11], is highly efficient in terms of chip area usage but reduces the switching time by only 50%, which may not be adequate for many demanding applications. A full-stacked switch can bypass the bias resistors and work efficiently but only for low-voltage RF switches, as reported in [2]. Lastly, the circuit demonstrated in [34] is limited when changing the state into off direction and introduces nonlinear distortion to the switch.

A conceptual circuit diagram of the implemented solution is demonstrated in Fig. 9.13b, where the elements of the bias resistive ladder are shorted by the switches SW_{sta} during change-state transient and are kept high-ohmic for RF operation. This allows substantial reduction of the switching time while achieving low power losses in RF operation.

The detailed circuit implementation of the switch is shown in Fig. 9.14a [10]. The topology comprises two auxiliary switch branches which short high-ohmic bias ladder during switching transient and significantly reduce the RC-time constant of the switch. These branches are controlled by the nodes V_N and V_P. The V_P stays normally positive, which corresponds to on-state of the switch MOSFET while V_N

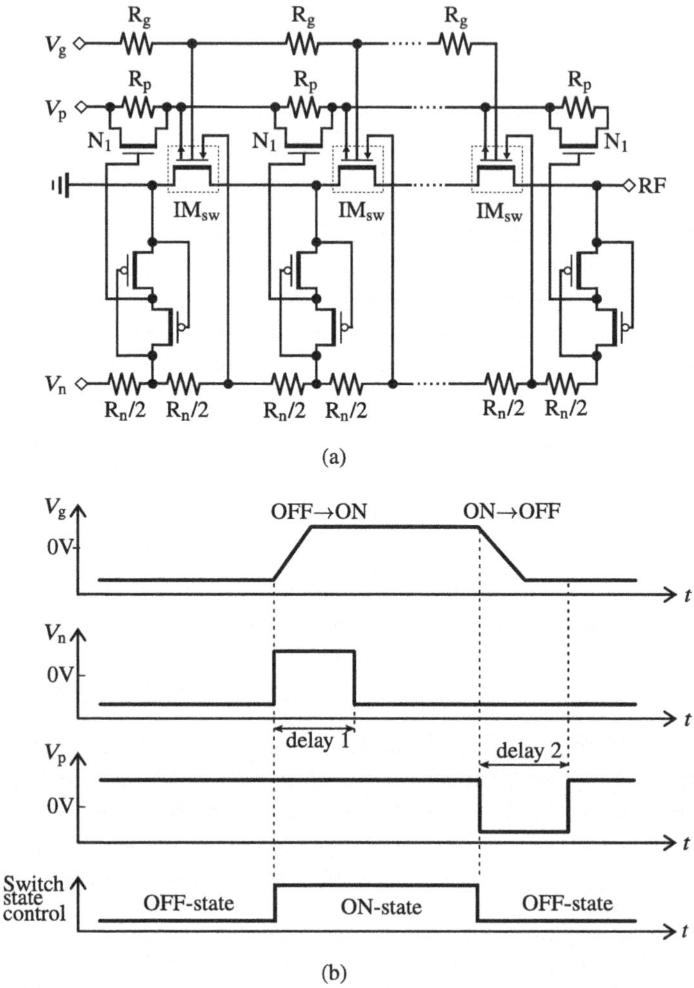

Fig. 9.14 (**a**) Schematic diagram of the switch with switching time acceleration, (**b**) corresponding time diagram

is biased negatively, which allows keeping all auxiliary devices in off-state. The V_P branch, along with all the PMOS devices connected to it, serves as the primary charge delivery path for the switch gates.

The time diagram shown in Fig. 9.14b showcases voltage waveforms at the nodes for corresponding switching events. During the switching from the off- to on- state, all bias resistors connected to the V_P node get shorted because of the positive V_N pulse and the NMOS auxiliary devices N_1 transition into the on-state. During the on $\rightarrow$ off switching phase, the negative pulse at V_P once again shorts all auxiliary gate resistors R_p with the help of N_1.

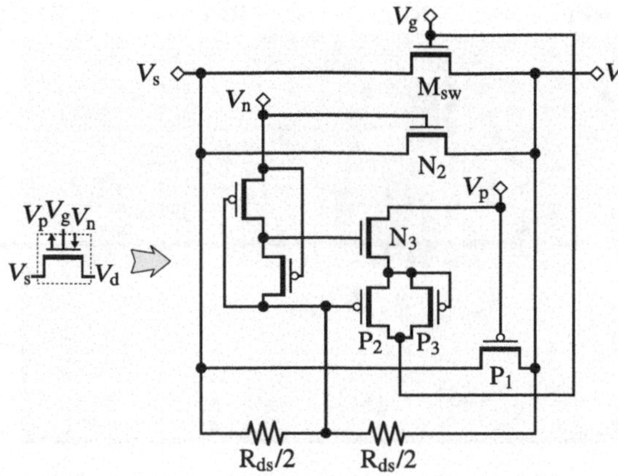

Fig. 9.15 Detailed schematic diagram of the cell IM_{sw} in Fig. 9.14a

Figure 9.15 depicts the circuit diagram of the subcells IM_{sw}. Despite the complexity of the circuit, the main switch transistor M_{sw} is the largest device by far. The other transistors are sized almost at minimum and make an insignificant contribution to the total switch silicon area. The gate recharging of M_{sw} is performed using N_3 and either P_2 (for switching on) or P_3 (for switching off). The recharging path is established by the positive pulse at V_N, which increases the gate–source voltage of N_3 and source–gate voltage of P_2. During the negative pulse at V_P, the main device switches to the off-state through N_3 and the diode-connected device P_3. The channel of M_{sw} is provided with a low-ohmic path through the transmission gate that is made up of N_2 and P_1. Both N_3 (controlled by the terminal V_n) and diode-connected P_3 devices allow unidirectional charging. Notably, the rising edge of V_P and falling edge of V_N do not discharge the main device M_{sw}.

The switch core is based on the conventional implementation described in Sect. 9.4.1. However, the drain–source and gate-to-gate resistors are increased to $R_g = 13\,\text{k}\Omega$ and $R_{ds} = 20\,\text{k}\Omega$ to minimize RF power losses in the circuit. The bias devices in the auxiliary branches are sized as $R_p = R_n = 20\,\text{k}\Omega$.

The performance-critical transistors in the switching time acceleration circuit comprise bypass devices N_1 sized as $W_{N1}/L_{N1} = 30\,\mu\text{m}/0.15\,\mu\text{m}$, P_1 sized as $W_{P1}/L_{P1} = 20\,\mu\text{m}/0.26\,\mu\text{m}$ and N_2 sized as $W_{N1}/L_{N1} = 10\,\mu\text{m}/0.15\,\mu\text{m}$. The width of said transistors defines the total resistance of the charge/discharge path for the gate oxide of M_{sw}. The other transistors are considered to be uncritical for the overall functionality and performance of the switch, therefore configured to have minimum or close to minimum channel size.

In Fig. 9.15b, the auxiliary control voltage pulses at V_P and V_N illustrated in the time diagram are generated by a delay cell based on CMOS inverters. These pulses have a pulse width of $1.5\,\mu\text{s}$ at the nominal PVT (process-voltage–temperature) corner.

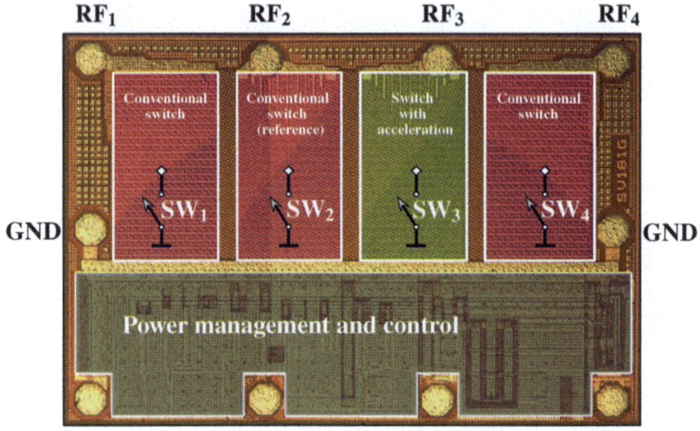

Fig. 9.16 Die photograph

9.4.3 *Top Level Design*

The floorplan of the hardware demonstrator is arranged in the same way as shown in the fabricated die photograph in Fig. 9.16. The chip contains four switch throws that are linked between the RF pins and a common ground, along with the power management and control unit, which provides the necessary control voltages for the switches. The device can be programmed using a 2-wire RFFE MIPI interface. The design details of the power management and control units are beyond the scope of this work and are not further discussed.

The acceleration circuit is incorporated into the shunt switch SW_3, which is connected to RF_3, while the other three switches are designed conventionally. The switch SW_2 located to the left of SW_3 is chosen as a measurement reference due to the layout symmetry. This way, any deviations in small- and large-signal characteristics caused by the placement of switch cells in top layout are minimized.

9.4.4 *Considerations for Backside Biasing*

In the used fabrication process, a common bulk silicon carrier for the SOI is utilized, meaning that the regions under the switches cannot be galvanically isolated from each other. Biasing the bulk silicon with a large positive or negative voltages can slightly improve the isolation in on-state or linearity in off-state, respectively. However, if the four switches are brought into different states, the common backside biasing would degrade the RF performance of switches in the unfavourable states. Also, the improvement in performance with backside bias is marginal due to large ratio (over 50) between the thickness of BOX and GOX. Therefore, the bulk silicon under the BOX is biased at 0 V.

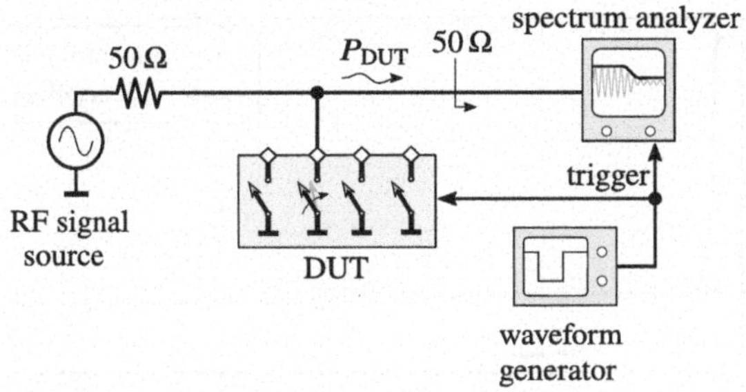

Fig. 9.17 Measurement setup for switching time evaluation

9.4.5 Measurement Results

The fabricated die shown in Fig. 9.16 has been flip-chip mounted into a plastic lead-less package with 10 pins and dimensions of 1.5 mm × 1.1 mm. The measurements have been conducted on a printed circuit board. The device under test (DUT) has been attached to a 50 Ω thru line on the board for the evaluation of switching time. The measurement setup can be seen in Fig. 9.17.

A continuous-wave RF excitation signal with a frequency of 824 MHz and a power of 0 dBm has been applied to the shunt DUT for switching time measurements. The output has been coupled to a spectrum analyzer. The spectrum analyzer has been set to zero-span mode, allowing it to record the power envelope of the signal over time. To synchronize the "change-state" command event and the start of the power envelope recording, both the DUT and the spectrum analyzer were triggered simultaneously by a waveform generator. An uncertainty related to control circuit reaction time, which is in the range of 10 ns, was not considered significant since it is well below the DUT switching time.

Figure 9.18 shows the envelope curves obtained from both switching directions. The "change-state" command event and the trigger signal for the spectrum analyzer are represented by the 0 second time point. When the switches are in the off-state, the recorded power is −0.75 dBm. This insertion loss can be attributed to power losses in the board thru line and mismatch losses caused by the DUT. In the off-state, the settled power is −19 dBm, primarily due to the strong mismatch loss in the thru line, which is shorted to ground by the enabled shunt switch.

The criterion for a switch to be in the off-state is when the signal power falls to a level that is 1 dB less than the settled off-state power, which is defined as −1.75 dBm in the measured DUT. Conversely, the switch is recognized in on-state when the power ascends 3 dB above the stable endpoint, reaching the −16 dBm mark. Specific reference points are established to signify the time it takes for both the conventional and the accelerated switches to change states. Transitioning from the off-to-on state, the switch with acceleration achieves the −16 dBm threshold in only 1.6 μs, a significant increase in speed compared to the 19.2 μs required by the conventional structure, enhancing the switching speed by a factor of 12. When

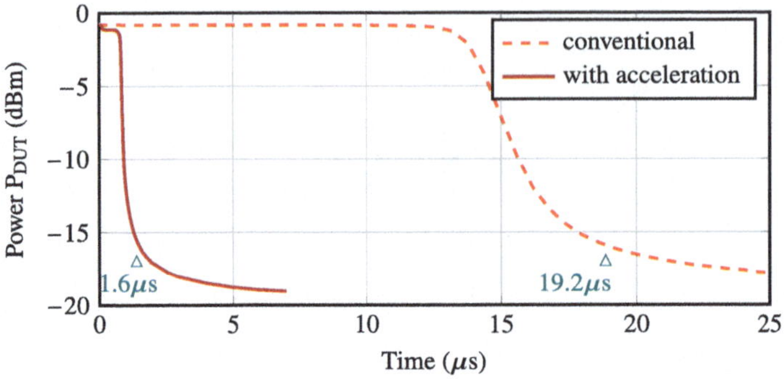

(a)

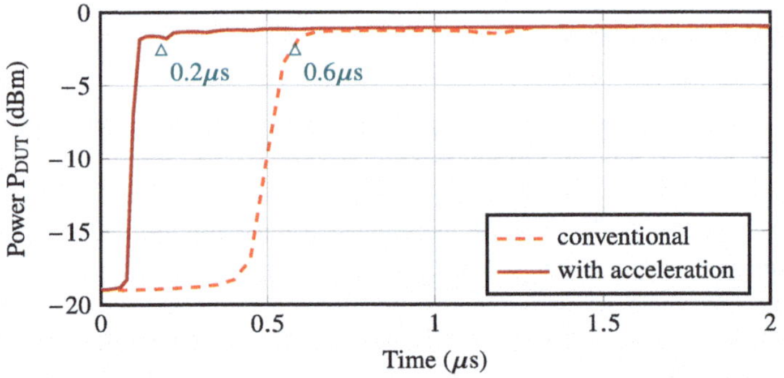

(b)

Fig. 9.18 Measured power envelope of RF signal P_{DUT} for (**a**) off →on switching transition and (**b**) on →off switching transition

moving from ON to OFF, the accelerated switch also shows superior performance, attaining the off-state in a mere 0.2 μs, which is three times faster than the 0.6 μs taken by the conventional design.

A single-tone harmonic distortion of the DUT has been tested to verify the impact of switching time acceleration circuit onto the linearity of the switch. The setup is similar to the one depicted in Fig. 9.17, wherein the spectrum analyzer is configured to detect the third harmonic generated by the DUT. For more details on the harmonics measurement setup, the reader is referred to [39].

The DUT in off-state has been excited with a 900 MHz signal, with power levels varied from 20 dBm to 48 dBm. At a load impedance of 50 ohms, the 48 dBm power translates to a peak RF voltage of 80 V across the DUT. The power of the third harmonic at 2700 MHz is plotted in Fig. 9.19. The analysis of the plotted data reveals that the two curves closely match, suggesting that the addition of the acceleration circuit to the switch does not negatively affect its linearity. Any small discrepancies observed between the curves are within the normal range of measurement and production variances for various samples of the DUT.

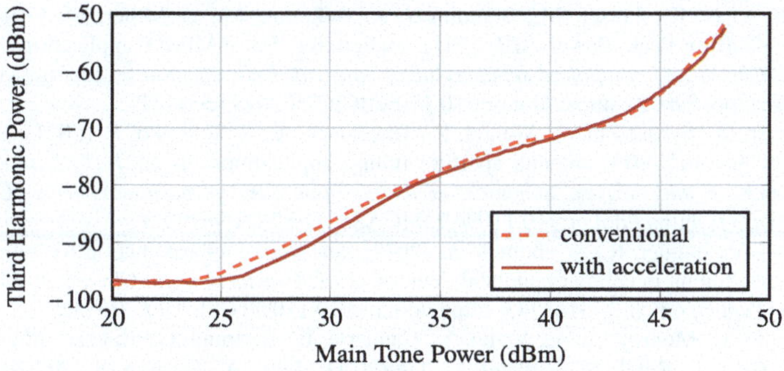

Fig. 9.19 Measured third harmonic of the off-state switch DUTs

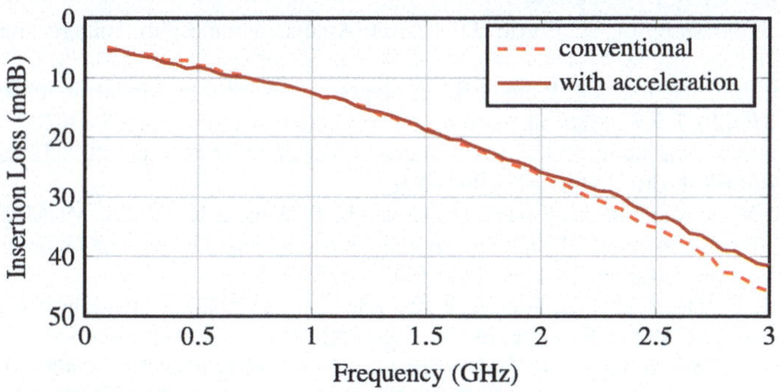

Fig. 9.20 Measured insertion loss of the off-state switches

The small-signal characteristics of the switches have been evaluated. Specifically, the insertion loss when the switches are in off-state is presented in Fig. 9.20. These characteristics are nearly identical, although the switch with the acceleration circuit shows marginally lower losses at elevated frequencies, which could be partially due to the increased overall shunt resistance from the biasing network. When in the on-state, both switches exhibit the same performance, with the resistance of 2.15 Ω at low frequencies.

References

1. Oezdamar, O., Weigel, R., Hagelauer, A., & Solomko, V. (2021). An integrated circuit for decoupling and tuning of inverted-F antennas in cellular user equipment. In *2021 IEEE International Symposium on Circuits and Systems (ISCAS)* (pp. 1–4). https://doi.org/10.1109/ISCAS51556.2021.9401633
2. Ilkov, N. (2008). Switch using an accelerating element. US7868683B2. .

3. Malladi, V. N. K., Fraser, M., Staudinger, J., Bokatius, M., & Miller, M. (2018). A 2–6 GHz, 45 dBm Peak Power T/R SPDT switch for 5G mMIMO applications. In *2018 IEEE BiCMOS and compound semiconductor integrated circuits and technology symposium (BCICTS)* (pp. 89–92). https://doi.org/10.1109/BCICTS.2018.8551151
4. Oezdamar, O., Syroiezhin, S., Weigel, R., Hagelauer, A., & Solomko, V. (2021). Hybrid C-Tuner IC for 40 V/80 V antenna aperture tuning applications. In *2021 IEEE international conference on microwaves, antennas, communications and electronic systems (COMCAS)* (pp. 334–337). https://doi.org/10.1109/COMCAS52219.2021.9629061
5. Solomko, V., Bauder, R., & Thomas, A. (2018). Successive approximation RF reflectometer for antenna tuning in cellular handheld devices. *IEEE Transactions on Circuits and Systems I: Regular Papers, 65*(5), 1731–1743. https://doi.org/10.1109/TCSI.2017.2765325
6. Gianesello, F., Monroy, A. A., Vialla, V., Canderle, E., Bertrand, G., Buczko, M., Coly, M., Nowakowski, J., Revil, N., Rolland, L., Gloria, D., Juge, A., Gachon, S., Aubert, J. P., & Granger, E. (2016). Highly linear and sub 120 fs Ron × Coff 130 nm RF SOI technology targeting 5G carrier aggregation RF switches and FEM SOC. In *2016 IEEE 16th topical meeting on silicon monolithic integrated circuits in RF systems (SiRF)* (pp. 9–12). https://doi.org/10.1109/SIRF.2016.7445454
7. Zhu, Y., Klimashov, O., & Bartle, D. (2014). Analytical model of voltage division inside stacked-FET switch. In *2014 Asia-Pacific microwave conference* (pp. 750–752).
8. Solomko, V., Oezdamar, O., Weigel, R., & Hagelauer, A. (2021). Model of substrate capacitance of MOSFET RF switch inspired by inverted microstrip line. In *ESSDERC 2021—IEEE 51st European solid-state device research conference (ESSDERC)* (pp. 207–210). https://doi.org/10.1109/ESSDERC53440.2021.9631806
9. Solomko, V., Syroiezhin, S., Tayari, D., Essel, J., & Weigel, R. (2023). MOSFETBased RF switch with active biasing. *IEEE Transactions on Microwave Theory and Techniques, 71*(12), 5295–5303. https://doi.org/10.1109/TMTT.2023.3279909
10. Syroiezhin, S., Oezdamar, O., Weigel, R., & Solomko, V. (2022). Switching time acceleration for high-voltage CMOS RF switch. In *ESSCIRC 2022-IEEE 48th European Solid State Circuits Conference (ESSCIRC)* (pp. 301–304). https://doi.org/10.1109/ESSCIRC55480.2022.9911368
11. Solomko, V., Oezdamar, O., Weigel, R., & Hagelauer, A. (2021). CMOS RF switch with fast discharge feature. *IEEE Solid-State Circuits Letters, 4*, 68–71. https://doi.org/10.1109/LSSC.2021.3059918
12. Rack, M., Nyssens, L., Wane, S., Bajon, D., & Raskin, J. P. (2020). DC-40 GHz SPDTs in 22 nm FD-SOI and back-gate impact study. In *2020 IEEE Radio Frequency Integrated Circuits Symposium (RFIC)* (pp. 67–70). https://doi.org/10.1109/RFIC49505.2020.9218317
13. Raskin, J. P., & Desbonnets, E. (2015). High resistivity SOI Wafer for Mainstream RF System-on-Chip. In *2015 IEEE 15th Topical Meeting on Silicon Monolithic Integrated Circuits in RF Systems*. https://doi.org/10.1109/sirf.2015.7119866
14. Le Meil, J.M., Aspar, B., Desbonnets, E., & Raskin, J. P. (2015). Engineered substrates: The foundation to meet current and future RF requirements. In *2015 International symposium on VLSI technology, systems and applications*. https://doi.org/10.1109/vlsi-tsa.2015.7117546
15. Raskin, J. P. (2012). SOI substrates for more than moore roadmap. In *2012 8th international caribbean conference on devices, circuits and systems (ICCDCS)*. https://doi.org/10.1109/iccdcs.2012.6188886
16. Raskin, J. P. (2014). SOI technology: An opportunity for RF designers?. In *Technical papers of 2014 international symposium on VLSI design, automation and test.*
17. Ali, K. B., Neve, C. R., Gharsallah, A., & Raskin, J.-P. (2012). RF SOI CMOS technology on commercial trap-rich high resistivity SOIWafer. In *2012 IEEE International SOI Conference (SOI)*. https://doi.org/10.1109/soi.2012.6404404
18. Jaffe, M., Abou-Khalil, M., Botula, A., Ellis-Monaghan, J., Gambino, J., Gross, J., He, Z. X., Joseph, A., Phelps, R., Shank, S., Slinkman, J., & Wolf, R. (2015). Improvements in SOI technology for RF switches. In *2015 IEEE 15th topical meeting on silicon monolithic integrated circuits in RF systems*. (pp. 30–32). https://doi.org/10.1109/SIRF.2015.7119865

19. *BGSA12GN10: Single Pole Dual Throw Antenna Tuning Switch*. Accessed: Feb. 01, 2024. Infineon Technologies AG. https://www.infineon.com/dgdl/Infineon-BGSA12GN10-DataSheet-v03_01-EN.pdf?fileId=5546d46255dd933d0155e9daddca09f6
20. Carusone, T. C., Johns, D., & Martin, K. (2011). *Analog integrated circuit design*. Wiley.
21. Niemeier, D., Feick, H., Bartels, M., Cattaneo, A., & Malkov, N. (2020). Modeling OFF-state harmonics in MOS transistors used as RF switches. In *Solid- State Electronics*. https://doi.org/10.1016/j.sse.2020.107872
22. Oezdamar, O. (2022). *High-Voltage RF switch-based antenna tuning circuits in CMOS technologies*. FAU Studien aus der Elektrotechnik Band 16. Erlangen: FAU University Press. https://doi.org/10.25593/978-3-96147-562-9
23. Fjeldly, T. A., & Shur, M. (1993). Threshold voltage modeling and the subthreshold regime of operation of short-channel MOSFETs. *IEEE Transactions on Electron Devices, 40*(1), 137–145. https://doi.org/10.1109/16.249436
24. Esfeh, B. K., Ali, K. B., & Raskin, J. P. (2014). RF Non-Linearities from Si-Based Substrates. In *2014 international workshop on integrated nonlinear microwave and millimetre-wave circuits (INMMiC)*. https://doi.org/10.1109/inmmic.2014.6815104
25. Neve, C. R., Alia, K. B., Malaquin, C., Allibert, F., Desbonnets, E., Bertrand, I., Van Den Daele, W., & Raskin, J. P. (2013). RF and linear performance of commercial 200 mm Trap-Rich HR-SOI Wafers for SoC applications. In *2013 IEEE 13th Topical Meeting on Silicon Monolithic Integrated Circuits in RF Systems*. https://doi.org/10.1109/sirf.2013.6489417
26. Roda Neve, C. (2012). Small- and large-signal characterization of Trap-Rich HRSi/HR-SOIWafers for SoC applications. PhD thesis. Catholic University of Louvain. http://hdl.handle.net/2078.1/109212.
27. Rack, M., & Raskin, J. P. (2017). 'RF harmonic distortion modeling in silicon-based substrates including non-equilibrium carrier dynamics. In *2017 IEEE MTT-S international microwave symposium (IMS)*. https://doi.org/10.1109/mwsym.2017.8058737
28. Xi, X., Dunga, M., He, J., Liu, W., Cao, K. M., Jin, X., Ou, J. J., Chan, M., Niknejad, A. M., & Hu, C. (2003). *BSIM4.3.0 MOSFET model*. Accessed: Feb. 01, 2024. http://ewh.ieee.org/r5/denver/sscs/References/2003_BSIM4v30_manual.pdf
29. Ohtou, T., Saraya, T., & Hiramoto, T. (2008). Variable-Body-Factor SOI MOSFET with ultrathin buried oxide for adaptive threshold voltage and leakage control. *IEEE Transactions on Electron Devices, 55*(1), 40–47. https://doi.org/10.1109/TED.2007.912612
30. Auerhammer, J., Hartig, C., Wendt, K., van Oostrum, R., Pfeiffer, G., Bayer, S., & Srocka, B. (2016). Silicon thickness variation of FD-SOI wafers investigated by differential reflective microscopy. In *2016 IEEE SOI-3D-Subthreshold microelectronics technology unified conference (S3S)* (pp. 1–3). https://doi.org/10.1109/S3S.2016.7804378
31. Rigato, M., Fleury, C., Schwarz, B., Mergens, M., Bychikhin, S., Simburger, W., & Pogany, D. (2018). Analysis ofESDBehavior of Stacked nMOSFET RF switches in bulk technology. *IEEE Transactions on Electron Devices, 65*, 829–837. https://doi.org/10.1109/ted.2018.2789941
32. Rigato, M. (2018). ESD experiments and simulations on RF CMOS switche. PhD thesis. TU Wien. https://repositum.tuwien.at/bitstream/20.500.12708/6998/2/Rigato%20Matteo%20-%202018%20-%20ESD%20experiments%20and%20simulations%20on%20RF%20CMOS%20switches.pdf
33. Katzin, P., Bedard, B. E., Shifrin, M. B., & Ayasli, Y. (1992). High-speed, 100+W RF switches using GaAs MMICs. *IEEE Transactions on Microwave Theory and Techniques, 40*(11), 1989–1996. https://doi.org/10.1109/22.168755
34. Levy, C. S., Asbeck, P. M., & Buckwalter, J. F. (2013). A CMOS SOI stacked shunt switch with Sub-500 ps time constant and 19-Vpp breakdown. In *2013 IEEE Compound Semiconductor Integrated Circuit Symposium (CSICS)* (pp. 1–4). https://doi.org/10.1109/CSICS.2013.6659196
35. Wheeler, H. A. (1977). Transmission-line properties of a strip on a dielectric sheet on a plane. *IEEE Transactions on Microwave Theory and Techniques, 25*(8), 631–647. https://doi.org/10.1109/tmtt.1977.1129179

36. Schneider, M. V. (1969). Microstrip lines for microwave integrated circuits. *Bell System Technical Journal, 48*(5), 1421–1444. https://doi.org/10.1002/j.1538-7305.1969.tb04274.x
37. Hammerstad, E., & Jensen, O.. Accurate models for microstrip computer-aided design. In *MTT-S international microwave symposium digest*. https://doi.org/10.1109/mwsym.1980.1124303
38. Tomar, R. S., & Bhartia, P. (1987). New Quasi-Static models for the computer-aided design of suspended and inverted microstrip lines (Short Paper). *IEEE Transactions on Microwave Theory and Techniques, 35*(4), 453–457. https://doi.org/10.1109/tmtt.1987.1133671
39. Oezdamar, O., Weigel, R., Hagelauer, A., & Solomko, V. (2020). Considerations for harmonics distribution in aperture-tuned inverted-F antenna for cellular handheld devices. *IEEE Transactions on Microwave Theory and Techniques, 68*(10), 4122–4130. https://doi.org/10.1109/TMTT.2020.2996195
40. Bloch, M., Lauterbauch, C., & Weber, W. (1998). High efficiency charge pump circuit for negative high voltage generation at 2 V supply voltage. In *Proceedings of the 24th European solid-state circuits conference* (pp. 100–103). https://doi.org/10.1109/ESSCIR.1998.186218
41. Solomko, V., Syroiezhin, S., & Weigel, R. (2023). High-Voltage analog switch inspired by stacked MOSFET-Based RF switches. *IEEE Transactions on Circuits and Systems II: Express Briefs, 70*(11), 4008–4012. https://doi.org/10.1109/TCSII.2023.3286328

Chapter 10
Variable Gain Control with Bulk Biasing in mmW Amplifier

Patrick James Artz and Friedel Gerfers

10.1 Introduction: Back-Gate Tuning for Gain Control

Variable gain control in amplifiers offers a wide range of possibilities. In a wireless receiver, tuning the gain in the Low-Noise Amplifier (LNA) allows to reduce the gain requirements for the baseband variable gain amplifier (VGA) or depending on the application, even omit it [4]. A reduction of gain in the LNA can extend the input power saturation and thereby relax the linearity requirements of the subsequent down-conversion mixer. If applied correctly, it can effectively increase the systems signal-to-noise and distortion ratio (SNDR) and thereby increase the dynamic range of the input power. Moreover, an adaptive gain LNA that adjusts its power consumption based on gain reduction not only enhances a transceiver's power efficiency but also facilitates sleep mode operation, which is crucial for a high-power D-band monolithic microwave integrated circuit (MMIC).

A unique gain-tuning technique involves modulating the bulk node, which in turn adjusts the threshold voltage, altering the transconductance and, consequently, the gain, as demonstrated in [1–3, 8]. The basic schematic for a differential common-source (CS) stage with bulk control is shown in Fig. 10.1. By using the bulk control of the 22 nm FDSOI technology, the operating point of an active device in a CS stage can be passively fine tuned without interfering in the biasing path of the active gate, thereby modulating the devices g_m and therefore the gain A. The amplifier's gain sensitivity respective to the applied back-gate voltage V_{bg} can be expressed as:

$$A \propto \left(V_{gs} - \left(V_{th,0} - 80\,\mathrm{mV} \cdot \frac{V_{bg}}{1\,V}\right)\right) \cdot Z_{load}, \tag{10.1}$$

P. J. Artz (✉) · F. Gerfers
Technische Universität Berlin, Berlin, Germany
e-mail: artz@tu-berlin.de

F. Gerfers (ed.), *Bulk-Driven Circuit Techniques for CMOS FDSOI Processes*,
https://doi.org/10.1007/978-3-031-85114-8_10

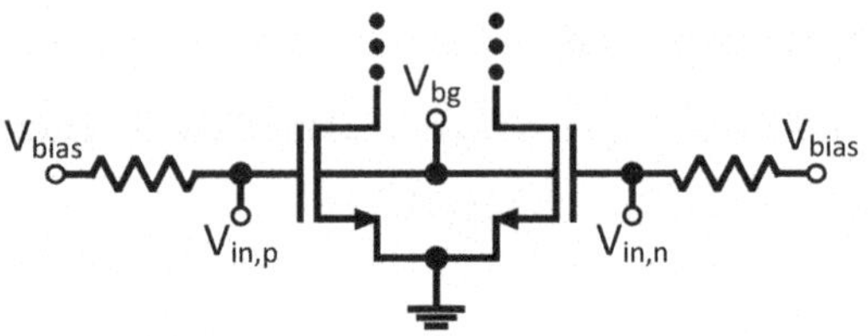

Fig. 10.1 Differential common source configuration with bulk control and resistive gate biasing

with V_{gs} being the gate–source voltage, $V_{th,0}$ the intrinsic threshold voltage at $V_{bg} = 0\,\text{V}$, and Z_{load} the load impedance seen at the drain node of the CS device. As can be seen, the voltage sensitivity of the gain tuning via the bulk is about $1\text{V}/80\,\text{mV} = 12.5x$ higher than tuning directly the gate making the gain control more resilient against noise and signal interference. Furthermore, the gain control is now completely isolated from the signal path. In RF amplifiers, the gate biasing is typically done with the use of a high-ohmic bias resistor R_{bias} to block RF signals. The advantage of the bulk control can be for applications where the gain control is used in active feedback loops with short response time. The high-ohmic bias resistor forms an RC time constant with the capacitive load of the RF devices, which is relaxed in the bulk gain control since the control voltage can be applied to the bulk via a low-impedance path. Therefore, a fast settling can be achieved.

10.2 D-band LNA Design Using Back-Gate Control

The following is an example of back-gate gain control for a 4-stage D-band LNA design operating around 150 GHz. Nevertheless, the described technique can be applied to a baseband amplifier design as well. Hereby, the design focuses on enabling gain control while reducing the impact on the noise figure (NF) degradation as well as the 1 dB compression point (P1dB). The gain control is only applied to the second and third stages, thereby enabling a measured peak gain-control between 9 and 18 dB while reducing the NF degradation below 1.5 dB and the output P1dB (OP1dB) degradation below 1 dB. This enables to further increase the signal-to-noise and distortion ratio (SNDR) by reducing the gain for higher input power saturation.

A simulated link-budget model [5] of a D-band IQ receiver including the measured data of the described LNA with variable gain control demonstrates this performance extension in Fig. 10.2. The simulated peak SNDR at the output of the analog–digital converter (ADC) can be significantly extended by reducing the gain in the LNA since its input compression increases inversely with gain.

$$F_{tot} = F_{St1} + \frac{F_{St2} - 1}{G_{St1}} + \frac{F_{St3} - 1}{G_{St1} \cdot G_{St2}} + \frac{F_{St4} - 1}{G_{St1} \cdot G_{St2} \cdot G_{St3}} \tag{10.2}$$

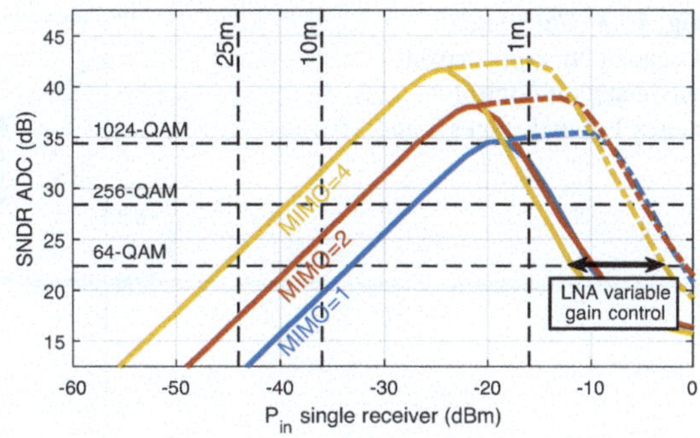

Fig. 10.2 Simulated SNDR at the ADC output of a D-band IQ receiver with different MIMO factors over the input power and distance visualizing the performance extension through gain control in the LNA © 2024 IEEE. Reprinted with permission from [5]

According to the cascaded linear NF (F) equation in 10.2, the noise and gain of the first stage have significant impact on the overall NF. Therefore, a reduction of gain is avoided since this will increase the noise contribution of the subsequent stages. Furthermore, adapting the biasing of the active devices will change the noise parameters including the minimum NF (NF_{min}) as well as the optimum noise impedance (R_{opt}, and Γ_{opt}). This eventually will reduce the overall NF_{tot} since the source admittance Y_{s} remains constant. Similarly, the gain compression (P1dB) and third-order intercept point (IP3) are dominated by the last stage as given in the equation for the cascaded IP3:

$$IP3_{\text{tot}} = \frac{1}{\frac{1}{IP3_{\text{St1-3}} \cdot G_{\text{St4}}} + \frac{1}{IP3_{\text{St4}}}}, \tag{10.3}$$

whereby $IP3_{\text{St1-3}}$ combines the nonlinear contribution of the previous stages. As evident from the equation, the gain of the last stage, G_{St4}, directly suppresses the influence of the preceding stages. Therefore, to maintain a constant output compression point (OP1dB) and output-referred third-order intercept point (OIP3), it is recommended to keep G_{St4} static.

In Fig. 10.3, the cascaded NF over the peak gain is calculated as a comparison for different gain-control options. Thereby a gain control of all four stages is compared with the control of only the second and third stages, indicating the improvement by having the first stage's gain being fixed. Nevertheless, applying gain control only in the second and third stages results in the compromise of a reduced gain-control range.

10.2.1 LNA Stage Design

The LNA stages are designed as differential CS stages with capacitive neutralization to improve the maximum available and maximum stable gain (MAG/MSG), which offers excellent performance at mmW frequencies in terms of high gain, reverse

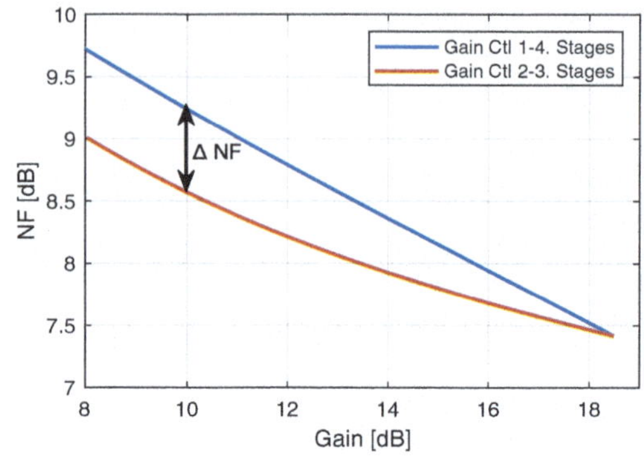

Fig. 10.3 Calculated cascaded NF comparison between gain-control in stages 1–4 and stages 2 and 3

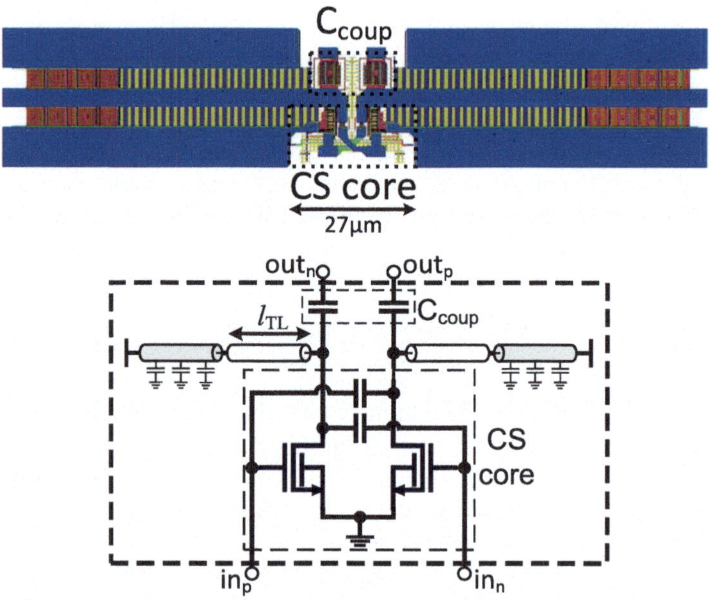

Fig. 10.4 Differential CS stage with capacitive neutralization and transmission line-based matching © 2024 IEEE. Reprinted with permission from [5]

isolation, and stability performance [6]. All stages use the same device geometries to enable an interstage matching with a single-stub combined with a Zero-Ohm transmission line (TL) [7] and a coupling capacitor enabling acceptable conjugated matched output reflection. The first stage is matched for noise performance and low losses, while the last stage is optimized for low losses and high IP3.

An exemplary stage with the corresponding layout is shown in Fig. 10.4. The gain in the controllable stages 2 and 3 is staggered by slightly tuning the TL by about $\pm 8\,\mu$m, thereby extending the effective bandwidth by slightly shifting each peak gain to lower and higher frequencies, respectively. A final schematic is shown in Fig. 10.5 with the chosen device geometries summarised in Table 10.1. Since the

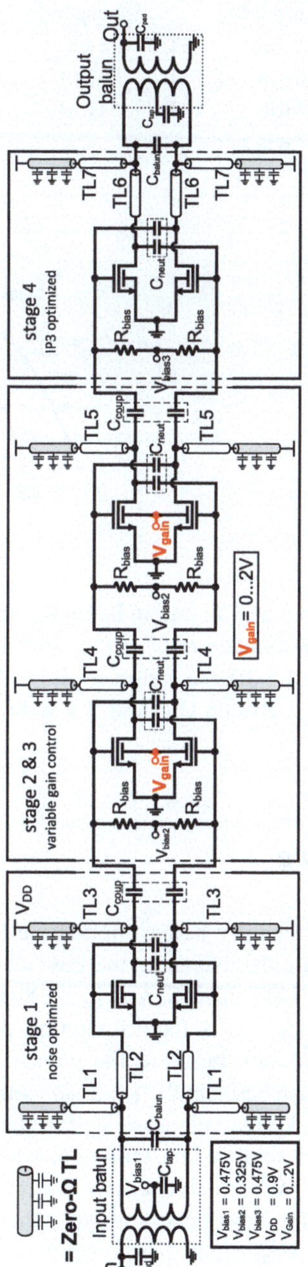

Fig. 10.5 Final schematic of the implemented 4-stage D-band LNA with gain control in the second and third stages via back gate © 2022 IEEE. Reprinted with permission from [8]

Table 10.1 Transmission line (TL) lengths of the final design in Fig. 10.5 [5]

	Length l_{TL}		Length l_{TL}
TL 1	54 μm	TL 5	71 μm
TL 2	70 μm	TL 6	80 μm
TL 3	63 μm	TL 7	54 μm
TL 4	54 μm		

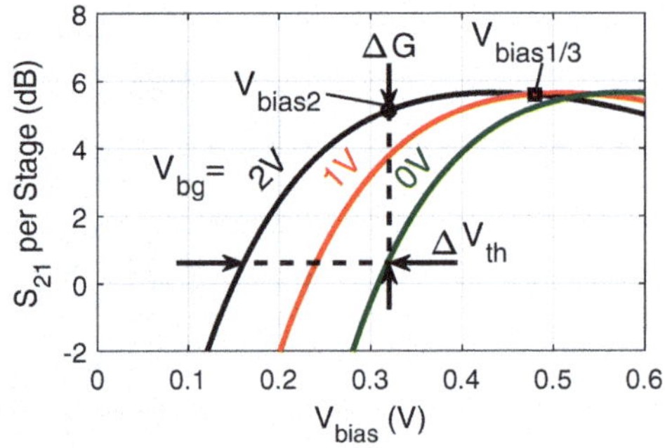

Fig. 10.6 Gain per matched stage over biasing for different back-gate voltages

LNA operates differently, an input and an output balun for measurement is required. A stacked spiral inductor with a diameter of 37 μm provides acceptable ± 3° phase imbalance and losses of about 1.5 dB per balun in the D-band. The gain can be controlled by the voltage V_{gain} between 0 V and 2 V achieving a gain control of about 9 dB.

10.2.2 LNA Stage Biasing

The 22 nm FDSOI technology enables active flipped well devices with the bulk being completely isolated from the channel, thereby allowing forward biasing with tuning the V_{th} by a sensitivity of $dV_{th}\ /\ dV_{bg} \approx 80\,\text{mV/1 V}$. By setting the direct biasing of the active devices to a fixed operating point allows to use this V_{th} dependency to tune the g_m and therefore the gain, as explained previously. In Fig. 10.6, the gain of a matched single stage is simulated over the direct biasing voltage for different back-gate voltages V_{bg}. For the first and fourth stages, $V_{bias1/3} = 0.475$ V is chosen to achieve constant gain, while the second and third stages are set to $V_{bias2} = 0.325$ V to enable a gain-control range of ≈ 5 dB with tuning $V_{bg} = 0 \ldots 2$ V. For the described prototype, these biasing voltages are directly applied off-chip. Later, V_{bias} needs to be generated with a current mirror fed by a reference current to enable robustness against process, voltage, and temperature variation (PVT).

10.3 Measurement Results

The fabricated prototype die for probing is shown in Fig. 10.7. The core LNA area is 0.0934 mm^2.

10.3.1 S-Parameter and Gain Compression

The small signal parameters are measured with a vector network analyzer (VNA) using R&S ZC170 frequency extender modules from 110 to 170 GHz. The measured S-parameters for different gain settings are shown in Fig. 10.8, achieving a controllable peak gain of 9–18 dB without reducing the 3 dB bandwidth of about 11 GHz. Furthermore, $S_{11/22}$ and S_{12} are negligibly affected by the reduction of the gain ($<$ 1 dB).

The analysis of the nonlinearity is done by measuring the input compression point P1dB. The measured peak gain over the input power is shown in Fig. 10.9, and the resulting input and output refereed P1dB along with the gain are plotted in Fig. 10.10. By reducing the back-gate control voltage V_{gain}, the peak gain decreases while the input refereed P1dB increases inversely. This concludes in a nearly constant output referred P1dB with degradation of less than 1 dB, which proves the reduced nonlinearity dependence on the gain control.

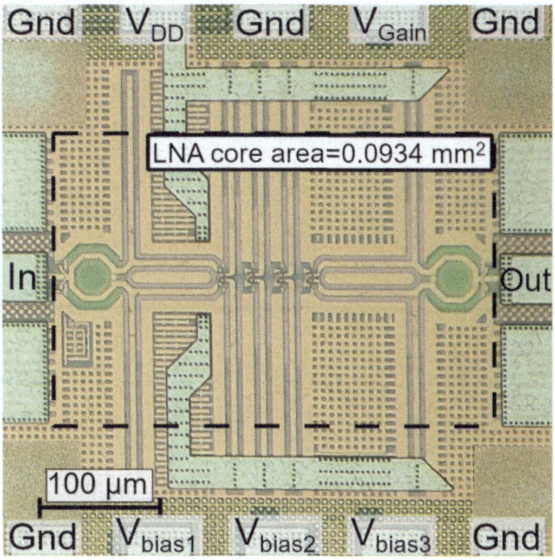

Fig. 10.7 Die photo of the implemented D-band LNA with gain control via back-gate tuning © 2022 IEEE. Reprinted with permission from [8]

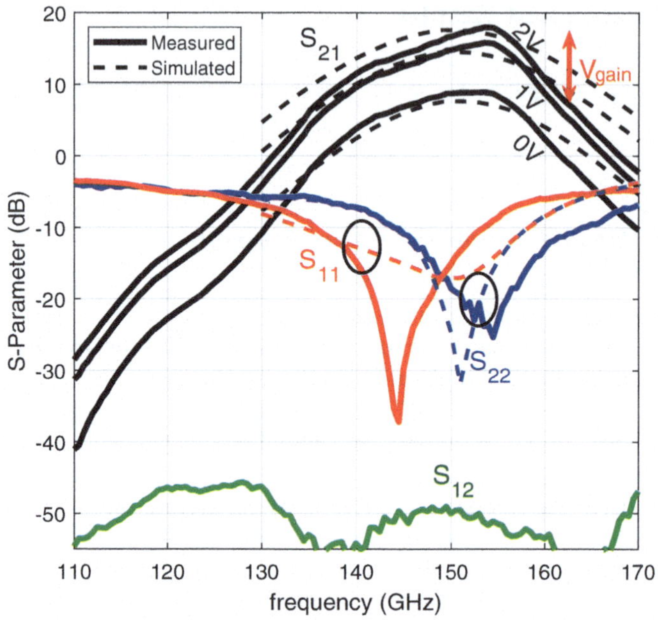

Fig. 10.8 S-Parameter measurement of 4-stage D-band LNA with different gain settings © 2024 IEEE. Reprinted with permission from [5]

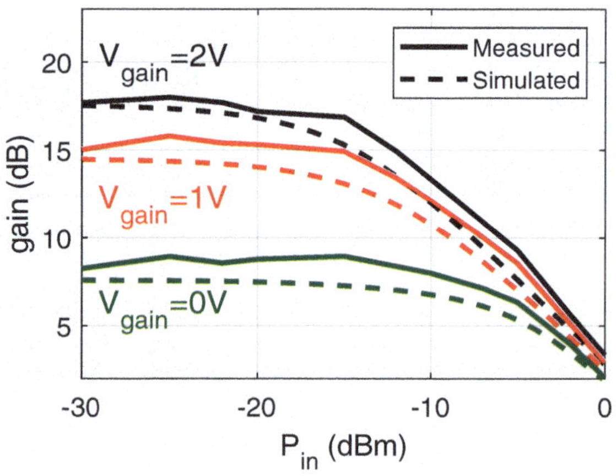

Fig. 10.9 Gain over input power for different gain settings © 2024 IEEE. Reprinted with permission from [5]

10.3.2 NF Measurement

The noise figure is measured using a tuner-based noise measurement system that employs the cold-source technique [5, 9], as illustrated in Fig. 10.11. This approach helps to address the large uncertainties typically encountered in most D-band NF measurements, which are largely caused by the mismatch-related constraints of the

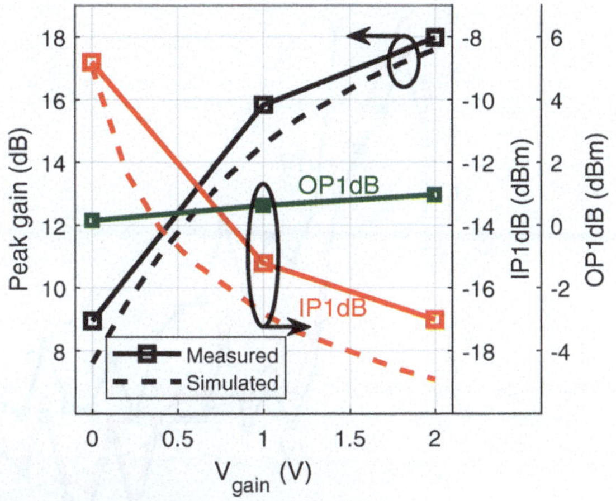

Fig. 10.10 Input/Output 1 dB compression point (IP1dB/OP1dB) with peak gain for different gain settings © 2024 IEEE. Reprinted with permission from [5]

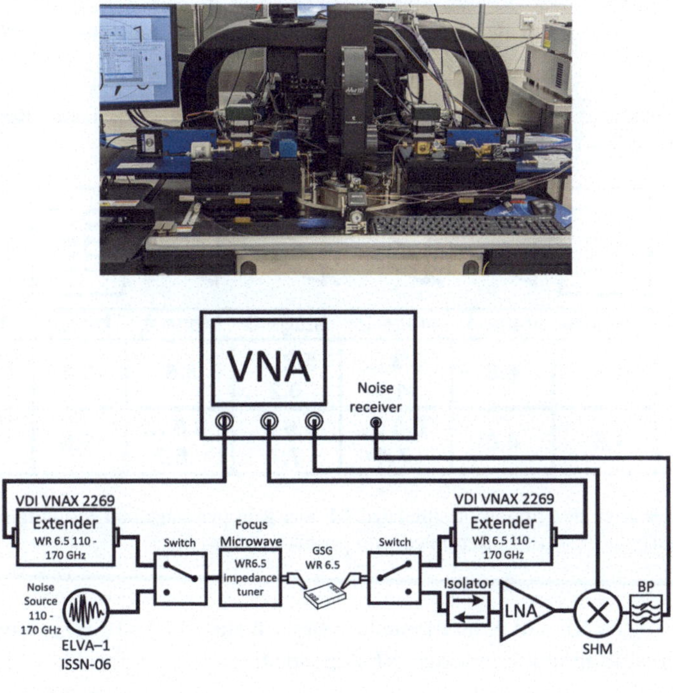

Fig. 10.11 Tuner-based noise measurement setup featuring the cold-source technique © 2024 IEEE. Reprinted with permission from [5]

Y-factor method [9, 10]. In Fig. 10.12, the measured NF is shown for the different gain settings in the 3 dB bandwidth with minimum NFs ranging between 7.9 dB for maximum gain and about 9.5 dB at minimum gain. Therefore, the NF degradation stays around 1.5–2 dB. This dependency can be further analyzed with the detailed

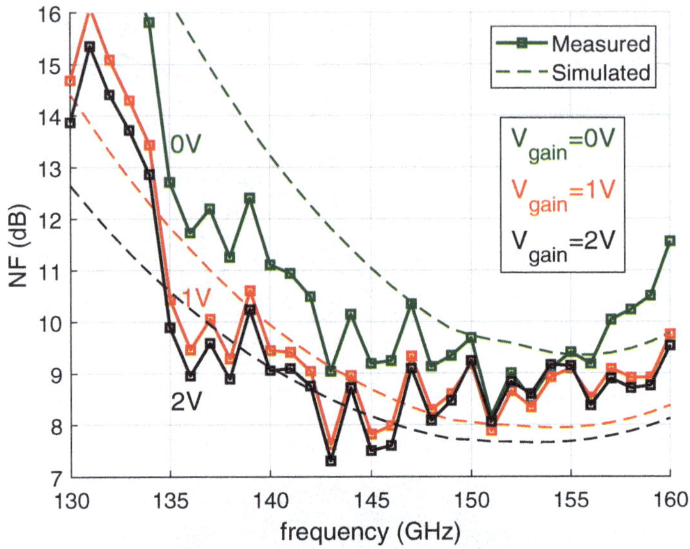

Fig. 10.12 Measured Noise figure for different gain settings © 2024 IEEE. Reprinted with permission from [5]

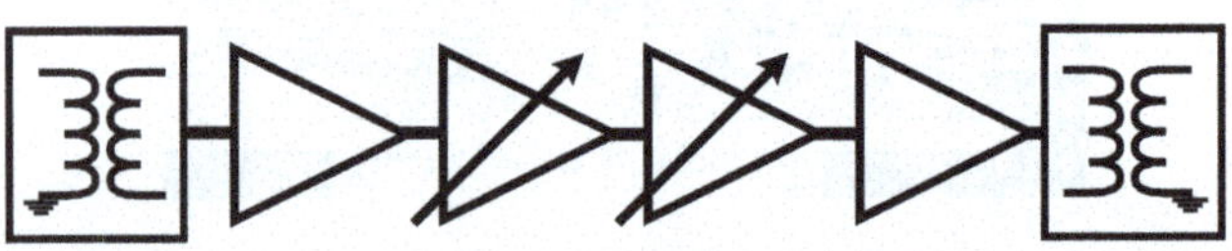

at 150 GHz:	balun	stage 1	stage 2	stage 3	stage 4	balun	total:
Gain / Atten.	-1.5	6.5	3.3 ... -1.7	5.2 ... 0.2	5.5	-1.5	17.5 ... 7.5
NF	1.5	4.6	6.5 ... 7.4	6.6 ... 7.7	6.0 ... 6.2	1.5	7.7 ... 9.7

Fig. 10.13 Detailed description of simulated NF and gain per stage and resulting total NF and gain at 150 GHz © 2024 IEEE. Reprinted with permission from [5]

simulation of the NF and gain/attenuation of each stage at 150 GHz, summarized in Fig. 10.13 resulting in the reported NF dependency.

10.3.3 Power Consumption

Since the operating points of the second and third stages are adjusted, the power consumption will be reduced as well, as shown in Fig. 10.14. From peak to

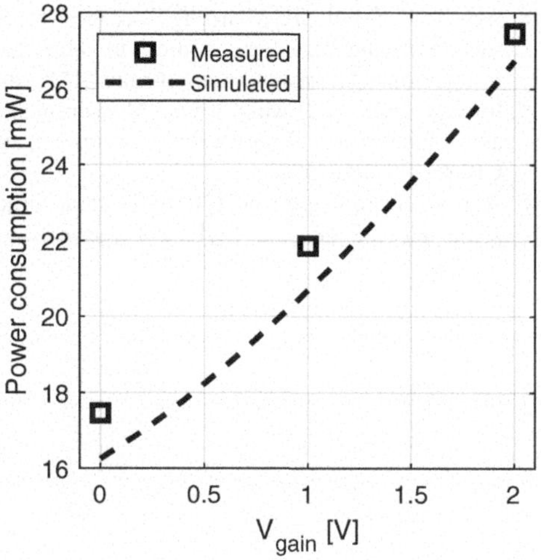

Fig. 10.14 Power consumption for different gain settings © 2022 IEEE. Reprinted with permission from [8]

lowest gain, the power consumption is reduced by about 36%. Therefore, this gain reduction might be used for a sleep mode operation to enhance system efficiency.

References

1. Rashtian, H., & Mirabbasi, S. (2014). Applications of body biasing in multistage CMOS low-noise amplifiers. *IEEE Transactions on Circuits and Systems I: Regular Papers, 61*(6), 1638–1647. https://doi.org/10.1109/TCSI.2013.2290848
2. Zaini, J., Hameau, F., Taris, T., Morche, D., Audebert, P., & Mercier, E. (2017). A tunable Ultra Low Power inductorless Low Noise Amplifier exploiting body biasing of 28 nm FDSOI technology. In *2017 IEEE/ACM international symposium on low power electronics and design (ISLPED)* (pp. 1–6). https://doi.org/10.1109/ISLPED.2017.8009161
3. Zandieh, A., Bonen, S., Dadash, M. S., Gong, M. J., Hasch, J., & Sorin P. (2021). Voinigescu. 155 GHz FMCW and stepped-frequency carrierOFDMradar sensor transceiver IC featuring a PLL with <30 ns settling time and 40 fs rms Jitter. *IEEE Transactions on Microwave Theory and Techniques, 69*(11), 4908–4924.
4. Ali, S. N. et al. (2019). A continually-stepped variable-gain LNA in 65-nm CMOS enabled by a tunable-transformer for mm-wave 5G communications. In 2019 IEEE MTT-S International Microwave Symposium (IMS) (pp. 926–929). https://doi.org/10.1109/MWSYM.2019.8701023
5. Artz, P. J., Le, Q. H., Huynh, D. K., Scholz, P., Kämpfe, T., Lehmann, S., Mausolf, T., & Gerfers, F. (2024). Design and characterization of a variable gain D-Band LNA for optimized link budgeting for a 6G receiver in 22FDX. *IEEE Transactions on Microwave Theory and Techniques, 72*(2), 1008–1017. https://doi.org/10.1109/TMTT.2023.3298197
6. Vigilante, M., & Reynaert, P. (2016). 20.10 A 68.1-to-96.4GHz variable-gain low-noise amplifier in 28 nm CMOS. In *2016 IEEE International Solid-State Circuits Conference (ISSCC)* (pp. 360–362). https://doi.org/10.1109/ISSCC.2016.7418056
7. Goda, R., et al. (2015). Characterization of wideband decoupling power line with extremely low characteristic impedance for millimeter-wave CMOS circuits. In *Proceedings of the 2015 International Conference on Microelectronic Test Structures* (pp. 220–223). https://doi.org/10.1109/ICMTS.2015.7106098

8. Artz, P. J., Scholz, P., Mausolf, T., & Gerfers, F. (2022). A fully-differential 146.6–157.4 GHz LNA utilizing back gate control to adjust gain in 22 nm FDSOI. In 2022 IEEE/MTT-S International Microwave Symposium—IMS 2022 (pp. 611–614).
9. Modern RF and Microwave Measurement Techniques. (2013). The Cambridge RF and Microwave Engineering Series. Cambridge University. https://doi.org/10.1017/CBO9781139567626
10. Agilent Technologies Inc. (2010). High-accuracy noise figure measurements using the PNA-X series network analyzer. In *Agilent Application Note 1408–20*.

Index

F. Gerfers (ed.), *Bulk-Driven Circuit Techniques for CMOS FDSOI Processes*, https://doi.org/10.1007/978-3-031-85114-8

The manufacturer's authorised representative in the EU is Springer Nature Customer Service Centre GmbH, Europaplatz 3, 69115 Heidelberg, Germany. If you have any concerns regarding our products, please contact ProductSafety@springernature.com

Printed and bound by CPI Group (UK) Ltd, Croydon, CR0 4YY

30/09/2026

02225564-0008